Biological Crystallization

Biological Crystallization

Special Issue Editors

Jaime Gómez-Morales
Giuseppe Falini
Juan Manuel García-Ruiz

MDPI • Basel • Beijing • Wuhan • Barcelona • Belgrade

Special Issue Editors
Jaime Gómez-Morales
Laboratorio de Estudios Cristalográficos,
Instituto Andaluz de Ciencias de la Tierra
Spain

Juan Manuel García-Ruiz
Laboratorio de Estudios Cristalográficos,
Instituto Andaluz de Ciencias de la Tierra
Spain

Giuseppe Falini
University of Bologna
Italy

Editorial Office
MDPI
St. Alban-Anlage 66
4052 Basel, Switzerland

This is a reprint of articles from the Special Issue published online in the open access journal *Crystals* (ISSN 2073-4352) from 2018 to 2019 (available at: https://www.mdpi.com/journal/crystals/special_issues/biological_crystallization)

For citation purposes, cite each article independently as indicated on the article page online and as indicated below:

LastName, A.A.; LastName, B.B.; LastName, C.C. Article Title. *Journal Name* **Year**, *Article Number*, Page Range.

ISBN 978-3-03921-403-7 (Pbk)
ISBN 978-3-03921-404-4 (PDF)

Cover image courtesy of Juan Manuel García-Ruiz.

Contents

About the Special Issue Editor vii

Preface to "Biological Crystallization" ix

Jaime Gómez-Morales, Giuseppe Falini and Juan Manuel García-Ruiz
Biological Crystallization
Reprinted from: *crystals* **2019**, *9*, 409, doi:10.3390/cryst9080409 1

Álvaro Esteban Torres-Aravena, Carla Duarte-Nass, Laura Azócar, Rodrigo Mella-Herrera, Mariella Rivas and David Jeison
Can Microbially Induced Calcite Precipitation (MICP) through a Ureolytic Pathway Be Successfully Applied for Removing Heavy Metals from Wastewaters?
Reprinted from: *crystals* **2018**, *8*, 438, doi:10.3390/cryst8110438 4

Julian Opel, Niklas Unglaube, Melissa Wörner, Matthias Kellermeier, Helmut Cölfen and Juan-Manuel García-Ruiz
Hybrid Biomimetic Materials from Silica/Carbonate Biomorphs
Reprinted from: *crystals* **2019**, *9*, 157, doi:10.3390/cryst9030157 17

Nives Matijaković, Giulia Magnabosco, Francesco Scarpino, Simona Fermani, Giuseppe Falini and Damir Kralj
Synthesis and Adsorbing Properties of Tabular {001} Calcite Crystals
Reprinted from: *crystals* **2019**, *9*, 16, doi:10.3390/cryst9010016 25

Zhong He, Zengzilu Xia, Mengying Zhang, Jinbo Wu and Weijia Wen
Calcium Carbonate Mineralization in a Surface-Tension-Confined Droplets Array
Reprinted from: *crystals* **2019**, *9*, 284, doi:10.3390/cryst9060284 37

Jaime Gómez-Morales, Luis Antonio González-Ramírez, Cristóbal Verdugo-Escamilla, Raquel Fernández Penas, Francesca Oltolina, Maria Prat and Giuseppe Falini
Induced Nucleation of Biomimetic Nanoapatites on Exfoliated Graphene Biomolecule Flakes by Vapor Diffusion in Microdroplets
Reprinted from: *crystals* **2019**, *9*, 341, doi:10.3390/cryst9070341 48

Adafih Blackburn, Shahla H. Partowmah, Haley M. Brennan, Kimberly E. Mestizo, Cristina D. Stivala, Julia Petreczky, Aleida Perez, Amanda Horn, Sean McSweeney and Alexei S. Soares
A Simple Technique to Improve Microcrystals Using Gel Exclusion of Nucleation Inducing Elements
Reprinted from: *crystals* **2018**, *8*, 464, doi:10.3390/cryst8120464 60

Christo N. Nanev
Recent Insights into Protein Crystal Nucleation
Reprinted from: *crystals* **2018**, *8*, 219, doi:10.3390/cryst8050219 73

Christo N. Nanev
Peculiarities of Protein Crystal Nucleation and Growth
Reprinted from: *crystals* **2018**, *8*, 422, doi:10.3390/cryst8110422 85

Alexander McPherson
pH and Redox Induced Color Changes in Protein Crystals Suffused with Dyes
Reprinted from: *crystals* **2019**, *9*, 126, doi:10.3390/cryst9030126 101

Katarina Koruza, Bénédicte Lafumat, Maria Nyblom, Wolfgang Knecht and Zoë Fisher
From Initial Hit to Crystal Optimization with Microseeding of Human Carbonic Anhydrase IX—A Case Study for Neutron Protein Crystallography
Reprinted from: *crystals* **2018**, *8*, 434, doi:10.3390/cryst8110434 . . . . **114**

Qian Sun, Sze Wan Cheng, Kelton Cheung, Marianne M. Lee and Michael K. Chan
Cry Protein Crystal-Immobilized Metallothioneins for Bioremediation of Heavy Metals from Water
Reprinted from: *crystals* **2019**, *9*, 287, doi:10.3390/cryst9060287 . . . . **125**

Jeong Kuk Park, Yeo Won Sim and SangYoun Park
Over-Expression, Secondary Structure Characterization, and Preliminary X-ray Crystallographic Analysis of *Xenopus tropicalis* Ependymin
Reprinted from: *crystals* **2018**, *8*, 284, doi:10.3390/cryst8070284 . . . . **133**

Zhao-Xin Liu, Zhenggang Han, Xiao-Li Yu, Guoyuan Wen and Chi Zeng
Crystal Structure of the Catalytic Domain of MCR-1 (cMCR-1) in Complex with D-Xylose
Reprinted from: *crystals* **2018**, *8*, 172, doi:10.3390/cryst8040172 . . . . **141**

Mohammad Mizanur Rahman, Bradley Goff, Li Zhang and Anna Roujeinikova
Refolding, Characterization, and Preliminary X-ray Crystallographic Studies on the *Campylobacter concisus* Plasmid-Encoded Secreted Protein Csep1^{P} Associated with Crohn's Disease
Reprinted from: *crystals* **2018**, *8*, 391, doi:10.3390/cryst8100391 . . . . **150**

Tatyana Prudnikova, Barbora Kascakova, Jeroen R. Mesters, Pavel Grinkevich, Petra Havlickova, Andrii Mazur, Anastasiia Shaposhnikova, Radka Chaloupkova, Jiri Damborsky, Michal Kuty and Ivana Kuta Smatanova
Crystallization and Crystallographic Analysis of a *Bradyrhizobium Elkanii* USDA94 Haloalkane Dehalogenase Variant with an Eliminated Halide-Binding Site
Reprinted from: *crystals* **2019**, *9*, 375, doi:10.3390/cryst9070375 . . . . **160**

About the Special Issue Editors

Jaime Gómez-Morales received his Ph.D. in Chemistry from the University of Barcelona in 1992. From 1992 until 2003 he was a researcher at the Institute of Materials Science of Barcelona, developing his activity in crystal growth, crystallization of inorganic monodisperse particles, and in the field of biomaterials. Since 2004, he has been a senior research scientist at the Laboratory of Crystallographic Studies of Andalusian Institute of Earth Sciences in Granada, Spain. His main research activities and interests are focused on the field of crystallization and crystallography of inorganic materials of natural interest, biological crystallization, nanocrystallization, and biomaterials science. In particular, his studies focus on nucleation, growth, polymorphism, aggregation, and the interaction of crystals with proteins and small molecules. In the field of biomaterials, his main interests include the preparation of biomimetic apatite nanoparticles, biohybrid composites, and 3D printing polymeric scaffolds.

Giuseppe Falini received his Ph.D. degree in Chemistry from the University of Bologna, Italy, in 1994, for studying the molecular recognition at the organic–inorganic interface in biomineralization processes. From 1994 to 1998, he spent several research and learning periods at the Weizmann Institute of Science (Israel) and the University of California Santa Barbara (USA). He was an associate professor at the University of Bologna from 2008 to 2018, when he became a full professor in General and Inorganic Chemistry at the same University. His research interests include the biomineralization process in marine calcifying organisms, synthesis of biomaterials inspired by biomineralization and reuse of waste byproducts from mussel aquaculture.

Juan Manuel García-Ruiz is Research Professor at the National Research Council (CSIC) of the University of Granada (Spain). He received his BS and Ph.D. from Complutense University (Madrid, Spain). Professor García-Ruiz is an expert in crystallization, and the founder of the Laboratory of Crystallographic Studies, and the Crystallization Factory in Granada. His main area of research is in the self-organization and self-assembly in biological and geological materials, applied to early life detection, the origin of life, and the synthesis of new materials. He is the author of the book The Mystery of the Giant Crystals, and the corresponding documentary film script. In 2014, he organized the exhibition "Crystals: A world to discover". Professor García-Ruiz is very much involved in activities to promote a citizen culture of science. http://www.garciaruiz.net/juanma/Inicio.html

Preface to "Biological Crystallization"

In September 2017, we were invited by the journal Crystals to Edit a Special Issue on "Biological Crystallization", a commitment that we accepted with enthusiasm and a desire to gather contributions from some of the best specialists on this subject.

Biological Crystallization deals with the formation of inorganic and organic crystals by living organisms and also with the crystallization of biological materials. In this Special Issue, we have intentionally widened the scope to also gather some articles of bioinspired and biomimetic crystallization, indirectly related to the core subject. The resulted Special Issue gathers 15 original manuscripts dealing with, among others, the precipitation of calcium carbonates, calcium phosphates, self-assembled silica/carbonate materials, and protein crystallization, and covering fundamental aspects such as nucleation and crystal growth, the characterization by different techniques including X-ray diffraction and the applications in fields as diverse as biomedicine, environment, materials science, and others. We hope this volume can be of interest and helpful for the wide readership of Crystals.

Jaime Gómez-Morales, Giuseppe Falini, Juan Manuel García-Ruiz
Special Issue Editors

crystals

MDPI

Editorial

Biological Crystallization

Jaime Gómez-Morales [1,*], Giuseppe Falini [2] and Juan Manuel García-Ruiz [1]

1 Laboratorio de Estudios Cristalográficos, IACT, CSIC-UGR. Avda. Las Palmeras 4, 18100 Armilla, Granada, Spain
2 Dipartimento di Chimica "Giacomo Ciamician", Alma Mater Studiorum Università di Bologna, via Selmi 2, 40126 Bologna, Italy
* Correspondence: jaime@lec.csic.es

Received: 2 August 2019; Accepted: 3 August 2019; Published: 6 August 2019

Keywords: biomineralization; biomimetic materials; biomorphs; calcium carbonate; nanoapatites; nucleation; growth; crystallization of macromolecules; bioremediation; materials science; biomedicine

---

"Biological Crystallization" is today a very wide topic that includes biomineralization, but also the laboratory crystallization of biological compounds such as macromolecules, carbohydrates or lipids, and the synthesis and fabrication of biomimetic materials by different routes. In this Special Issue, special attention is paid to the fundamental phenomena of crystallization (nucleation and growth), and the potential applications of the crystals in environmental science, materials science and biomedicine.

This issue collects 15 contributions, starting with the paper of Torres-Aravena et al. [1]. This paper reviews the main characteristics of a microbially induced precipitation process (MICP), which promotes calcium carbonate (calcite) precipitation. The authors propose to consider this method for heavy metal removal from wastewater/waters.

In the second article, Opel et al. [2] present a method to convert silica/carbonate biomorphs into hybrid organic/carbonate composite materials similar to biominerals. It is worth highlighting that silica/carbonate biomorphs are a class of biomimetic materials named so since they resemble primitive living organisms and their inner textures mimic biominerals. However, compared to biominerals, which are hybrid inorganic-organic materials, the biomorphs are purely inorganic composite materials, the structuring role of organic compounds being taken over by amorphous silica.

Calcium carbonate ($CaCO_3$) is considered a key mineral by many organisms to build its exoskeletons for protecting and supporting purposes. The most common crystal habit of the thermodynamically stable polymorph of calcium carbonate, calcite, is the rhombohedral one, which exposes {10.4} faces. However, in presence of Li^+ the tabular {00.1} faces appear together with the {10.4}, thus generating truncated rhombohedrons. The paper of Matijaković et al. [3] explores the morphological aspects and adsorbing properties of model organic substances of the {10.4} versus {00.1} faces, which are relevant for the understanding of biomineralization processes, in which the {00.1} faces often interact with organic macromolecules and open new routes for the usage of calcite as adsorbing substrate with applications in the environment.

In biomineralization the interactions between organic macromolecules and the nascent inorganic solids play a pivotal role in controlling the shape, size distribution, polymorphism, orientation and even assembly of the formed crystals. At the laboratory scale, it is not easy to carry out high-throughput experiments with only a few macromolecule reagents using conventional experimental methods. In the fourth paper, He et al. [4] explore the surface-tension-confined droplet arrays technique to fabricate $CaCO_3$ using polyacrylic acid as a modified organic molecule control. These authors prove the possibility of performing biomimetic crystallization and biomineralization experiments using this technique.

Nanocrystalline calcium phosphates apatites are a class of biomimetic materials displaying morphological and crystalline properties close to those of bone and dentine apatites. Due to their excellent biocompatibility, osteoconductivity and osteoinductivity, these nanoparticles find application in the field of biomaterials. In the fifth paper, Gómez Morales et al. [5] explore the nucleation of apatite nanoparticles on exfoliated graphene flakes to yield graphene/apatite nanocomposites with applications as bone grafts.

Crystallization of biological macromolecules is the largest part of this Special Issue (papers 6–15 fall in this section). Crystallization is a crucial step in the pathway to determine the three-dimensional structure of macromolecules by X-ray diffraction techniques, and also to obtain crystals for specific applications in environment, industry or medicine. Blackburn et al. [6] present a simple technique, based on gel exclusion of nucleation inducing elements, for generating large well-diffracting crystals from conditions that yield microcrystals when using other techniques. This method is successfully applied to generate diffraction quality crystals of lysozyme, cubic insulin, proteinase k, and ferritin.

Nanev tackles the fundamental aspects of nucleation and growth of protein in the successive articles n° 7 and 8. In his first paper [7], he presents a study that establishes the supersaturation dependence of the protein crystal nucleus size of arbitrary lattice structures. His approach is compared to the classical one of Stranski and Kaischew, which is applied merely for the so-called Kossel-crystal and vapor grown crystals. In his second paper [8], Nanev reviews investigations on protein crystallization and aims to present a comprehensive rather than complete account of recent studies and efforts to elucidate the mechanisms of protein crystal nucleation, and the importance of both physical and biochemical factors in these mechanisms.

An interesting feature of protein crystals is that they are usually colorless. However, they can be stained a variety of hues by saturating them with dyes or by co-crystallization. The colors assumed by dyes are a function of chemical factors, particularly pH and redox potential. In paper n° 9, McPherson [9] presents a number of experiments using pH or redox sensitive dye-saturated protein crystals, and some experiments using double dye, sequential redox–pH changes.

In this book, membrane proteins are also represented. Human carbonic anhydrase IX is a multi-domain membrane protein that is, therefore, difficult to express or crystallize. In paper n° 10, Koruza et al. [10] present successful crystallization results of the catalytic domain SV of the human carbonic anhydrase IX by using the microseed matrix screening technique. The crystals were employed as a case-study for neutron protein crystallography.

In paper n° 11, Sun et al. [11] report a new strategy that allows for the removal of cadmium and chromium from wastewater by using fusion crystals of a Cry protein and a low molecular weight cysteine-rich protein (SmtA) known to bind heavy metals. These fusion crystals were microbially grown on *Bacillus thuringiensis* (Bt). The authors suggest the potential uses of these types of crystals for bioremediation of heavy metals.

Park et al. [12] authored paper n° 12. They report crystallization and preliminary X-ray crystallographic data of frog (*Xenopus tropicalis*) ependymin, obtained in a synchrotron facility. Ependymin is a glycoprotein of the extracellular fluid of brain fish and it has been suggested to have various roles related to learning behavior. Ependymin-related proteins also exist in other animals such as sea urchins, frogs, and even mammals. In the same line of crystal structure determination, paper n° 13, authored by Liu et al. [13], presents results of the catalytic domain of the phosphoethanolamine transferase MCR-1 (cMCR-1) co-crystallized with d-Xylose. This study is of great interest to fight drug-resistant enterobacteria. Paper n° 14, authored by Rahman and coworkers, reports the purification, crystallization by hanging drop vapor diffusion, and preliminary X-ray crystallographic studies on plasmid-encoded *Campylobacter concisus*-secreted protein 1 (Csep1p) [14]. This plasmid was recently identified as a putative pathogenicity marker associated with active Crohn's disease, a clinical form of the inflammatory bowel disease. Finally, in paper n° 15 Prudnikova and coworkers report the crystallization and crystal structure determination of the double mutant (Ile44Leu + Gln102His) of the haloalkane dehalogenase DbeA from *Bradyrhizobium elkanii* USDA94 (DbeAΔCl [15]. Haloalkane dehalogenases are a very important class of microbial enzymes for environmental detoxification of halogenated pollutants.

In summary, the articles presented in this Special Issue are representative of some of the lines of a topic as broad as biological crystallization as well as of its importance in different scientific fields, and cover aspects ranging from biomineralization and biomimetic crystallization to crystallization of biological macromolecules and its applications in bioremediation and biomedicine.

Funding: Grant number PGC2018-102047-B-I00 (MCIU/AEI/FEDER, UE).

Acknowledgments: The Guest Editors thank all the authors contributing in this Special Issue and the Editorial staff of *Crystals* for their priceless support.

References

1. Torres-Aravena, Á.E.; Duarte-Nass, C.; Azócar, L.; Mella-Herrera, R.; Rivas, M.; Jeison, D. Can microbially induced calcite precipitation (MICP) through a ureolytic pathway be successfully applied for removing heavy metals from wastewaters? *Crystals* **2018**, *8*, 438. [CrossRef]
2. Opel, J.; Unglaube, N.; Wörner, M.; Kellermeier, M.; Cölfen, H.; García-Ruiz, J.M. Hybrid biomimetic materials from silica/carbonate biomorphs. *Crystals* **2019**, *9*, 157. [CrossRef]
3. Matijaković, N.; Magnabosco, G.; Scarpino, F.; Fermani, S.; Falini, G.; Kralj, D. Synthesis and adsorbing properties of tabular {001} calcite crystals. *Crystals* **2019**, *9*, 16. [CrossRef]
4. He, Z.; Xia, Z.; Zhang, M.; Wu, J.; Wen, W. Calcium carbonate mineralization in a surface-tension-confined droplets array. *Crystals* **2019**, *9*, 284. [CrossRef]
5. Gómez-Morales, J.; González-Ramírez, L.A.; Verdugo-Escamilla, C.; Fernández Penas, R.; Oltolina, F.; Prat, M.; Falini, G. Induced nucleation of biomimetic nanoapatites on exfoliated graphene biomolecule flakes by vapor diffusion in microdroplets. *Crystals* **2019**, *9*, 341. [CrossRef]
6. Blackburn, A.; Partowmah, S.H.; Brennan, H.M.; Mestizo, K.E.; Stivala, C.D.; Petreczky, J.; Perez, A.; Horn, A.; McSweeney, S.; Soares, A.S. A simple technique to improve microcrystals using gel exclusion of nucleation inducing elements. *Crystals* **2018**, *8*, 464. [CrossRef]
7. Nanev, C.N. Recent insights into protein crystal nucleation. *Crystals* **2018**, *8*, 219. [CrossRef]
8. Nanev, C.N. Peculiarities of protein crystal nucleation and growth. *Crystals* **2018**, *8*, 422. [CrossRef]
9. McPherson, A. pH and redox induced color changes in protein crystals suffused with dyes. *Crystals* **2019**, *9*, 126. [CrossRef]
10. Koruza, K.; Lafumat, B.; Nyblom, M.; Knecht, W.; Fisher, Z. From initial hit to crystal optimization with microseeding of human carbonic anhydrase IX—A case study for neutron protein crystallography. *Crystals* **2018**, *8*, 434. [CrossRef]
11. Sun, Q.; Cheng, S.W.; Cheung, K.; Lee, M.M.; Chan, M.K. Cry protein crystal-immobilized metallothioneins for bioremediation of heavy metals from water. *Crystals* **2019**, *9*, 287. [CrossRef]
12. Park, J.K.; Sim, Y.W.; Park, S. Over-expression, secondary structure characterization, and preliminary X-ray crystallographic analysis of xenopus tropicalis ependymin. *Crystals* **2018**, *8*, 284. [CrossRef]
13. Liu, Z.-X.; Han, Z.; Yu, X.-L.; Wen, G.; Zeng, C. Crystal structure of the catalytic domain of MCR-1 (cMCR-1) in complex with d-Xylose. *Crystals* **2018**, *8*, 172. [CrossRef]
14. Rahman, M.M.; Goff, B.; Zhang, L.; Roujeinikova, A. Refolding, characterization, and preliminary X-ray crystallographic studies on the campylobacter concisus plasmid-encoded secreted protein Csep1p associated with Crohn's disease. *Crystals* **2018**, *8*, 391. [CrossRef]
15. Prudnikova, T.; Kascakova, B.; Mesters, J.R.; Grinkevich, P.; Havlickova, P.; Mazur, A.; Shaposhnikova, A.; Chaloupkova, R.; Damborsky, J.; Kuty, M.; et al. Crystallization and crystallographic analysis of a bradyrhizobium elkanii USDA94 haloalkane dehalogenase variant with an eliminated halide-binding site. *Crystals* **2019**, *9*, 375. [CrossRef]

crystals

MDPI

Review

Can Microbially Induced Calcite Precipitation (MICP) through a Ureolytic Pathway Be Successfully Applied for Removing Heavy Metals from Wastewaters?

Álvaro Esteban Torres-Aravena [1,2,*], Carla Duarte-Nass [2], Laura Azócar [3], Rodrigo Mella-Herrera [2], Mariella Rivas [4,5] and David Jeison [6]

1 Departamento de Ingeniería Química, Universidad de La Frontera, Temuco 4780000, Chile
2 Núcleo Científico y Tecnológico de Biorecursos (BIOREN), Universidad de La Frontera, Temuco 4780000, Chile; carla.duarte@ufrontera.cl (C.D.-N.); rodrigo.mella@ufrontera.cl (R.M.-H.)
3 Departamento de Química Ambiental, Facultad de Ciencias, Universidad Católica de la Santísima Concepción, Concepción 4090541, Chile; lazocar@ucsc.cl
4 Centro de Investigación Científico Tecnológico para la Minería CICITEM, Antofagasta 1240000, Chile; mariella.rivas@uantof.cl
5 Laboratorio de Biotecnología Algal y Sustentabilidad, Facultad de Ciencias del Mar y recursos Biológicos, Universidad de Antofagasta, Antofagasta 1240000, Chile
6 Escuela de Ingeniería Bioquímica, Facultad de Ingeniería, Pontificia Universidad Católica de Valparaíso, Valparaíso 2362803, Chile; david.jeison@pucv.cl
* Correspondence: alvaro.torres@ufrontera.cl

Received: 11 October 2018; Accepted: 2 November 2018; Published: 21 November 2018

Abstract: Microbially induced calcite precipitation (MICP) through a ureolytic pathway is a process that promotes calcite precipitation as a result of the urease enzymatic activity of several microorganisms. It has been studied for different technological applications, such as soil bio-consolidation, bio-cementation, CO_2 sequestration, among others. Recently, this process has been proposed as a possible process for removing heavy metals from contaminated soils. However, no research has been reported dealing with the MICP process for heavy metal removal from wastewater/waters. This (re)view proposes to consider to such possibility. The main characteristics of MICP are presented and discussed. The precipitation of heavy metals contained in wastewaters/waters via MICP is exanimated based on process characteristics. Moreover, challenges for its successful implementation are discussed, such as the heavy metal tolerance of inoculum, ammonium release as product of urea hydrolysis, and so on. A semi-continuous operation in two steps (cell growth and bio-precipitation) is proposed. Finally, the wastewater from some typical industries releasing heavy metals are examined, discussing the technical barriers and feasibility.

Keywords: microbially induced calcite precipitation (MICP); heavy metals; wastewater treatment; bioprecipitation; calcium carbonate

1. Heavy Metals and Environmental Problems

The contamination of watercourses by heavy metals is a serious environmental problem that has increased because of rapid industrial development. Indeed, several economic activities, such as metal plating, galvanization, the extraction and processing of minerals, tanning, battery production, paper manufacture, and pesticide synthesis, generate wastewaters that can contain pollutants [1]. Many of these metals are micronutrients, that is, they are essential for cell growth [2]. However, at high concentrations, they may turn toxic or carcinogenic, causing serious health problems. Moreover, when entering the food chain, these can accumulate in the human body [3]. In this sense, special attention has been paid to the most hazardous pollutants, such as zinc, copper, nickel, mercury, cadmium, lead, and chromium [4]. Their adverse effect on natural and human environments demands the development of efficient and cost-effective technologies in order to ensure the removal of these heavy metals from the environment.

2. Conventional Treatment of Wastewaters Containing Heavy Metals

Nowadays, several processes such as flotation, chemical precipitation, adsorption, ion exchange, membrane filtration, coagulation, and electrochemical deposition are available to treat metal-containing water [1,3]. Although these processes can remove metals efficiently (metal removal exceeding 90%) [1], their main constraints are associated with high energy requirements, the use of chemicals, and the production of toxic metal sludge. All of these characteristics contribute to increase overall costs [5,6]. In recent years, biological processes have been proposed and developed for metal removal from waters and wastewaters, for example bio-sorption, bio-accumulation, phytoremediation, bio-coagulation, bio-leaching, and application of sulfate-reducing bacteria (SRB) [6,7]. These biological processes have relevant advantages when compared with their conventional counterparts, such as reduced energy and material consumption, possibilities for metal recycling or recovery, and lower sludge production [8]. These characteristics transform them into potentially more eco-friendly alternatives. However, there are disadvantages mainly related to the final waste disposal, which is often a metal-containing biomass [9]. Moreover, the bacterial immobilization of metals may not constitute a long-term solution, when it is the result of changes in the redox state, as conditions in the environment may re-mobilize them [10].

3. Microbially Induced Calcite Precipitation (MICP) Process

3.1. Precipitation by Ureolytic MICP Process

MICP is a biological process in which calcite ($CaCO_3$) formation is achieved as a result of the active metabolism of bacteria, which generates a favorable micro-environment for precipitation [11,12]. In calcite formation in the MICP through ureolytic pathway, bacteria catalyze urea hydrolysis into carbonate and ammonium. The latter produces an alkalization of the micro-environment, favoring the binding of calcium and carbonate, and furthermore, calcite precipitation [13,14]. Moreover, bacteria also provide nucleation sites in which the calcite precipitation takes place. Four steps can be identified for biological calcite precipitation, as illustrated in Figure 1.

(a) Urea hydrolysis: The urease enzyme hydrolyzes urea into carbamic acid and ammonia (Equation (1)). Furthermore, a spontaneous chemical equilibrium takes place and carbamic acid is converted into carbonic acid and ammonia (Equation (2)) [10,15].

$$CO(NH_2)_2 + H_2O \overset{\text{Urease}}{\rightarrow} NH_2COOH + NH_3 \tag{1}$$

$$NH_2COOH + H_2O \overset{\text{Spontaneous}}{\rightarrow} H_2CO_3 + NH_3 \tag{2}$$

(b) Chemical equilibrium: Ammonia from the urea hydrolysis turns into ammonium, releasing hydroxide ions and increasing the micro-environmental pH, which generates favorable conditions for further precipitation [11] (Equation (3)). Hydroxide ions induce carbonate formation from carbonic acid (Equations (4) and (5)).

$$2NH_3 + 2H_2O \leftrightarrow 2NH_4^+ + 2OH^- \tag{3}$$

$$H_2CO_3 \leftrightarrow HCO_3^- + H^+ \tag{4}$$

$$HCO_3^- + H^+ + 2OH^- \leftrightarrow CO_3^{2-} + 2H_2O \tag{5}$$

(c) Heterogeneous nucleation: Calcium ions are bound to the external cell surface because of the negatively charged functional groups in the cell wall (Equation (6)). Then, calcite formation occurs in the cell surface, once the calcium ion activity is sufficient and the saturation conditions are favorable for $CaCO_3$ precipitation (Equation (7)) [16].

$$Ca^{2+} + Cell \rightarrow Ca^{2+} - Cell \tag{6}$$

$$Ca^{2+} - Cell + CO_3^{2-} \rightarrow Cell - CaCO_3 \tag{7}$$

(d) Successive stratification: Successive calcite layers are developed on the external cell surface (stratification) [17]. The nutrients transfer is limited, and the cells get embedded by calcite crystals, provoking cellular death.

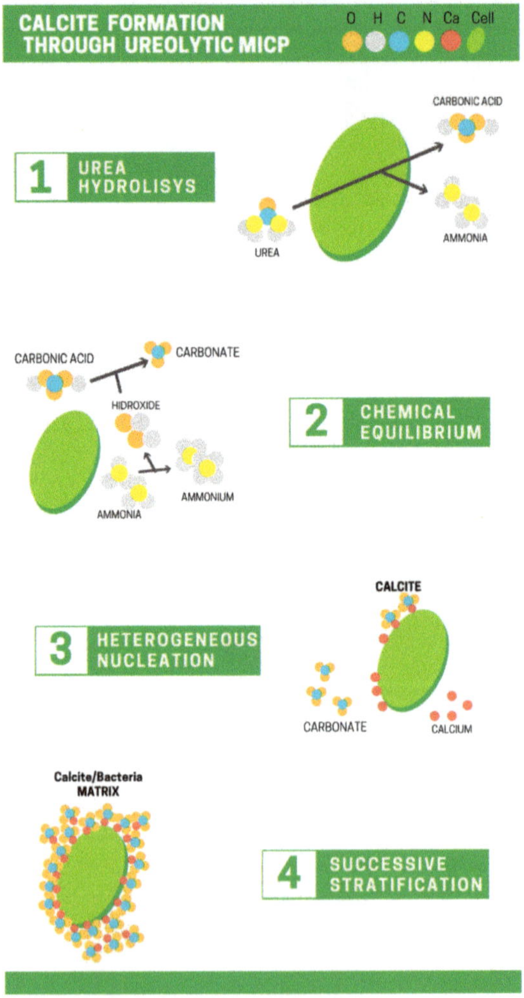

Figure 1. Schematic illustration of calcite production through the microbially induced calcite precipitation (MICP) process: 1—urea hydrolysis; 2—chemical equilibrium; 3—heterogeneous nucleation; 4—successive stratification.

Calcite precipitation through MICP has been regarded as a promising technology for different applications, such as for the improvement of mechanical properties in soils (bio-consolidation) [18,19], bio-cementation [14,20], crack repair in concrete structures [21], CO_2 sequestration [22], bio-composites [23], hydraulic control [10], and so on. For a detailed review about the engineering applications of the MICP process, the authors recommend viewing the following literature [10,19].

3.2. Factors Governing the Ureolytic MICP Process

Several factors govern the ureolytic MICP process, such as the bacteria type, urea and Ca^{2+} concentrations, nucleation sites, pH, and temperature. Several microorganisms presenting high levels of urease activity have been identified, such as *Sporosarcina pasteurii, Bacillus spCR2, Lysinibacillus sphaericus CH5, Bacillus pasteurii NCIM 2477, Kocuria flava CR1, Bacillus megaterium SS3, Bacillus thuringiensis*, and *Halomonas ssp.* [24]. *S. pasteurii* is a non-pathogenic bacterium that is able to tolerate extreme conditions [24], and is probably the most common species used for testing and studying the MICP process [13,25–27].

As already mentioned, the bacteria provide nucleation sites where the precipitation is enhanced. Specifically, the cell surface in the bacteria has negatively charged groups (negative zeta potential) [28], providing binding sites for the Ca^{2+} ions, where carbonate and calcium can react, forming calcite [24,29]. In this sense, the research carried out by Stocks-Fischer et al. (1999) showed the beneficial effect of bacteria as nucleation sites. They observed that the removal of calcium with chemical precipitation (adding carbonate) was a 34–54%, but this value was increased to 98% when MICP bacteria were used.

Another factor is the pH, which determines the acid-based chemical equilibria, and thus defines the presence of carbonate and the precipitation processes. Then, it is considered a key parameter, affecting ureolytic MICP. Moreover, the pH affects the activity of the urease enzyme [26]. It has been reported that the urease activity increases when the pH rises from 6 to 10, but decreases when the pH exceeds 10 [15]. As is the case with all enzymes, the urea catalysis is temperature-dependent. The optimal temperature is in the range of 20 to 37 °C [30,31]. Moreover, the temperature affects the chemical equilibrium and thus the solubility of the $CaCO_3$ in the media.

4. The Ureolytic MICP Process as Treatment for Heavy Metal Removal from Wastewaters

4.1. How Can the Ureolytic MICP Process Remove Heavy Metals?

Heavy metals may be removed through direct precipitation, where metal carbonate is formed, or by co-precipitation, in which metals such Cu^{2+}, Cd^{2+}, Co^{2+}, Ni^{2+}, Zn^{2+}, Pb^{2+}, and Fe^{2+} can be incorporated in the lattice structure of calcite via the substitution of Ca^{2+} [32]. To date, research conducted on ureolyitic MICP for metal removal has been directed toward soil remediation. So far, the reported research has dealt with the isolation of bacteria, presenting important urease activity and metal resistance, the morphological evaluation of metal precipitates, and the removal efficiency of different metals. High removal efficiencies have been reported for different metals (Table 1), when working with species isolated from contaminated soils, as follows: 89–97% for copper, 98–100% for lead, 96–100% for cadmium, ca. 90% for nickel, 93–100% for zinc, and 90–94% for cobalt (Table 1). Unlike the high extent of metal removal mentioned above, the research carried out by Mugwar and Harbottle [28] showed that the metal removal for *S. pasteurii* decreased when the copper and zinc concentrations were increased to 0.5 and 2.0 mM, respectively, but the lead and cadmium removal was not affected by concentration (Table 1). Therefore, from the results summarized in Table 1, we can see that the ureolytic MICP process is more effective for cadmium and lead than for copper, which may be associated with the toxicity of copper. Moreover, the results in Table 1 confirm that the metal tolerance of a particular strain is a key parameter for an efficient ureolytic MICP process. Additionally, it is worth noting that previous research has dealt with the removal of heavy metals from soil by washing soil samples, and then applying MICP to the resulting aqueous solution. Thus, it is expected that the application of the ureolytic MICP to water/wastewater may be feasible from a technical point of view.

Table 1. Summary of assay for heavy metal removal through the ureolytic microbially induced calcite precipitation (MICP) process.

	Bacteria Strain	Medium Conditions				Assay Conditions				Reference
		Medium	Calcium	Urea	Metal	Temperature	Initial pH	Assay Time	Metal Removal	
			mM	mM	mM	°C		d	%	
Copper	Kocuria flava CR1	Nutrient broth	25	333	4.0	30	8	5	97	[33]
	Sporosarcina pasteurii *Terrabacter tumescens* *Sporosarcina sp.* R-31323 (UR31) *Bacillus lentus* (UR41) *Sporosarcina koreensis* (UR47) *Sporosarcina globispora* (UR53)	NH_4–YE medium	0	500	14.88	30	n.r.	2	90.5 [a] 90 [a] 90 [a] 89.5 [a] 93 [a] 89.5 [a]	[27]
	Sporosarcina pasteurii	Oxoid CM0001 nutrient broth/NH_4Cl/sodium bicarbonate	50	333	0.01 0.5 5.0	30	6.5	7	100 30 10	[28]
	Terrabacter tumescens A12	NH_4–YE medium	0	500	14.88	Room T°	n.r.	4	90	[34]
Nickel	*Sporosarcina pasteurii* *Terrabacter tumescens* *Sporosarcina sp.* R-31323 (UR31) *Bacillus lentus* (UR41) *Sporosarcina koreensis* (UR47) *Sporosarcina globispora* (UR53)	NH_4–YE medium	0	500	15.43	30	n.r.	2	90 [a] 90.5 [a] 90 [a] 89.5 [a] 89.5 [a] 89.5 [a]	[27]
	Terrabacter tumescens A12	NH_4–YE medium	0	500	15.43	Room T°	n.r.	4	90	[34]
Lead	*Sporosarcina pasteurii* *Terrabacter tumescens* *Sporosarcina sp.* R-31323 (UR31) *Bacillus lentus* (UR41) *Sporosarcina koreensis* (UR47) *Sporosarcina globispora* (UR53)	NH_4–YE medium	0	500	7.19	30	n.r.	2	100 100 100 100 100 100	[27]
	Enterobacter cloacae KJ46 *Enterobacter cloacae* KJ47	NH_4–YE medium	0	500	0.035 0.028	30	7	2	68.1 54.2	[35]
	Sporosarcina pasteurii	Oxoid CM0001 nutrient broth/NH_4Cl/sodium bicarbonate	50	333	0.05 0.5 5.0	30	6.5	7	100 100 100	[28]
	Pararhodobacter sp.	ZoBell marine broth 2216	500	500	5.0	30	7.6–7.8	0.25	100	[36]
	Penicillium chrysogenum CS1	modified martin broth	40	333	0.48 0.96	27	6.5	12	98.7 98.8	[37]
	Terrabacter tumescens A12	NH_4–YE medium	0	500	7.19	Room T°	n.r.	4	100	[34]

Table 1. *Cont.*

	Bacteria Strain	Medium Conditions				Assay Conditions				Reference
		Medium	Calcium	Urea	Metal	Temperature	Initial pH	Assay Time	Metal Removal	
			mM	mM	mM	°C		d	%	
Cobalt	*Sporosarcina pasteurii* *Terrabacter tumescens* *Sporosarcina sp.* R-31323 (UR31) *Bacillus lentus* (UR41) *Sporosarcina koreensis* (UR47) *Sporosarcina globispora* (UR53)	NH_4–YE medium	0	500	15.4	30	n.r.	2	92 [a] 91.5 [a] 94 [a] 90 [a] 93 [a] 90 [a]	[27]
	Terrabacter tumescens A12	NH_4–YE medium	0	500	15.4	Room T°	n.r.	4	91 [a]	[34]
Zinc	*Sporosarcina pasteurii* *Terrabacter tumescens* *Sporosarcina sp.* R-31323 (UR31) *Bacillus lentus* (UR41) *Sporosarcina koreensis* (UR47) *Sporosarcina globispora* (UR53)	NH_4–YE medium	0	500	14.68	n.r.	n.r.	2	95.5 97 99.5 93 99 96.5	[27]
	Sporosarcina pasteurii	Oxoid CM0001 nutrient broth/NH_4Cl/sodium bicarbonate	50	333	0.1 2.0 10.0	30	6.5	7	100 70 65	[28]
	Terrabacter tumescens A12	NH_4–YE medium	0	500	14.68	Room T°	n.r.	4	97	[34]
Cadmium	*Sporosarcina pasteurii* *Terrabacter tumescens* *Sporosarcina sp.* R-31323 (UR31) *Bacillus lentus* (UR41) *Sporosarcina koreensis* (UR47) *Sporosarcina globispora* (UR53)	NH_4–YE medium	0	500	10.91	30	n.r.	2	99.5 100 100 97.5 100 98	[27]
	Lysinibacillus sphaericus CH-5	Beef extract–peptone broth	0	333	7.32	30	8.3	2	99.95	[38]
	Exiguobacterium undae YR10	Soil mixed with bacterial suspension at a ratio of 2:1 (w/w) in nutrient broth	25	333	39.3	10 25	7.5	14	96.7 97.2	[39]
	Nostoc calcicola	Fed-batch reactor	2.5	-	0.0025	25	8	60	98.8	[40]
	Sporosarcina pasteurii	Oxoid CM0001 nutrient broth/NH_4Cl/sodium bicarbonate	50	333	0.015 0.15 1.5	30	6.5	7	100 100 100	[28]
	Sporosarcina ginsengisoli *Neurospora crassa* *Terrabacter tumescens* A12	Nutrient broth modified AP1 medium NH_4–YE medium	25 n.r. 0	333 40 500	0.05 500 10.91	n.r. 25 Room T°	8 6.09 n.r.	7 n.r. 4	96.3 51.2 100	[41] [42] [34]
Chromium	*Penicillium chrysogenum* CS1	Modified martin broth	40	333	0.96 1.92	27	6.5	12	65.2 39.4	[37]

[a] Values were computed from graphic data shown from an article. n.r. = not reported.

4.2. Heavy Metals Precipitation through Ureolytic MICP Process for Wastewaters

To date, there is incipient research considering the ureolytic MICP for heavy metal removal from water/wastewater. However, researchers have tested this technology in aqueous solutions for removing radionuclides from groundwater [31], for the removal of ions Ca^{2+} and Mg^{2+} ions from wastewater [43], and for phosphate precipitation from anaerobic effluents [44]. Thus, the possibility of removing heavy metals through the ureolytic MICP process provides an interesting and less studied field of research [10,19,45]. Moreover, the ureolytic MICP process can provide an effective way to sequester metals as mineral precipitates for long periods [46]. Formed precipitates can resist an acid pH, reaching acid resistance values as low as 2 [27].

4.3. Key Aspects for the Application of Ureolytic MICP Process for Wastewater Treatment

Subsequently, it seems clear why the ureolytic MICP process may turn into an effective treatment alternative for heavy metal removal from water/wastewaters. However, its full-scale application will require taking several technical issues into consideration, which are discussed below. Potential solutions to overcome these challenges are proposed.

4.3.1. Inoculum (Soil/Wastewater)

Obviously, bacteria presenting relevant urease activity are required as inoculum for MICP. Moreover, inoculum should also show resistance to high concentrations of heavy metals. The isolation of bacteria strains from soil contaminated with the metal target for remediation has been the most common way to face this challenge. Several species have been isolated from contaminated soils in mining areas, showing tolerance to high concentrations of heavy metals, which may be useful when considering the remediation of contaminated sites [27,33,35,47–49]. In this sense, even though *S. pasteurii* is the most used strain for MICP applications, some soil isolated bacteria with lower urease activity have attained higher metal removal, as a result of their metal tolerance [27]. Moreover, the water/wastewater objective of the treatment could be an appropriated source for isolation.

An important factor for a potential operation of the MICP process will be the kind of culture that is used—mono or mixed culture are two possible strategies for operation. When (waste)water treatment is considered, the use of axenic cultures may not be possible, as a result of their scale. Then, the conditions for the proliferation of ureolytic bacteria could be considered. Furthermore, Kang et al. [50] reported that mixed cultures present advantages over mono cultures, such as a major metal tolerance and a higher degree of removal of heavy metals. Moreover, the presence of non-ureolytic bacteria in mono or mixed culture can contribute to MICP; even though no contribution for urea hydrolysis is expected, an increase in the nucleation sites may increase the precipitation rates. This was observed by Gat et al. [51], who cultivated *S. pasteurii* and *N. subtilis* (non-ureolytic bacteria) for $CaCO_3$ precipitation. Thus, culture contamination may not be as inconvenient, as long as a minimum urease activity can be provided by the existing ureolytic bacteria.

4.3.2. Substrates

The ureolytic MICP process requires a source of urea and calcium. The application to wastewater treatment also imposes the condition that the substrate source should have a low cost or no cost at all. Then, the use of wastewater containing urea seems to be the simplest solution. Typical urea containing wastes can come from fertilizer plants, urea synthesis industry, or human activities (sewage) [52,53]. In relation to calcium, this substrate could be obtained from calcium-rich wastewater, such as from citric acid production, landfill leachates, paper recycling, and bone processing [54]. However, the presence of organic matter, recalcitrant compounds, inhibitors of bacterial activity, or other ions may potentially interfere with the bio-precipitation. Then, the characterization of wastewater is certainly relevant.

4.3.3. pH of Metal Containing Wastewaters

The ureolytic MICP process has an optimal pH range between 6 and 10 [15]. Thus, it is expected that bio-precipitation in acidic wastewaters may not be feasible, or may require a pH adjustment. However, it is important to highlight that, as a result of urease activity, the pH will increase, because of the ammonium release. Then, the eventual requirements of the pH adjustment will depend on the wastewater pH and on the alkalinity level. This will be certainly a key factor when considering MICP technology for metal removal from mining wastewaters, as most have a low pH, as is the case with acid mine drainage [55,56]. On the other hand, the treated wastewater will contain an alkaline pH, which have to be corrected in order to avoid any environmental impact in watercourses where it is being discharged. In this sense, an additional study and the evaluation of adequate neutralization treatment for a feasible solution will be necessary.

4.3.4. Bacterial Re-Use after Precipitation/Re-Inoculation

A key challenge arises when the ureolytic MICP process for metal treatment is considered—bacteria provide nucleation sites for precipitation, and once the bio-precipitation takes place, the bacteria are embedded in carbonate crystals. Then, the activity is reduced, as a result of mass transfer limitations [10]. This means that the ureolytic bacteria is not available for future precipitation, as they are trapped in the precipitates, and in order to promote additional precipitation in the process, bacterial re-inoculation will be required. In spite of this, the bacteria embedded in the precipitates would provide a free-biomass effluent, which means that no further treatment for biomass removal will be necessary.

Thus, the continuous operation of a MICP system based on a single reactor, where both cell growth and bio-precipitation occur, may not be feasible. Under this scenario, a MICP process in two steps would be needed. In the first step, the bacteria are grown, and in a second step, bio-precipitation takes place (Figure 2). Thus, the bacteria from the first step are supplied to the second step, where urea, calcium, and wastewater meet, providing the conditions for bio-precipitation.

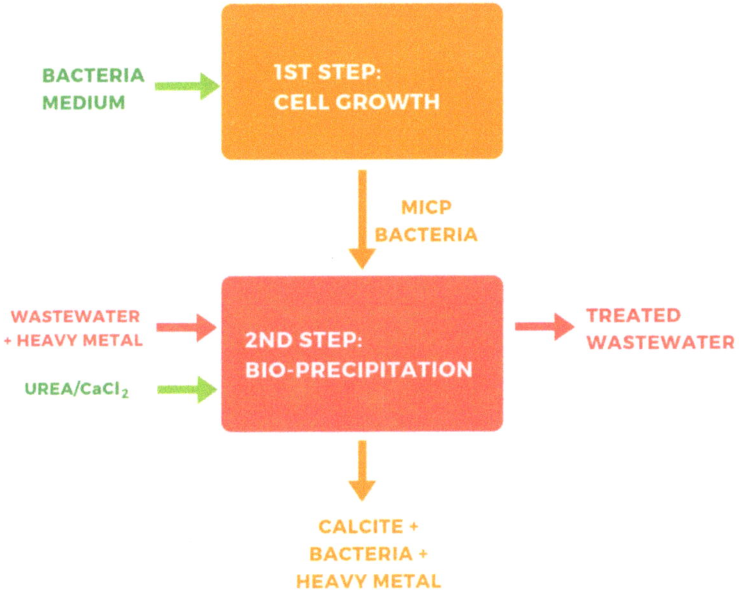

Figure 2. Potential approach for treating heavy metal containing wastewater through the ureolytic MICP process.

4.3.5. Ammonium Release

When ureolytic MICP for heavy metal removal from wastewater is considered, probably one of the greatest challenges to be faced is ammonium release. Two moles of ammonium are released per one mole of hydrolyzed urea (Equations (1) and (2)). This will produce an ammonium-rich effluent, which must be managed. Moreover, if the pH rises over 8–9, ammonia volatilization may become relevant, depending on the gas/liquid mass transfer coefficient. Indeed, Gat et al. [57] reported that ammonia volatilization affected the long-term sustainability of the calcite precipitate, so that ammonia volatilization caused a pH decrease, which showed a 30% calcite dissolution. It is clear that the first step to approaching this issue must be the optimization of urea dosages, in order to minimize the ammonia release. One alternative to managing an ammonia-rich effluent, is to couple MICP to a process for the re-utilization of ammonium for urea synthesis, based on the Basarov reaction described by the authors of [58]. A second alternative is to use one of the processes for ammonia removal and for the conversion normally used in full-scale wastewater treatment, such as the anammox process [59].

5. Wastewaters That May Be Potentially Treated by the Ureolytic MICP Process

Several economic activities release wastewaters containing heavy metals that could be potentially treated through the MICP process. Some of them are the refining and mining of ores, pesticide industries, batteries production, paper industries, tanneries, fertilizer production facilities, solid wastes disposal including sewage sludge, and wastewater irrigation [60].

5.1. Tannery

Tannery and electroplating industries are the major sources of chromium contamination. Hexavalent chromium is genotoxic and carcinogenic, and the reaction of Cr(VI) with ascorbate and hydrogen peroxide results in the accumulation of hydroxyl radicals, thereby causing damage to the DNA [61]. MICP could be an alternative for chromium removal from tannery wastewaters, as calcium carbonate precipitates at pH levels where these effluents are normally present (neutral to alkaline). Moreover, calcium carbonate can be deposited around chromium (VI) compounds [10,62], making stable precipitates for long time periods [63]. Lead is another metal that can be present in tannery wastewater [64]. Because of its valence, Pb (II) could act in a similar way to calcium, and could bind carbonate ions to form stable lead carbonate that precipitates through the MICP process.

5.2. Mining and Metal Refinery

The main environmental impacts of the mining industry include pollution as a result of the discharge of liquid effluents into rivers, the infiltration of soils, the contamination of underground waters, residue disposal, and threats from tailings dams [65]. Processed tailings dam overflow, acid wastewaters, and acid mine drainage (AMD) are some of the most common liquid industrial residues of this sector [56,65].

AMD has pH levels under 4 [55], and they are usually around pH 2 [66]. At such a pH, the urease positive microbes cannot hydrolyze urea, because of the inactivation of this enzyme. Then, a previous step to neutralize the AMD would be necessary before a bio-precipitation treatment. A second option would be to find a urease positive bacterium adapted to these acidic conditions.

On the other hand, mine tailing in the active process has high pH levels [67]. Then, in such situations, the MICP process could be used as a treatment alternative for heavy metal precipitation, leading to recovering those metals or to their disposition in a safe way. The challenge with this wastewater is the high content of heavy metals. It would be necessary to work with bacteria with a high tolerance to those elements.

5.3. Electroplating

In general, wastewaters from electroplating contain a low concentration of organic matter, but are of high toxicity because of heavy metals. Rinse waters are continuously produced and contain relatively low concentrations of contaminants, depending on the electroplating process, such as cadmium, chromium, copper, nickel, lead, and zinc [68]. Bath solutions, in contrast, contain markedly higher concentrations of metals (of the order of hundreds of g/L) and are replaced every several weeks or months, depending on the process [68]. The treatment of wastewater and sludge containing heavy metals is one of the main ecological problems induced by the electroplating industry [69].

In general, solutions with a high content of chromium (VI) are acidic [68,70]. Therefore, the chromium (VI) bio-precipitation from these wastewaters will need a pH increase. However, the other metals, in the pH range, are suitable for the MICP process. On the other hand, bacteria with very high tolerance to heavy metal will be required, because of the high concentration of metal in these wastewaters.

6. Concluding Remarks

The ureolytic MICP is a promising process for the removal of heavy metals. Even though so far it has only been tested for soils, it is expected that it could also represent a potential tool for heavy metal removal from wastewaters. Several relevant economic activities produce wastewater containing heavy metals that could benefit from this process. However, some challenges need to be overcome, such as pH adjustment, ammonium release, and calcium/urea source. Then, future research should be focused on such challenges. Considering that the process requires the production of a biomass presenting urease activity, as well as its subsequent use for bio-precipitation, a potential process configuration is proposed, based on these two steps.

Funding: This publication was funded by CRHIAM center CONICYT FONDAP 1513001.

Acknowledgments: This publication was supported by CONICYT FONDECYT/POSTDOCTORADO 2018 3180648, CONICYT PAI/CONCURSO NACIONAL TESIS DE DOCTORADO EN EL SECTOR PRODUCTIVO, CONVOCATORIA 2017 T7817110005 and by CRHIAM center CONICYT FONDAP 1513001.

Conflicts of Interest: The authors declare no conflict of interest.

References

1. Fu, F.; Wang, Q. Removal of heavy metal ions from wastewaters: A review. *J. Environ. Manag.* **2011**, *92*, 407–418. [CrossRef] [PubMed]
2. Ilhan, S.; Nourbaksh, M.; Kilicarslan, S.; Ozdag, H. Removal of chromium, lead and copper ions from industrial waste waters by *Staphylococcus saprophyticus*. *Turk. Electron. J. Biotechnol.* **2004**, *2*, 50–57.
3. Barakat, M.A. New trends in removing heavy metals from industrial wastewater. *Arab. J. Chem.* **2011**, *4*, 361–377. [CrossRef]
4. Zhao, M.; Xu, Y.; Zhang, C.; Rong, H.; Zeng, G. New trends in removing heavy metals from wastewater. *Appl. Microbiol. Biotechnol.* **2016**, *100*, 6509–6518. [CrossRef] [PubMed]
5. Arbabi, M.; Hemati, S.; Amiri, M. Removal of lead ions from industrial wastewater: A review of Removal methods. *Int. J. Epidemiol. Res.* **2015**, *2*, 105–109.
6. Ahluwalia, S.S.; Goyal, D. Microbial and plant derived biomass for removal of heavy metals from wastewater. *Bioresour. Technol.* **2007**, *98*, 2243–2257. [CrossRef] [PubMed]
7. Nancharaiah, Y.V.; Mohan, S.V.; Lens, P.N.L. Biological and Bioelectrochemical Recovery of Critical and Scarce Metals. *Trends Biotechnol.* **2016**, *34*, 137–155. [CrossRef] [PubMed]
8. Tsezos, M.; Hatzikioseyian, A.; Remoudaki, E. Biofilm Reactors in Mining and Metallurgical Effluent Treatment: Biosorption, Bioprecipitation, Bioreduction Processes. 2012. Available online: http://www.metal.ntua.gr/uploads (accessed on 2 September 2018).
9. Lesmana, S.O.; Febriana, N.; Soetaredjo, F.E.; Sunarso, J.; Ismadji, S. Studies on potential applications of biomass for the separation of heavy metals from water and wastewater. *Biochem. Eng. J.* **2009**, *44*, 19–41. [CrossRef]

10. Phillips, A.J.; Gerlach, R.; Lauchnor, E.; Mitchell, A.C.; Cunningham, A.B.; Spangler, L. Engineered applications of ureolytic biomineralization: A review. *Biofouling* **2013**, *29*, 715–733. [CrossRef] [PubMed]
11. Hammes, F.; Verstraete, W. Key roles of pH and calcium metabolism in microbial carbonate precipitation. *Rev. Environ. Sci. Biotechnol.* **2002**, *1*, 3–7. [CrossRef]
12. Seifan, M.; Samani, A.K.; Berenjian, A. New insights into the role of pH and aeration in the bacterial production of calcium carbonate ($CaCO_3$). *Appl. Microbiol. Biotechnol.* **2017**, *101*, 3131–3142. [CrossRef] [PubMed]
13. Whiffin, V.S.; van Paassen, L.A.; Harkes, M.P. Microbial Carbonate Precipitation as a Soil Improvement Technique. *Geomicrobiol. J.* **2007**, *24*, 417–423. [CrossRef]
14. Seifan, M.; Samani, A.K.; Berenjian, A. Bioconcrete: Next generation of self-healing concrete. *Appl. Microbiol. Biotechnol.* **2016**, *100*, 2591–2602. [CrossRef] [PubMed]
15. Stocks-Fischer, S.; Galinat, J.K.; Bang, S.S. Microbiological precipitation of $CaCO_3$. *Soil Biol. Biochem.* **1999**, *31*, 1563–1571. [CrossRef]
16. Kumari, D.; Qian, X.-Y.; Pan, X.; Achal, V.; Li, Q.; Gadd, G.M. Microbially-induced Carbonate Precipitation for Immobilization of Toxic Metals. *Adv. Appl. Microbiol.* **2016**, *94*, 79–108. [CrossRef] [PubMed]
17. Castanier, S.; Le Métayer-Levrel, G.; Perthuisot, J.-P. Ca-carbonates precipitation and limestone genesis–The microbiogeologist point of view. *Sediment. Geol.* **1999**, *126*, 9–23. [CrossRef]
18. DeJong, J.T.; Mortensen, B.M.; Martinez, B.C.; Nelson, D.C. Bio-mediated soil improvement. *Ecol. Eng.* **2010**, *36*, 197–210. [CrossRef]
19. Dhami, N.K.; Reddy, M.S.; Mukherjee, M.S. Biomineralization of calcium carbonates and their engineered applications: A review. *Front. Microbiol.* **2013**, *4*, 314. [CrossRef] [PubMed]
20. Ramachandran, S.K.; Ramakrishnan, V.; Bang, S.S. Remediation of concrete using micro-organisms. *ACI Mater. J.* **2001**, *98*, 3–9.
21. Achal, V.; Mukherjee, A. A review of microbial precipitation for sustainable construction. *Constr. Build. Mater.* **2015**. [CrossRef]
22. Mitchell, A.C.; Dideriksen, K.; Spangler, L.H.; Cunningham, A.B.; Gerlach, R. Microbially enhanced carbon capture and storage by mineral-trapping and solubility-trapping. *Environ. Sci. Technol.* **2010**, *44*, 5270–5276. [CrossRef] [PubMed]
23. Declet, A.; Reyes, E.; Suárez, O.M. Calcium Carbonate Precipitation: A Review of the Carbonate Crystallization Process and Applications in Bioinspired Composites. *Rev. Adv. Mater. Sci.* **2016**, *44*, 87–107.
24. Anbu, P.; Kang, C.-H.; Shin, Y.-J.; So, J.-S. Formations of calcium carbonate minerals by bacteria and its multiple applications. *Springerplus* **2016**, *5*, 250. [CrossRef] [PubMed]
25. Bhaduri, S.; Debnath, N.; Mitra, S.; Liu, Y.; Kumar, A. Microbiologically Induced Calcite Precipitation Mediated by *Sporosarcina pasteurii*. *J. Vis. Exp.* **2016**, e53253. [CrossRef] [PubMed]
26. Gorospe, C.M.; Han, S.H.; Kim, S.G.; Park, J.Y.; Kang, C.H.; Jeong, J.H.; So, J.S. Effects of different calcium salts on calcium carbonate crystal formation by Sporosarcina pasteurii KCTC 3558. *Biotechnol. Bioprocess Eng.* **2013**, *18*, 903–908. [CrossRef]
27. Li, M.; Cheng, X.; Guo, H. Heavy metal removal by biomineralization of urease producing bacteria isolated from soil. *Int. Biodeterior. Biodegrad.* **2013**, *76*, 81–85. [CrossRef]
28. Mugwar, A.J.; Harbottle, M.J. Toxicity effects on metal sequestration by microbially-induced carbonate precipitation. *J. Hazard. Mater.* **2016**, *314*, 237–248. [CrossRef] [PubMed]
29. De Muynck, W.; De Belie, N.; Verstraete, W. Microbial carbonate precipitation in construction materials: A review. *Ecol. Eng.* **2010**, *36*, 118–136. [CrossRef]
30. Okwadha, G.D.O.; Li, J. Optimum conditions for microbial carbonate precipitation. *Chemosphere* **2010**, *81*, 1143–1148. [CrossRef] [PubMed]
31. Mitchell, A.C.; Ferris, F.G. The coprecipitation of Sr into calcite precipitates induced by bacterial ureolysis in artificial groundwater: Temperature and kinetic dependence. *Geochim. Cosmochim. Acta* **2005**, *69*, 4199–4210. [CrossRef]
32. Krajewska, B. Urease-aided calcium carbonate mineralization for engineering applications: A review. *J. Adv. Res.* **2017**. [CrossRef] [PubMed]
33. Achal, V.; Pan, X.; Zhang, D. Remediation of copper-contaminated soil by Kocuria flava CR1, based on microbially induced calcite precipitation. *Ecol. Eng.* **2011**, *37*, 1601–1605. [CrossRef]

34. Li, M.; Fu, Q.L.; Zhang, Q.; Achal, V.; Kawasaki, S. Bio-grout based on microbially induced sand solidification by means of asparaginase activity. *Sci. Rep.* **2015**, *5*, 16128. [CrossRef] [PubMed]
35. Kang, C.-H.; Oh, S.J.; Shin, Y.; Han, S.-H.; Nam, I.-H.; So, J.-S. Bioremediation of lead by ureolytic bacteria isolated from soil at abandoned metal mines in South Korea. *Ecol. Eng.* **2015**, *74*, 402–407. [CrossRef]
36. Mwandira, W.; Nakashima, K.; Kawasaki, S. Bioremediation of lead-contaminated mine waste by *Pararhodobacter* sp. based on the microbially induced calcium carbonate precipitation technique and its effects on strength of coarse and fine grained sand. *Ecol. Eng.* **2017**, *109*, 57–64. [CrossRef]
37. Qian, X.; Fang, C.; Huang, M.; Achal, V. Characterization of fungal-mediated carbonate precipitation in the biomineralization of chromate and lead from an aqueous solution and soil. *J. Clean. Prod.* **2017**, *164*, 198–208. [CrossRef]
38. Kang, C.H.; Han, S.H.; Shin, Y.; Oh, S.J.; So, J.S. Bioremediation of Cd by microbially induced calcite precipitation. *Appl. Biochem. Biotechnol.* **2014**, *172*, 2907–2915. [CrossRef] [PubMed]
39. Kumari, D.; Pan, X.; Lee, D.J.; Achal, V. Immobilization of cadmium in soil by microbially induced carbonate precipitation with Exiguobacterium undae at low temperature. *Int. Biodeterior. Biodegrad.* **2014**, *94*, 98–102. [CrossRef]
40. Zhao, C.; Fu, Q.; Song, W.; Zhang, D.; Ahati, J.; Pan, X.; Al-Misned, F.A.; Mortuza, M.G. Calcifying cyanobacterium (*Nostoc calcicola*) reactor as a promising way to remove cadmium from water. *Ecol. Eng.* **2015**, *81*, 107–114. [CrossRef]
41. Achal, V.; Pan, X.; Fu, Q.; Zhang, D. Biomineralization based remediation of As(III) contaminated soil by Sporosarcina ginsengisoli. *J. Hazard. Mater.* **2012**, *201–202*, 178–184. [CrossRef] [PubMed]
42. Li, Q.; Csetenyi, L.; Gadd, G.M. Biomineralization of metal carbonates by Neurospora crassa. *Environ. Sci. Technol.* **2014**, *48*, 14409–14416. [CrossRef] [PubMed]
43. Hammes, F.; Seka, A.; Van Hege, K.; Van De Wiele, T.; Vanderdeelen, J.; Siciliano, S.D.; Verstraete, W. Calcium removal from industrial wastewater by bio-catalytic $CaCO_3$ precipitation. *J. Chem. Technol. Biotechnol.* **2003**, *78*, 670–677. [CrossRef]
44. Carballa, M.; Moerman, W.; De Windt, W.; Grootaerd, H.; Verstraete, W. Strategies to optimize phosphate removal from industrial anaerobic effluents by magnesium ammonium phosphate (MAP) production. *J. Chem. Technol. Biotechnol.* **2009**, *84*, 63–68. [CrossRef]
45. Arias, D.; Cisternas, L.; Rivas, M. Biomineralization Mediated by Ureolytic Bacteria Applied to Water Treatment: A Review. *Crystals* **2017**, *7*, 345. [CrossRef]
46. Fujita, Y.; Ferris, F.; Lawson, R.; Colwell, F.; Smith, R. Calcium Carbonate Precipitation by Ureolytic Subsurface Bacteria. *Geomicrobiol. J.* **2000**, *17*, 305–318. [CrossRef]
47. Zhao, Y.; Yao, J.; Yuan, Z.; Wang, T.; Zhang, Y.; Wang, F. Bioremediation of Cd by strain GZ-22 isolated from mine soil based on biosorption and microbially induced carbonate precipitation. *Environ. Sci. Pollut. Res.* **2017**, *24*. [CrossRef] [PubMed]
48. Yang, J.; Pan, X.; Zhao, C.; Mou, S.; Achal, V.; Al-Misned, F.A.; Mortuza, M.G.; Gadd, G.M. Bioimmobilization of Heavy Metals in Acidic Copper Mine Tailings Soil. *Geomicrobiol. J.* **2016**, *33*, 261–266. [CrossRef]
49. Arias, D.; Valdes, P.; Cisternas, L.A.; Rivas, M. Isolation and Selection of Halophilic Ureolytic Bacteria for Biocementation of Calcium and Magnesium from Seawater. *Adv. Mater. Res.* **2015**, *1130*, 489–492. [CrossRef]
50. Kang, C.; Kwon, Y.; So, J. Bioremediation of heavy metals by using bacterial mixtures. *Ecol. Eng.* **2016**, *89*, 64–69. [CrossRef]
51. Gat, D.; Tsesarsky, M.; Shamir, D.; Ronen, Z. Accelerated microbial-induced $CaCO_3$ precipitation in a defined coculture of ureolytic and non-ureolytic bacteria. *Biogeosciences* **2014**, *11*, 2561–2569. [CrossRef]
52. Matijašević, L.; Dejanović, I.; Lisac, H. Treatment of wastewater generated by urea production. *Resour. Conserv. Recycl.* **2010**, *54*, 149–154. [CrossRef]
53. Barmaki, M.M.; Rahimpour, M.R.; Jahanmiri, A. Treatment of wastewater polluted with urea by counter-current thermal hydrolysis in an industrial urea plant. *Sep. Purif. Technol.* **2009**, *66*, 492–503. [CrossRef]
54. Van Langerak, E.P.A.; Hamelers, H.V.M.; Lettinga, G. Influent calcium removal by crystallization reusing anaerobic effluent alkalinity. *Water Sci. Technol.* **1997**, *36*, 341–348. [CrossRef]
55. Akcil, A.; Koldas, S. Acid Mine Drainage (AMD): Causes, treatment and case studies. *J. Clean. Prod.* **2006**, *14*, 1139–1145. [CrossRef]

56. Gaikwad, R.W.; Gupta, D.V. Review on Removal of Heavy Metals From Acid Mine Drainage. *Appl. Ecol. Environ. Res.* **2008**, *6*, 81–98. [CrossRef]
57. Gat, D.; Ronen, Z.; Tsesarsky, M. Long-term sustainability of microbial-induced $CaCO_3$ precipitation in aqueous media. *Chemosphere* **2017**, *184*, 524–531. [CrossRef] [PubMed]
58. Meessen, J.; Wo, F. Urea synthesis. **2014**, 2180–2189. [CrossRef]
59. Mao, N.; Ren, H.; Geng, J.; Ding, L.; Xu, K. Engineering application of anaerobic ammonium oxidation process in wastewater treatment. *World J. Microbiol. Biotechnol.* **2017**, *33*, 153. [CrossRef] [PubMed]
60. Khalid, S.; Shahid, M.; Niazi, N.K.; Murtaza, B.; Bibi, I.; Dumat, C. A comparison of technologies for remediation of heavy metal contaminated soils. *J. Geochem. Explor.* **2016**, *182*, 247–268. [CrossRef]
61. Kalidhasan, S.; Kumar, A.S.K.; Rajesh, V.; Rajesh, N. The journey traversed in the remediation of hexavalent chromium and the road ahead toward greener alternatives—A perspective. *Coord. Chem. Rev.* **2016**, *317*, 157–166. [CrossRef]
62. Achal, V.; Pan, X.; Lee, D.J.; Kumari, D.; Zhang, D. Remediation of Cr(VI) from chromium slag by biocementation. *Chemosphere* **2013**, *93*. [CrossRef] [PubMed]
63. Ferris, F.G.; Phoenix, V.; Fujita, Y.; Smith, R.W. Kinetics of calcite precipitation induced by ureolytic bacteria at 10 to 20 °C in artificial groundwater. *Geochim. Cosmochim. Acta* **2004**, *68*, 1701–1722. [CrossRef]
64. Rehman, A.; Shakoori, F.R.; Shakoori, A.R. Heavy metal resistant freshwater ciliate, Euplotes mutabilis, isolated from industrial effluents has potential to decontaminate wastewater of toxic metals. *Bioresour. Technol.* **2008**, *99*, 3890–3895. [CrossRef] [PubMed]
65. Castro, S.H.; Sa, M. Environmental viewpoint on small-scale copper, gold and silver mining in Chile. *J. Clean. Prod.* **2003**, *11*, 207–213. [CrossRef]
66. Obreque-Contreras, J.; Pérez-Flores, D.; Gutiérrez, P.; Chávez-Crooker, P. Acid Mine Drainage in Chile: An Opportunity to Apply Bioremediation Technology. *J. Forensic Res.* **2015**, *6*, 1–8. [CrossRef]
67. Wang, L.; Ji, B.; Hu, Y.; Liu, R.; Sun, W. A review on in situ phytoremediation of mine tailings. *Chemosphere* **2017**, *184*, 594–600. [CrossRef] [PubMed]
68. Martín-Lara, M.A.; Blázquez, G.; Trujillo, M.C.; Pérez, A.; Calero, M. New treatment of real electroplating wastewater containing heavy metal ions by adsorption onto olive stone. *J. Clean. Prod.* **2014**, *81*, 120–129. [CrossRef]
69. Panayotova, T.; Dimova-Todorova, M.; Dobrevsky, I. Purification and reuse of heavy metals containing wastewaters from electroplating plants. *Desalination* **2007**, *206*, 135–140. [CrossRef]
70. Kurniawan, T.A.; Chan, G.Y.S.; Lo, W.H.; Babel, S. Physico-chemical treatment techniques for wastewater laden with heavy metals. *Chem. Eng. J.* **2006**, *118*, 83–98. [CrossRef]

crystals

MDPI

Communication

Hybrid Biomimetic Materials from Silica/Carbonate Biomorphs

Julian Opel [1,2], Niklas Unglaube [1], Melissa Wörner [1], Matthias Kellermeier [3], Helmut Cölfen [1,*] and Juan-Manuel García-Ruiz [2,*]

1 Physical Chemistry, University of Konstanz, Universitätsstrasse 10, D-78457 Konstanz, Germany; julian.opel@uni-konstanz.de (J.O.); niklas.unglaube@uni.kn (N.U.); melissa.woerner@uni.kn (M.W.)
2 Laboratorio de Estudios Cristalográficos, Instituto Andaluz de Ciencias de la Tierra (CSIC-UGR), Avenida de las Palmeras 4, E-18100 Armilla, Granada, Spain
3 Material Physics, BASF SE, RAA/OS–B007, Carl-Bosch-Strasse 38, D-67056 Ludwigshafen, Germany; matthias.kellermeier@basf.com
* Correspondence: helmut.coelfen@uni-konstanz.de (H.C.); jmgruiz@ugr.es (J.-M.G.-R.); Tel.: +49-7531-884063 (H.C.); +34-669434700 (J.-M.G.-R.)

Received: 28 February 2019; Accepted: 14 March 2019; Published: 18 March 2019

Abstract: The formation of a polymer protection layer around fragile mineral architectures ensures that structures stay intact even after treatments that would normally destroy them going along with a total loss of textural information. Here we present a strategy to preserve the shape of silica-carbonate biomorphs with polymers. This method converts non-hybrid inorganic-inorganic composite materials such a silica/carbonate biomorphs into hybrid organic/carbonate composite materials similar to biominerals.

Keywords: biomorphs; barium carbonate; silica; PCDA; pyrrole

1. Introduction

Silica-earth-alkaline carbonate composites show exceptional shapes which is so far a one-of-a-kind appearance within the field of pure inorganic composites [1,2]. They were named silica-biomorphs due to their morphology, which resembles primitive living organisms and their inner textures mimic biominerals [3]. Compared to biominerals, which are normally hierarchically ordered hybrid composites consisting of inorganic minerals and structure conducting organic matter [4–6], biomorphs are purely inorganic composite materials, the structuring role of organic compounds being taken over by amorphous silica. They are self-organized structures that forms upon the coupled co-precipitation of silica (SiO_2) and alkaline -earth metal carbonates, namely witherite ($BaCO_3$), strontianite ($SrCO_3$) or either aragonite or monohydrocalcite ($CaCO_3$) [7–10]. The formation of biomorphs can be described in three stages which are related to pH [11]. In the first stage, the initial single crystal of alkaline -earth metal carbonate experiences splitting provoked by selective adsorption of silica. Iterative splitting triggers fractal growth and eventually leads to primary globular particles [12–14]. The precipitation of the carbonate induces a local and bulk decrease of the pH that can be monitored even at this very early stage [11,15,16]. At this stage nearly no silica is adsorbed within the structure [11]. The second stage starts after some time (depending on the initial pH) and more complex structures form by determining the final shape of the biomorph. During this stage a polycrystalline growth of a myriad of elongated witherite nanorods with a typical size of 200–400 nm in length and 30–50 nm in thickness can be observed [11]. Once the pH drops far enough, the biomorph formation enters the last stage where only secondary precipitation processes occur. Due to the inverse solubility of silica with respect to pH [17,18], the structures become embedded in a thick silica shell, which grows bigger if the structures mature in the mother solution.

Once extracted from the mother solution, the biomorphs can be further treated as shown in Figure 1. The composite is hollowed by acidic treatment. A diluted hydrochloric or acetic acid can be used to dissolve the inner part within minutes [19]. The result is a hollow structure, which is called a biomorph "ghost." Alternatively, an alkaline treatment with sodium hydroxide solution allows the removal of the silica shell and excavates the so-called "naked" biomorph (*cf.* Figure 1) [20]. A naked biomorph is useful for attachment of molecules or particles with carboxylate groups. One example for an attachment of a monomeric carboxylate species is 10,12-pentacosadiynoic acid (PCDA), a light-polymerizable diacetylen [20]. Poly-PCDA (pPCDA), as a member of the polydiacetylene family, comes with thermochromic properties and can be reversibly or irreversibly switched from a blue into a red state [21,22]. Besides PCDA, conductive polymers like polypyrrole (pPy) or poly thiophenes generated a growing interest in the field of biomineral preservation and replication [23–25]. Furthermore they provide some additional functionality and are used as chemical sensors [26,27], in drug delivery [27], as electronic devices like fuel cells [28] or electro-catalysts [29] and in combination with silica for chromatographic applications [30]. Choi et al. presented hierarchically structured pPy in helicoidal shapes and consequently, a pPy replica of a biomorph helix should also be usable as stretchable supercapacitors [31].

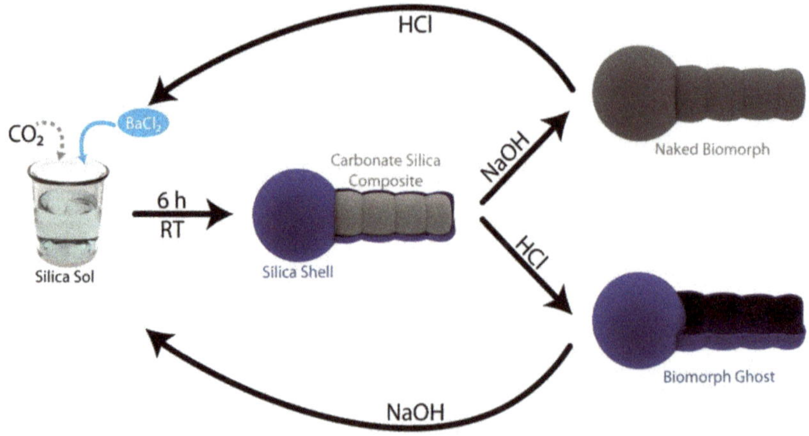

Figure 1. Schematic illustration of biomorph formation at room temperature (RT) and pathways to dissolve each part of composite selectively and return to its precursor state. As intermediate a naked biomorph with an excavated carbonate surface or a hollow biomorph ghost can be obtained.

So far, two known strategies exist to transfer hierarchical structures of a given material into a new one of different composition. The first method is to demineralize a biomineral to obtain an organic template and re-infiltrate the matrix with the new compound [32]. The second strategy consists of replacing the organic matrix with a new material, as shown by Imai and co-workers with dyes or polymers between nanocrystals [33,34]. We show here that a polymer layer around a biomorphic structure sustains the morphology after additional treatment with alkaline and/or acidic solutions while an unpreserved structure vanishes into its precursor state. In total, the following work demonstrates a successful preservation of micro-sculptures with two chosen straightforward pathways and elaborate a third strategy to transfer hierarchical structures into functional organic materials.

2. Materials and Methods

Barium chloride dihydrate (>99%), sodium hydroxide (reagent grade, >98%), sodium silicate solution (commercial water glass, containing 10.6% Na_2O and 26.5% SiO_2, reagent grade, density 1.39 g/ml), copper chloride (97%), 10,12-pentacosadiynoic acid (>97%) and pyrrole (reagent grade, 98%) were purchased from Sigma-Aldrich and used without further purification. Aqueous solutions were prepared using MilliQ water with a conductivity of 18 μS/cm.

2.1. Biomorph Formation

Silica-biomorphs were grown from alkaline silica sol which was prepared by 1.39 g of silica solution diluted with 349 mL of water. The pH of the solution was adjusted to 11.3 with aliquots of 0.1 M NaOH solution. The crystallization occurs in a 6-wells linbro plate by mixing 4 mL of the silica solution with 4 mL of a 10 mM $BaCl_2$ solution. After 8-16 h the structures were extracted from the wells and washed several times with water and ethanol (EtOH). The outer silica layer was removed by incubating 20 mg biomorph powder in 1 mL of a 1 M NaOH solution for 6 h.

2.2. pPCDA Functionalization

PCDA solution was prepared by dissolving 50 mg PCDA in 10 mL tetrahydrofuran (THF). The solution was filtered through a syringe filter (pore size: 0.22 μm) and diluted with 9 mL water and 1 mL 0.1 M NaOH solution. 10 mg of naked biomorphs were incubated for 8 h inside a PCDA solution with different concentrations and stored in the dark. After extracting and washing, the structures can be stored under UV-light for 5 min inducing the polymerization to pPCDA. The colour of the biomorph PCDA powder changes to blue.

2.3. pPy Functionalization

10 mg naked biomorphs were added to 1 mL of a 0.01 M $CuCl_2$ solution in isopropanol (iPrOH). After 1 h, additional 750 μl were added. After 1 h the structures were centrifuged and washed twice with 200 μl iso-propanol (iPrOH). After drying the structures were stored in a desiccator containing 10 mL pyrrole. After 2 h the structures begin to darken and turned completely black after 6 h. After the polymerization the inner core of the structures was removed with 0.05 M acetic acid. The residue was washed several times with water and dried under reduced pressure.

2.4. Analytical Methods

The pH of the solutions was measured with a pH meter (Eutech pH 510, Eutech Instruments, Singapore). Scanning electron microscopy (SEM) images were recorded on a Hitachi Tabletop SEM TM3000 (Hitachi-Hightech, Krefeld, Germany) with a backscatter detector and a Zeiss Crossbeam 1540XB (Zeiss, Oberkochen, Germany) with a secondary electron (SE2) and an inlens detector. Energy dispersive X-ray spectroscopy (EDX) was also performed on the Hitachi TM3000. Light microscopy was performed on a Zeiss Imager m2m (Zeiss, Jena, Germany) and Axio Zoom (Zeiss, Jena, Germany). Confocal laser scanning microscopy (cLSM) was performed on a Zeiss LSM700 (Zeiss, Jena, Germany). Attenuated total reflection Fourier transformed infrared (ATR-FTIR) spectroscopy was performed on a Perkin Elmer Spectrum 100 (PerkinElmer, Waltham, MA, USA).

3. Results and Discussion

3.1. *Preservation with pPCDA*

The first preservation route focuses on a light-polymerizable surfactant named PCDA. This molecule is immobilized on the witherite surface by incubating naked biomorphs (without an outer silica shell) in a mixture of sodium hydroxide, THF and water with different PCDA concentrations (120–12,000 ppm). The attachment process is schematically shown in Figure 2A. To make sure that only the monomer is attached on the surface the solution was pushed through a syringe filter to remove polymer particles from the solution. The naked biomorphs were immersed in the colourless solution for several hours. To ensure that the PCDA does not polymerize during the attachment process, the samples were stored in darkness. After the chosen incubation time (4–16 h) the solution was removed and the modified biomorphs were dried under reduced pressure. The obtained modified structures were investigated with scanning electron microscopy (SEM). Observation of the structures with the electron beam induces polymerization of the attached monomers to the bluish pPCDA (*cf.* inlet photos of the SEM stub before and after the SEM investigation in Figure 2B). To demonstrate the attachment of the PCDA on the structure, the process was observed by FTIR-spectroscopy. The transformation of the silica-biomorphs to the pPCDA-biomorphs is shown in Figure 2C. The black spectrum shows the characteristic carbonate (witherite) vibrations as well as the most abundant signal for amorphous silica (peak at 1000 cm^{-1}). After treatment with NaOH and the transformation to the naked biomorphs, the broad peak vanishes and only the witherite signals remain (red spectrum). The modification with PCDA is observable as lines at 2870 and 2965 cm^{-1}, which are the symmetric and asymmetric C-H vibrations. The carboxyl signal of the PCDA appears at 1540 cm^{-1} which is in good accordance to the literature [22]. Choosing a higher concentration of PCDA in the functionalization solution leads to the full coverage of the biomorph worm with pPCDA flakes, which preserve and display the outline of the structure beneath quite well. A detailed replica of the topography is not achievable with these high monomer concentrations. In previous work, we have demonstrated that the shape of the biomorph structure was preserved in acidic solutions even after removing the outer silica layer [20]. To ensure that a full coverage was achieved confocal laser scanning microscopy (cLSM) was used due to the strong fluorescence of pPCDA [20,35] (*cf.* Figure 2D,E). Smoothening the reprint might be possible by reducing the amount of monomer but a full coverage is not sustainable because an enrichment of PCDA was found inside of the notches of the biomorph worm (*cf.* Figure 2F,G). In Figure 2G at higher magnification the witherite nanorods and pPCDA flakes (false coloured in blue) can be visualized alongside each other. Excavated carbonate rods are not able to resist acidic solutions and therefore, the structures vanish completely after acid treatment. Also, the pPCDA flakes do not stick together and a polymer replica of the structures does not remain. The pPCDA coverage is useful to passivate bigger carbonate structures but a smooth and detailed replica cannot be obtained. Nevertheless, it still remains a readily applicable strategy to produce inverse biominerals.

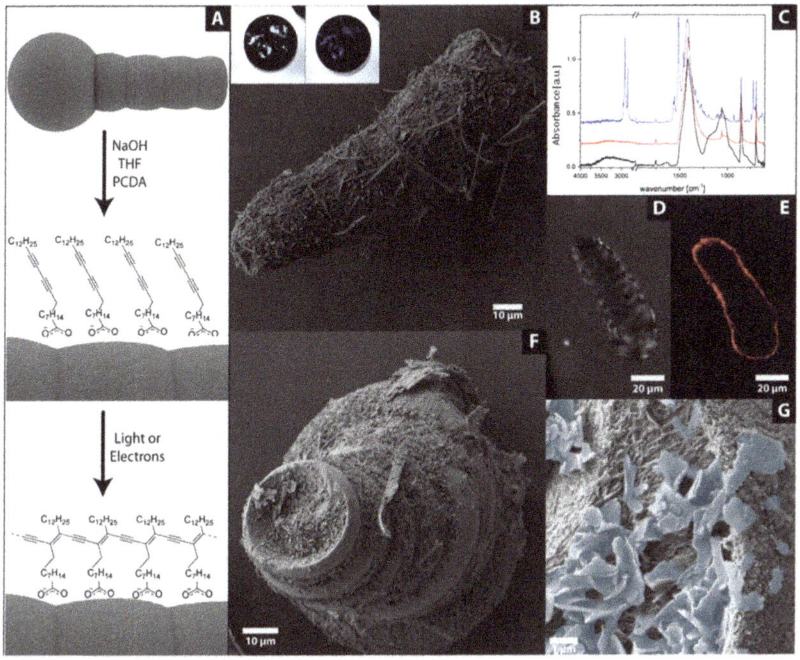

Figure 2. (**A**) Scheme of biomorph coverage with 10,12-pentacosadiynoic acid (PCDA) and the polymerization to pPCDA. (**B**) Scanning electron microscope (SEM) image of a fully covered biomorph worm (inlet: colour transformation of the powder after SEM investigation; c(PCDA) = 12,000 ppm). (**C**) Attenuated total reflection Fourier transformed infrared (ATR-FTIR) spectra of an untreated biomorph (black) a naked biomorph (red) and a pPCDA covered biomorph (blue). (D/E) Confocal Laser scanning microscopy (cLSM) images of a fully covered biomorph worm in the transmission (**D**) and fluorescence (**E**) channel. (**F**) SEM image of a partially covered biomorph worm (c(PCDA) = 2000 ppm) with the corresponding image at higher magnification (**G**), the blue coloured flakes indicate the pPCDA).

3.2. Preservation with pPy

Polypyrrole has gained attention due to its ability to produce polymer replicas of biominerals like sea urchin spines [25]. Recent breakthroughs in the preparation of polymer replicas on the micro scale are a huge improvement to preserve fragile structures and it now seems transferable to silica-biomorphs [23]. To obtain a full pPy coating around a biomorph we followed the route presented in Figure 3A. At first $CuCl_2$ is brought onto the structure as a catalyst. Due to the rough surface of the naked biomorph, many small crystals of $CuCl_2$ attach on the surface. These small crystals do not affect the shape as can be seen in Figure 3B. After drying, the structures were transferred into a desiccator. On the bottom of the desiccator, pyrrole was deposited creating a pyrrole saturated atmosphere. The pyrrole molecules diffuse to the catalyst and begin to polymerize. During this process, a colour change of the biomorphs can be observed by the naked eye. After 3 h the white powder turns grey and after 6 h a black powder was obtained. To ensure that the polymer layer is neither growing too big nor crosslinks the structures, the incubation time was kept at 6 h. After extracting the structures from the desiccator, the shapes of the naked biomorphs used as starting material remained nearly unchanged. Elemental analysis via energy dispersive X-ray spectroscopy (EDX) showed Ba, Cu, Cl, C, O, N and traces of Si. Except the higher amount of N and the signals from the catalyst, the spectrum looks identical to an EDX of the naked biomorphs. The most meaningful proof of the full coverage and subsequent preservation of the biomorph form is the dissolution of the inner core which was achieved with 0.05 M acetic acid. The core dissolves much slower compared to normal silica-biomorphs and as a result hollow structures were obtained and shown in the SEM image in Figure 3C. Note that the SEM images were recorded with a backscatter detector, which gives a good material contrast. The structures in Figure 3B show a higher contrast compared to the hollow structures without $BaCO_3$ in C. The EDX spectrum of these polymer replicas as well as their preliminary stages are shown in Figure 3D. The black spectrum shows the silica-biomorphs, the red spectrum the naked biomorphs with $CuCl_2$, which can be found at 8.02 eV. Here we can see that the outer silica layer has vanished and no Si signal was measured at 1.75 eV. In the upper blue spectrum, the Ba signals (green rectangular outline) have vanished and only traces of the raw material (Ba: 0.24; Si: 0.25 at.%) and the catalyst (0.4 at.%) are detected. The main component consists of carbon and nitrogen. Therefore, the pPy route is most suitable to preserve micro-sculptures on the microscale and to obtain conductive inverse biominerals.

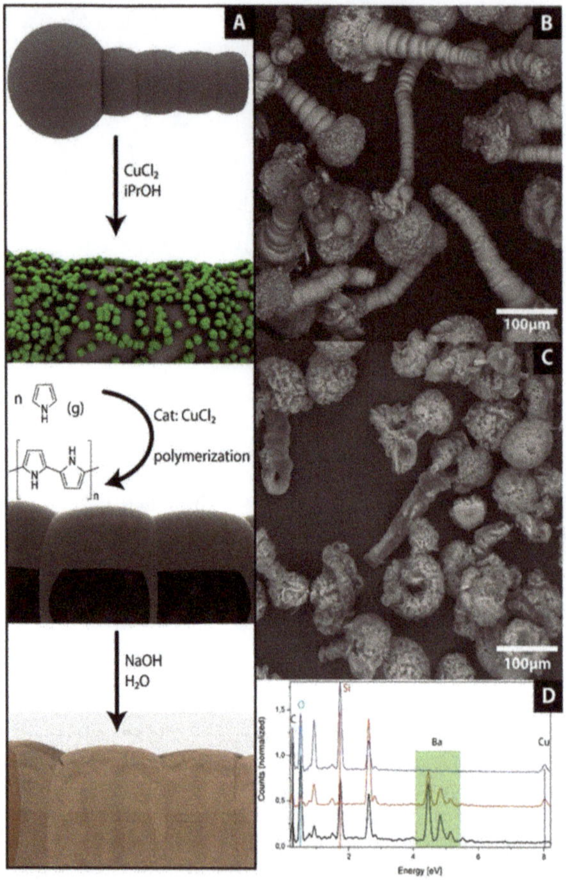

Figure 3. (**A**) Schematic illustration of the pPy route to preserve biomorphic structures. $CuCl_2$ crystals are attached from iPrOH solution on the surface of a naked biomorph. In the next step pyrrole diffuses via the gas phase onto the surface and is polymerized by $CuCl_2$. After the polymerization the inner parts can be removed by sodium hydroxide solution. (**B**) SEM image of catalyst crystals attached on biomorph worms. (**C**) Inverse biomorphs after Polymerization of pyrrole (Py) and removal of the inner part. (**D**) Energy dispersive X-ray (EDX) spectra of the different steps to plastic biomorphs. The untreated biomorphs are shown in black. Naked biomorphs with $CuCl_2$ in red and the final inverse biomorphs without inorganics are shown in blue.

4. Conclusions

We have achieved a full coverage of silica-carbonate composites by using PCDA and pyrrole to form an organic layer around the inorganic motifs. Thus, we have converted purely inorganic-inorganic composites into hybrids than can be named inverse biomorphs. The pathways to obtain these organic or hybrid structural motifs are described in detail and the methodology can be adapted to other microscopic biominerals or synthetic biomimetic architectures. Furthermore, the inorganic and structuring part of the biomorphs can be removed at ambient conditions resulting in a structured organic replica of the biomorph. With this straight-forward method, micro-structured functional materials can be formed and used as stretchable supercapacitors.

Author Contributions: J.O., M.K., H.C. and J.-M.G.-R. conceived and designed the experiments. The experiments where performed by J.O., N.U. (pPy part) and M.W. (pPCDA part). Analytical measurements where performed by J.O., N.U. and M.W., J.O. wrote the paper under supervision of J.-M.G.-R., H.C. and M.K.

Funding: The authors thank the European Research Council under the European Union's seventh Framework Program (FP7/2007-2013)/ERC grant agreement no. 340863.

Acknowledgments: The authors thank the Particle Analysis Center of the University of Konstanz (SFB 1214), the Nanostructure Laboratory and the Bioimaging center of the University of Konstanz for access to their instruments and Andra-Lisa Hoyt for corrections.

Conflicts of Interest: The authors declare no conflict of interest.

References

1. Kellermeier, M.; Colfen, H.; Garcia-Ruiz, J.M. Silica biomorphs: Complex biomimetic hybrid materials from "sand and chalk". *Eur. J. Inorg. Chem.* **2012**, *2012*, 5123–5144. [CrossRef]
2. Nakouzi, E.; Steinbock, O. Self-organization in precipitation reactions far from the equilibrium. *Sci. Adv.* **2016**, *2*, e1601144. [CrossRef]
3. Garcia-Ruiz, J.M.; Hyde, S.T.; Carnerup, A.M.; Christy, A.G.; van Kranendonk, M.J.; Welham, N.J. Self-assembled silica-carbonate structures and detection of ancient microfossils. *Science* **2003**, *302*, 1194–1197. [CrossRef] [PubMed]
4. Mann, S. The chemistry of form. *Angew. Chem. Int. Ed.* **2000**, *39*, 3392–3406. [CrossRef]
5. Addadi, L.; Weiner, S. Control and design principles in biological mineralization. *Angew. Chem. Int. Ed.* **1992**, *31*, 153–169. [CrossRef]
6. Lowenstam, H.A.; Weiner, S.; Weiner, S. *On Biomineralization*; Oxford University Press: New York, NY, USA, 1989; Volume 324.
7. Kellermeier, M.; Melero-García, E.; Kunz, W.; García-Ruiz, J.M. Local autocatalytic co-precipitation phenomena in self-assembled silica-carbonate materials. *J. Colloid Interface Sci.* **2012**, *380*, 1–7. [CrossRef] [PubMed]
8. Zhang, G.; Delgado-Lopez, J.M.; Choquesillo-Lazarte, D.; Garcia-Ruiz, J.M. Growth behavior of monohydrocalcite (caco3 center dot h2o) in silica-rich alkaline solution. *Cryst. Growth Des.* **2015**, *15*, 564–572. [CrossRef]
9. Zhang, G.; Garcia-Ruiz, J.M.; Sanchez-Migallon, J.M. Growth behaviour of silica/carbonate nanocrystalline composites of calcite and aragonite. *J. Mater. Chem. B* **2017**, *5*, 1658–1663. [CrossRef]
10. Zhang, G.; Verdugo-Escamilla, C.; Choquesillo-Lazarte, D.; Garcia-Ruiz, J.M. Thermal assisted self-organization of calcium carbonate. *Nat. Commun.* **2018**, *9*, 5221. [CrossRef]
11. Kellermeier, M.; Melero-Garcia, E.; Glaab, F.; Eiblmeier, J.; Kienle, L.; Rachel, R.; Kunz, W.; Garcia-Ruiz, J.M. Growth behavior and kinetics of self-assembled silica-carbonate biomorphs. *Chem. Eur. J.* **2012**, *18*, 2272–2282. [CrossRef] [PubMed]
12. García-Ruiz, J.M.; Melero-García, E.; Hyde, S.T. Morphogenesis of self-assembled nanocrystalline materials of barium carbonate and silica. *Science* **2009**, *323*, 362–365. [CrossRef] [PubMed]
13. Opel, J.; Kellermeier, M.; Sickinger, A.; Morales, J.; Cölfen, H.; García-Ruiz, J.-M. Structural transition of inorganic silica–carbonate composites towards curved lifelike morphologies. *Minerals* **2018**, *8*, 75. [CrossRef]
14. Bittarello, E.; Roberto Massaro, F.; Aquilano, D. The epitaxial role of silica groups in promoting the formation of silica/carbonate biomorphs: A first hypothesis. *J. Cryst. Growth* **2010**, *312*, 402–412. [CrossRef]

15. Opel, J.; Hecht, M.; Rurack, K.; Eiblmeier, J.; Kunz, W.; Colfen, H.; Kellermeier, M. Probing local ph-based precipitation processes in self-assembled silica-carbonate hybrid materials. *Nanoscale* **2015**, *7*, 17434–17440. [CrossRef]
16. Montalti, M.; Zhang, G.; Genovese, D.; Morales, J.; Kellermeier, M.; García-Ruiz, J.M. Local ph oscillations witness autocatalytic self-organization of biomorphic nanostructures. *Nat. Commun.* **2017**, *8*, 14427. [CrossRef]
17. Alexander, G.B.; Heston, W.; Iler, R.K. The solubility of amorphous silica in water. *J. Phys. Chem.* **1954**, *58*, 453–455. [CrossRef]
18. Iler, K.R. *The Chemistry of Silica. Solubility, Polymerization, Colloid and Surface Properties and Biochemistry of Silica*; John Wiley & Sons: Hoboken, NJ, USA, 1979.
19. Garcia Ruiz, J.M.; Carnerup, A.; Christy, A.G.; Welham, N.J.; Hyde, S.T. Morphology: An ambiguous indicator of biogenicity. *Astrobiology* **2002**, *2*, 353–369. [CrossRef] [PubMed]
20. Opel, J.; Wimmer, F.P.; Kellermeier, M.; Cölfen, H. Functionalisation of silica–carbonate biomorphs. *Nanoscale Horiz.* **2016**, *1*, 144–149. [CrossRef]
21. Pang, J.B.; Yang, L.; McCaughey, B.F.; Peng, H.S.; Ashbaugh, H.S.; Brinker, C.J.; Lu, Y.F. Thermochromatism and structural evolution of metastable polydiacetylenic crystals. *J. Phys. Chem. B* **2006**, *110*, 7221–7225. [CrossRef] [PubMed]
22. Patlolla, A.; Zunino, J.; Frenkel, A.I.; Iqbal, Z. Thermochromism in polydiacetylene-metal oxide nanocomposites. *J. Mater. Chem.* **2012**, *22*, 7028–7035. [CrossRef]
23. Göppert, A.; Cölfen, H. Infiltration of biomineral templates for nanostructured polypyrrole. *Rsc. Adv.* **2018**, *8*, 33748–33752. [CrossRef]
24. Munekawa, Y.; Oaki, Y.; Imai, H. An experimental study on the processes of hierarchical morphology replication by means of a mesocrystal: A case study of poly (3, 4-ethylenedioxythiophene). *Langmuir* **2014**, *30*, 3236–3242. [CrossRef]
25. Oaki, Y.; Kijima, M.; Imai, H. Synthesis and morphogenesis of organic polymer materials with hierarchical structures in biominerals. *J. Am. Chem. Soc.* **2011**, *133*, 8594–8599. [CrossRef]
26. Janata, J.; Josowicz, M. Conducting polymers in electronic chemical sensors. *Nat. Mater.* **2003**, *2*, 19–24. [CrossRef] [PubMed]
27. Geetha, S.; Rao, C.R.K.; Vijayan, M.; Trivedi, D.C. Biosensing and drug delivery by polypyrrole. *Anal. Chim. Acta* **2006**, *568*, 119–125. [CrossRef]
28. Unni, S.M.; Dhavale, V.M.; Pillai, V.K.; Kurungot, S. High pt utilization electrodes for polymer electrolyte membrane fuel cells by dispersing pt particles formed by a preprecipitation method on carbon "polished" with polypyrrole. *J. Phys. Chem. C* **2010**, *114*, 14654–14661. [CrossRef]
29. Olson, T.S.; Pylypenko, S.; Atanassov, P.; Asazawa, K.; Yamada, K.; Tanaka, H. Anion-exchange membrane fuel cells: Dual-site mechanism of oxygen reduction reaction in alkaline media on cobalt-polypyrrole electrocatalysts. *J. Phys. Chem. C* **2010**, *114*, 5049–5059. [CrossRef]
30. Ge, H.; Wallace, G.G. High-performance liquid chromatography on polypyrrole-modified silica. *J. Chromatogr. A* **1991**, *588*, 25–31. [CrossRef]
31. Choi, C.; Kim, J.H.; Sim, H.J.; Di, J.; Baughman, R.H.; Kim, S.J. Microscopically buckled and macroscopically coiled fibers for ultra-stretchable supercapacitors. *Adv. Energy Mater.* **2017**, *7*, 1602021. [CrossRef]
32. Siglreitmeier, M.; Wu, B.; Kollmann, T.; Neubauer, M.; Nagy, G.; Schwahn, D.; Pipich, V.; Faivre, D.; Zahn, D.; Fery, A.; et al. Multifunctional layered magnetic composites. *Beilstein J. Nanotechnol.* **2015**, *6*, 134–148. [CrossRef]
33. Oaki, Y.; Imai, H. The hierarchical architecture of nacre and its mimetic material. *Angew. Chem. Int. Ed.* **2005**, *44*, 6571–6575. [CrossRef] [PubMed]
34. Oaki, Y.; Imai, H. Nanoengineering in echinoderms: The emergence of morphology from nanobricks. *Small* **2006**, *2*, 66–70. [CrossRef] [PubMed]
35. Dei, S.; Matsumoto, A.; Matsumoto, A. Thermochromism of polydiacetylenes in the solid state and in solution by the self-organization of polymer chains containing no polar group. *Macromolecules* **2008**, *41*, 2467–2473. [CrossRef]

crystals

MDPI

Article

Synthesis and Adsorbing Properties of Tabular {001} Calcite Crystals

Nives Matijaković [1], Giulia Magnabosco [2], Francesco Scarpino [2], Simona Fermani [2], Giuseppe Falini [2] and Damir Kralj [1,*]

1 Laboratory for Precipitation Processes, Division of Materials Chemistry, Ruđer Bošković Institute, P. O. Box 180, HR-10002 Zagreb, Croatia; nives.matijakovic@irb.hr
2 Dipartimento di Chimica "Giacomo Ciamician", Alma Mater Studiorum-Universitá di Bologna, via Selmi 2, 40126 Bologna, Italy; giulia.magnabosco3@unibo.it (G.M.); francesco.scarpino3@studio.unibo.it (F.S.); simona.fermani@unibo.it (S.F.); giuseppe.falini@unibo.it (G.F.)
* Correspondence: kralj@irb.hr; Tel.: +385-1-468-0207

Received: 12 November 2018; Accepted: 22 December 2018; Published: 27 December 2018

Abstract: One of the most common crystal habits of the thermodynamically stable polymorph of calcium carbonate, calcite, is the rhombohedral one, which exposes {10.4} faces. When calcite is precipitated in the presence of Li^+ ions, dominantly {00.1} faces appear together with the {10.4}, thus generating truncated rhombohedrons. This well-known phenomenon is explored in this work, with the aim of obtaining calcite crystals with smooth {00.1} faces. In order to achieve this objective, the formation of calcite was examined in precipitation systems with different $c(Ca^{2+})/c(Li^+)$ ratios and by performing an initial high-power sonication. At the optimal conditions, a precipitate consisting of thin, tabular {001} calcite crystals and very low content of incorporated Li^+ has been obtained. The adsorption properties of the tabular crystals, in which the energetically unstable {00.1} faces represent almost all of the exposed surface, were tested with model dye molecules, calcein and crystal violet, and compared to predominantly rhombohedral crystals. It was found that the {00.1} crystals showed a lower adsorption capability when compared to the {10.4} crystals for calcein, while the adsorption of crystal violet was similar for both crystal morphologies. The obtained results open new routes for the usage of calcite as adsorbing substrates and are relevant for the understanding of biomineralization processes in which the {00.1} faces often interact with organic macromolecules.

Keywords: calcium carbonate; {00.1} calcite; lithium ions; ultrasonic irradiation; vaterite transformation; adsorption; calcein; crystal violet

1. Introduction

Investigation of precipitation mechanisms of calcium carbonates ($CaCO_3$) and their interactions with additives and surface chemistry, has attracted growing interest, due to promising technological applications [1–3] and the crucial role of this mineral in biomineralization [4–6]. Among $CaCO_3$ polymorphs, calcite is the most frequently studied, since it is in a thermodynamically-favored phase at ambient conditions and in the absence of additives. Calcite is an important product in the pharmaceutical, chemical, paper, and glass industries, and is also used as a sorbent for exhaust gasses and for the determination of the quality of drinking water [7–9]. Moreover, several organisms possess tissues mineralized with calcite that perform vital functions, like skeletal protection and support, light perception, or storage of calcium ions [10–13].

The morphology of calcite crystals produced in vitro is largely controlled by experimental conditions, such as pH, temperature, and supersaturation, as well as by the presence of impurities or additives that typically exert significant influence on the crystallization process [7,14–17]. The addition of inorganic additives such as Mg^{2+} and Li^+ causes selective stabilization of calcite crystals with different morphologies [18,19]. Thus, when precipitated in the presence of a Li^+-containing solution undersaturated with respect to Li_2CO_3, calcite shows {00.1} faces in addition to the {10.4} ones, which are typically obtained without additives. The extension of {00.1} faces depends on the $c(Ca^{2+})/c(Li^+)$ ratio [19–28]. In silico experiments on the calcite surface energy have shown that the rhombohedral {10.4} faces are the most stable, while the {00.1} faces have lower stability in the absence of additives [29]. The substitution of the surface Ca^{2+} with Li^+ destabilizes {10.4} faces, making the {00.1} ones the most stable and causing the formation of tabular crystals with {00.1} basal and {10.4} side faces [23,29].

Calcite has been used in experimental and theoretical studies as a model substrate for the investigation of the interactions occurring at organic-inorganic interfaces. Different classes of compounds have been tested, ranging from those with small molecular mass, such as water [30,31], alcohols [32], carboxylic [33,34] or amino acids [35,36], to bigger ones, such as polypeptides [37–39], proteins [40], or polysaccharides [41]. Most of these studies focus on the interaction of molecules on the calcite {10.4} surface, as this face is typically exposed by synthetic crystals, while the investigations of adsorption processes on other faces are relatively unexplored. Indeed, the structure of the monoatomic growing step on the {10.4} surface corresponds to {00.1} faces. Therefore, the study of the adsorption process on the well exposed and dominant {00.1} faces is also relevant. because the diffusion of additives along the {10.4} steps of calcite crystals occurs on them. Furthermore, this is a position of strong interfacial interactions with non-constituent molecules and organic matrix in biomineralization.

This study reports an optimized procedure for the preparation of size-uniform, thin, tabular calcite crystals with well-developed {00.1} faces ({00.1} calcite). In addition, the adsorption of model organic molecules, calcein and crystal violet, on {00.1} calcite is examined and compared to their adsorption on stable {10.4} faces. Besides the basic understanding of the specific {00.1} interactions with organic molecules, the knowledge gained from this research may be relevant to technology in which the presence of different crystal faces may tune the adsorbing capacities of a substrate. In addition, these data can contribute to the understanding of the processes of biomineralization, since the mineral interactions with intra-mineral macromolecules often occur exactly at {00.1} faces.

2. Materials and Methods

The reactant solutions were prepared by using the analytical grade chemicals, $CaCl_2 \cdot 2H_2O$, LiCl, and $NaHCO_3$ and deionized water (conductivity <0.055 $\mu S \cdot cm^{-1}$). Calcium carbonate precipitation was initiated by mixing equal volumes (200 cm^3) of $CaCl_2/LiCl$ solution with $NaHCO_3$ solution. All solutions were freshly prepared and the initial concentrations of $CaCl_2$ and $NaHCO_3$ were identical, $c = 0.1$ $mol \cdot dm^{-3}$, while the concentration of LiCl varied in the range from 0 to 1.0 $mol \cdot dm^{-3}$. The initial supersaturation, expressed with respect to calcite, was $S_c \approx 23$. Calcite supersaturation, S, was defined as a square root of the activity product quotient, $S = [(a_{Ca} \times a_{CO3})/K_{sp}]^{1/2}$, where a indicates the species activity and the K_{sp} is the calcite solubility product. The reference experiments were performed in the absence of LiCl (reference system).

The sonication was initiated in $NaHCO_3$ solution before mixing with $CaCl_2/LiCl$ solution. The systems were sonicated for 10 min and the pH was continuously measured. After that period, the mixture was left without stirring for 5 days. The ultrasonic irradiation was performed by using the Branson Sonifier 250 (20 kHz frequency). The ultrasound power output used in the experiments and set at 20% was 40 W. In order to avoid overheating the system, the pulsed mode was applied (60 cycles per minute, $t = 0.2$ s). The horn diameter was 5 mm and it was immersed in the center of the reaction vessel, 5 cm above the bottom. The crystal samples were collected by filtering 10 cm^3 of suspension through a 0.22 μm membrane filter. The precipitates were washed with small portions of water and dried at 90 °C for 2 h.

The polymorphic composition of the dried samples was determined by FT-IR spectroscopy (FT-IR TENSOR II, Bruker, Billerica, MA, USA), using the KBr pellets technique. X-ray powder diffraction, XRD (Philips X'Celerator diffractometer, Philips, Amsterdam, The Netherlands) was made in the angular range $20^\circ \leq 2\theta \leq 70^\circ$, setting a step size $2\theta = 0.05^\circ$ and measuring time of 120 s per step. The crystal morphology was determined by scanning electron microscopy (FEG SEM Hitachi 6400 (Hitachi, Tokyo, Japan), JEOL JSM-7000F (JEOL, Tokyo, Japan) and Phenom model G2, (Phenom-World BV, Eindhoven, The Netherlands)), operating in low-voltage mode. The SEM samples were placed on carbon tape without any coating. The calcium and Li^+ ion content in the crystals was determined by an ion chromatography system (ICS-1100, Dionex, Sunnyvale, CA, USA) fitted with a SC16 Analytical Column and using 30 mM MSA eluent. The specific surface area was determined by the multiple BET method (Micromeritics, Gemini), using liquid nitrogen.

For adsorption measurements the solutions of respective dye ($c = 10\ \mu mol \cdot dm^{-3}$) were diluted in HEPES buffer ($c = 0.1\ mol \cdot dm^{-3}$, pH 8.0). The buffer was pre-saturated with excess of rhombohedral calcite crystals, by mixing the suspension for 1 h and filtering by means of 0.2 μm membrane filter. Exactly 4 mg of calcite crystals (tabular or rhombohedral) were suspended in 2 mL of dye solution and stirred for 24 h, after which the crystals were separated by centrifugation. The quantity of residual dye in the solution was determined by UV-Vis spectroscopy (Cary UV-Vis 300Bio, Agilent Technologies, Santa Clara, CA, USA), measuring the absorbance at $\lambda = 495$ nm (calcein) and $\lambda = 594$ nm (crystal violet). The kinetics of adsorption was determined by the same protocol, by sampling at predetermined time intervals.

3. Results and Discussion

3.1. Synthesis and Characterization of Thin Tabular {00.1} Calcite Crystals

The first objective of this research was to identify the optimal conditions for the preparation of thin tabular calcite crystals, exposing predominantly the {00.1}, over the {10.4} faces. A series of $CaCO_3$ precipitation systems having high initial supersaturation, $c_i(CaCl_2) = c_i(NaHCO_3) = 0.1\ mol \cdot dm^{-3}$; $S_c \approx 23$, and different $c(Li^+)$, were initially irradiated by high-power ultrasound for 10 minutes and the precipitates were left to age in the mother solutions without stirring for 5 days (Table 1).

The precipitates were sampled and characterized immediately after the sonication process, continuously during the aging of precipitate, and finally after the equilibrating of the process, which typically lasted for 5 days (Figures SI1–SI6, Table SI1). The addition of Li^+ was supposed to stabilize {00.1} calcite crystals, while the ultrasound irradiation was applied in order to influence the nucleation and crystal growth processes, which may cause the changes of polymorphic composition, the morphology, or the size distribution of the precipitate. Indeed, in order to test a critical role of high-power ultrasound irradiation in the synthesis of {00.1} calcite, different types of stirring of the solutions during the nucleation period have been applied. Thus, either a mechanical propeller, Teflon-coated magnetic stirrer, or ultrasonic irradiation has been used during the first 10 minutes of the precipitation process. In Figure SI7, typical progress curves, pH vs time, of the systems ($c_i(CaCl_2) = c_i(NaHCO_3) = 0.1\ mol \cdot dm^{-3}$; $c(Li^+) = 0.3\ mol \cdot dm^{-3}$, or $c(Li^+) = 0$), stirred by different devices, are shown. Since the initial supersaturation was relatively high in all systems, precipitation started immediately after mixing the reactants (pH drop from 8.3 to 7.3, not shown). However, a shortening of the induction time for vaterite formation, caused by applying more intensive agitation and seen as a pH drop, can be observed (ultrasonic < magnetic < mechanical). At that, the most likely mechanism of the promotion of nucleation in the case of sonication is a collapse of cavitation bubbles, which caused localized high pressure and temperature spots, while the literature data indicated that the energetic collisions among particles may enhance the transition of metastable vaterite to calcite as well [42–45]. In addition, the SEMs of the respective samples isolated after 5 days of aging (Figure SI8) showed significantly different morphologies of calcite crystals. Thus, only a partial truncation of {10.4} calcite faces occurred when magnetic or mechanical stirring were applied, while in the case of ultrasonication, relatively uniform, thin, and hexagonal tabular crystals could be observed. Consequently, it can be concluded that in this particular system, the morphology of calcite can be assigned to the both parameters, initial mode of mixing and the LiCl addition.

The results of analysis of the polymorphic distribution of the precipitates from the different systems are reported in Table 1 and showed in Figure 1.

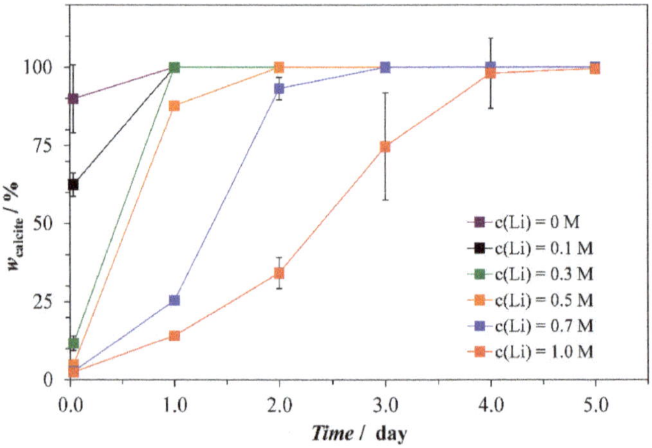

Figure 1. Content of calcite in a mixture with vaterite during the transformation process in the precipitation systems, $c_i(CaCl_2) = c_i(NaHCO_3) = 0.1$ mol·dm^{-3}, with different initial concentrations of Li^+. In the reference system, $c(Li^+) = 0$. The initial concentrations of Li^+ are indicated: M correspond to mol·dm^{-3}.

It can be seen that during the early stages of the precipitation process, vaterite always co-precipitates with calcite, and its relative amount increases by increasing the $c(Li^+)$ in solution. Since vaterite is thermodynamically metastable, it converts into calcite during the time, with a transition time proportional to the $c(Li^+)$. Indeed, after 5 days of aging, vaterite completely transformed into calcite at all examined conditions, with differences obtained in the system of the highest $c(Li^+)$ applied, in which just 1.89 wt % of vaterite still persists. Aquilano et al. [27] showed that Li^+ is sporadically adsorbed on growing calcite crystals and that Li^+ is not homogeneously distributed in the crystals. On the other hand, no data reporting Li^+ distribution into the vaterite crystals are available, to the best of our knowledge. Therefore, we hypothesize that in the systems with higher Li^+ content, the increasing adsorption onto the vaterite surface cause its slower dissolution and kinetic stabilization, which consequently postpone the formation of calcite, as already reported for organic molecules [46]. However, the adsorption of Li^+ onto the calcite can also inhibit its growth, which influences the overall course of transformation.

The analysis of X-ray powder diffraction XRPD patterns during the transformation process in all systems showed only diffraction peaks of vaterite and calcite (Figure 2 and Figures SI9–SI14, Table SI2), indicating that the formation of Li_2CO_3 does not occur at conditions studied in this work. On the contrary, in the systems described in the literature, which were supersaturated with respect to Li_2CO_3, the precipitation of this phase has been detected on calcite surface [25].

Shape and morphology of crystals were monitored during the progress of precipitation (Figure 3 and Figures SI15–SI17) and the obtained morphometric data are reported in Table 1. Thus, in the absence of Li^+, a mixture of hollow spherical vaterite and rhombohedral calcite crystals was obtained after 10 min of ultrasonic irradiation (Figure SI15). The comparison of higher-magnification SEM of the vaterite samples, prepared in the presence (0.3 mol·dm^{-3}) or in the absence of Li^+ (Figure SI17), indicates that the lithium ions do not significantly change the morphology of spherulites. However, after 5 days, only calcite crystals were detected, homogeneous in size and with surface pits, truncations, and ragged edges (Figure 3a and Figure SI15), which can be attributed to the etching caused by the atmospheric CO_2 dissolved in the mother solution during the aging [47,48].

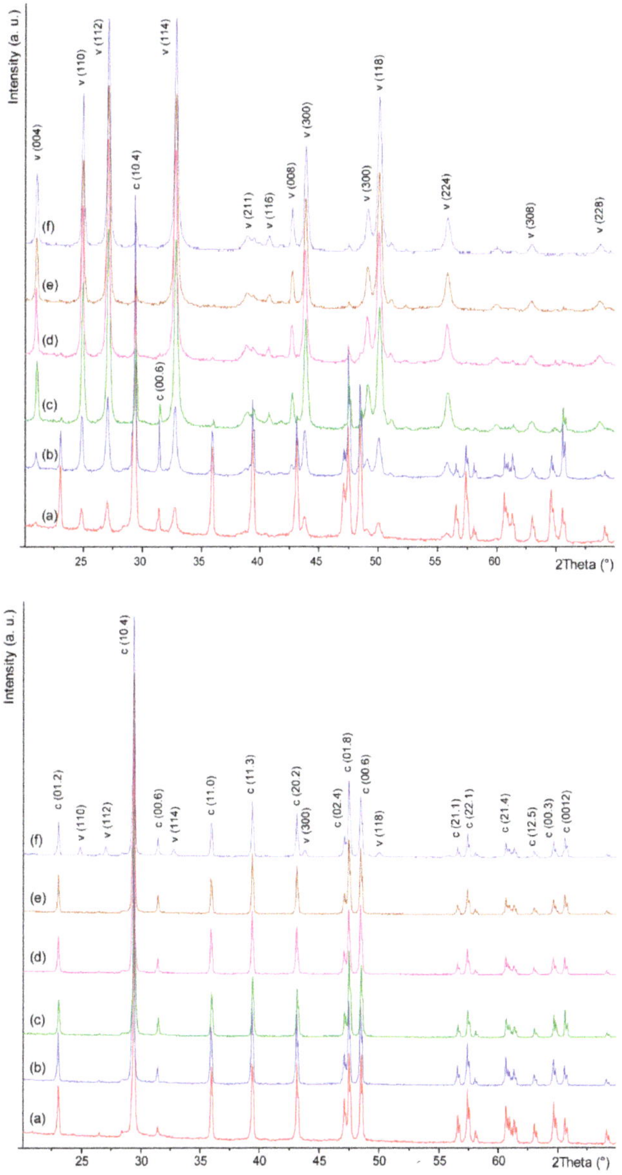

Figure 2. X-ray powder diffraction patterns of the precipitates obtained in the precipitation systems, $c_i(CaCl_2) = c_i(NaHCO_3) = 0.1$ mol·dm^{-3}, after 10 min (**top**) and 5 days of aging (**bottom**), in the presence of Li^+ at the concentration of (a) 0.0 mol·dm^{-3}, (b) 0.1 mol·dm^{-3}, (c) 0.3 mol·dm^{-3}, (d) 0.5 mol·dm^{-3}, (e) 0.7 mol·dm^{-3} or (f) 1.0 mol·dm^{-3}. The diffraction patterns were indexed accordingly to the PDF 00-005-0586 for calcite and PDF 00-033-0268 for vaterite PDF. In all systems only calcite and vaterite have been detected.

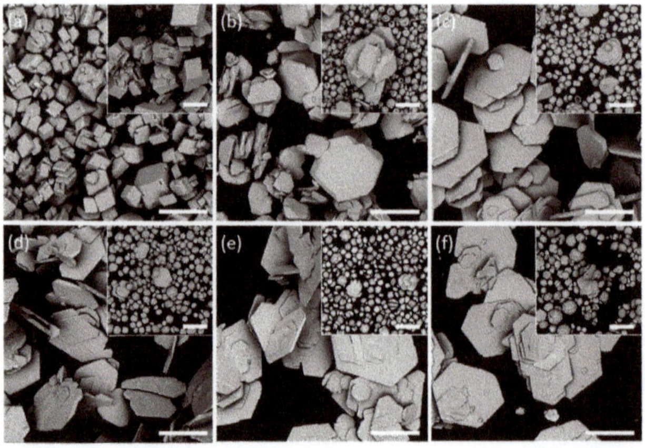

Figure 3. Scanning electron images of calcium carbonate samples obtained after 5 days of aging in the reference system, $c(Li^+) = 0$ (**a**) and in the systems with $c(Li^+) = 0.1$ mol·dm^{-3} (**b**), $c(Li^+) = 0.3$ mol·dm^{-3} (**c**), $c(Li^+) = 0.5$ mol·dm^{-3} (**d**), $c(Li^+) = 0.7$ mol·dm^{-3} (**e**) and $c(Li^+) = 1.0$ mol·dm^{-3} (**f**). The inset shows the precipitates obtained after 10 min of ultrasonic irradiation at $P = 40$ W and at room temperature. Scale bars: 20 μm.

In the system with $c(Li^+) = 0.1$ mol·dm^{-3}, after 10 min of sonication, hollow spherulitic vaterite and rhombohedral calcite were observed as well. However, after 5 days the calcite crystals exhibited triangular faces, intergrowth, and crystal aggregation (Figure 3b and Figure SI15). In addition, the smooth {10.4} crystal faces were truncated with {001} faces. Rajam and Mann [19] showed that calcite crystals truncation depends on Li^+/Ca^{2+} molar ratio, and that at low ratios the truncations are triangular, owing to growth of {00.1} and partial truncation of three {10.4} faces.

The system with $c(Li^+) = 0.3$ mol·dm^{-3}, after 10 min of ultrasonic irradiation contained mainly spherical vaterite particles agglomerated into 3 μm structures, as well as tabular calcite crystals and calcite crystals with triangular truncations (Figure 3c inset and Figure SI15). After 5 days of aging, only thin hexagonal tabular calcite crystals, with well-developed basal {00.1} faces and {10.4} side faces, were obtained (Figure 3c and Figure SI15). The hexagonal crystal truncations, formed by partial truncation of all six {10.4} calcite faces, are promoted by the adsorption of Li^+ and followed by absorption of Li_2CO_3 into growing calcite crystals, as showed by Aquilano [27].

At the highest Li^+ concentrations applied, $c(Li^+) = 0.5$ mol·dm^{-3} and 1.0 mol·dm^{-3}, and after 10 min of ultrasonic irradiation a mixture of vaterite particles, together with small amounts of tabular calcite with well-developed {00.1} faces (Figure 3d–f insets and Figure SI16), was obtained. The vaterite particles were aggregated in structures larger than those in the systems containing lower content of Li^+ (6–10 μm). At the end of the process, the remained vaterite particles converted into hollow particles (Figure 3f inset and Figure SI16). Tabular hexagonal crystals obtained in the same systems, contained macro-steps on the {00.1} faces, which were not observed at lower $c(Li^+)$ (Figure 3d–f and Figure SI16).

The aspect ratios of the calcite crystal were evaluated by analyzing the respective SEM images of the samples prepared in the systems with different Li^+ concentrations and isolated after 5 days of aging (Figures SI15 and SI16). At the same time, the ratios of relative intensities, $I_{10.4}/I_{00.1}$, were estimated from the XRPD patterns (Figures SI9–SI14). The data shown in the Table 1 and in Figure SI18 indicates that the aspect ratio increased with increasing the Li^+ content for each crystal population obtained in the respective system. Also, the ratios of relative intensities, $I_{10.4}/I_{00.1}$, decrease with increasing Li^+ content. Indeed, the aspect ratio values of two classes of the crystals obtained in the final precipitate pointed to two nucleation events, which imply that the conditions in the solutions during these two events (S, Li/Ca) were not completely identical. However, these findings are consistent with SEM observations

(Figure 3, Figures SI15 and SI16) which showed that in the system containing $c(Li^+) = 0.3$ mol·dm^{-3}, crystals were plate-like with well-developed {001} faces and without macro-steps. On other hand, in the systems with higher $c(Li^+)$, the crystals were larger, but contained macro-steps with expressed {104} side faces.

The content of Li^+ incorporated into the $CaCO_3$ samples prepared in different chemical environments (different additions of Li^+) and isolated during the aging of precipitate was determined by ion chromatography. Thus, Figure 4 shows that the Li^+ content in the precipitate continuously decreased during the progress of transformation in each precipitation system.

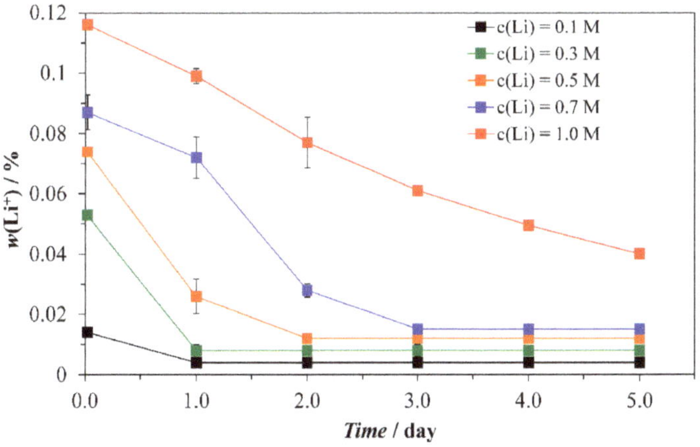

Figure 4. The content of Li^+ incorporated into the crystals during the transformation process in the precipitation systems, $c_i(CaCl_2) = c_i(NaHCO_3) = 0.1$ mol·dm^{-3}, with different initial concentration of Li^+. The initial concentrations of Li^+ in solution are indicated: M correspond to mol·dm^{-3}.

The highest incorporation determined immediately after 10 minutes of sonication ($t = 0$ days) was found in the system with $c(Li^+) = 1.0$ mol·dm^{-3} ($w(Li^+) = 0.116$ wt %), while in the systems with lower initial Li^+ concentrations, the content was in the range from 0.087 to 0.014 wt % (SI19). The amount of Li^+ incorporated into the precipitate and determined at the end of the aging process was found to vary in the very narrow range between 0.004 and 0.04 wt %. Again, the Li^+ incorporation increased almost linearly with increasing the initial Li^+ up to the $c = 0.7$ mol·dm^{-3} (SI19), so the obtained values may be considered as typical for distribution between solution and tabular calcite crystals with dominant expression of {00.1} crystal planes. On the other hand, in the systems in which vaterite was still present, the measured incorporation of Li^+ was significantly higher, which implicate different mechanisms of Li^+ uptake. Thus, it can be concluded that, besides the adsorption, the calcite incorporates the Li^+ in the crystal structure, while vaterite particles can additionally entrap the Li-containing solution into the pores. Indeed, at the applied supersaturation level, vaterite spherulites are most probably formed by fast spherulitic growth, which is characterized exactly by the increased trapping of dissolved impurities, like Li^+ in this particular case [45,49,50].

The results of structural, morphological, and chemical analyses of calcium carbonate precipitate formed in the Li^+ containing solutions pointed out that the calcite crystals obtained in the system with $c(Li^+) = 0.3$ mol·dm^{-3} are appropriate for use in adsorption experiments, in which the {00.1} crystal faces should be investigated. Namely, the prepared crystals are relatively uniform in size, without expressed macro-steps and {10.4} faces, as well as the content of incorporated Li^+, is relatively low (Figure SI19)

Table 1. Properties of the precipitate and the calcite crystal obtained after initial sonication and 5 days of aging in mother solutions, containing different concentrations of Li^+ ions. In the system with $c(Li^+) = 1.0$ mol·dm^{-3}, about 2 wt % of vaterite has been obtained as well ($d \approx 3$ μm).

$c(Li^+)$/ mol·dm^{-3}	w(calcite)/ wt %	Particle Size */ μm	Aspect Ratio #	$I_{(10.4)}/I_{(00.1)}$	w(Li)/wt %
0	100	9	-	68	0
0.1	100	11/27	11:1 27:1	39	0.004 ± 0.00017
0.3	100	12/32	12:1 32:1	25	0.008 ± 0.00011
0.5	100	23/43	23:1 43:1	30	0.012 ± 0.00015
0.7	100	23/40	23:1 40:1	23	0.015 ± 0.00023
1.0	98 ± 0.055	30/48	30:1 48:1	21	0.049 ± 0.00037

* Two values indicate two different population of crystals. # The aspect ratio indicates the ratio between the longest axis of (00.1) planes and the thickness of the crystals: values of two different populations are shown.

3.2. Adsorption of Model Molecules

The adsorption proprieties of {00.1} calcite prepared in the system $c(Li^+) = 0.3$ mol·dm^{-3} were compared with the {104} calcite obtained by the same procedure but in the absence of Li^+. The specific surface area of the {00.1} calcite used for the adsorption experiments was, $s = 0.38$ m^2·g^{-1}, while of the {10.4} calcite was, $s = 0.34$ m^2·g^{-1}. At that, calcein and crystal violet (CV) have been selected as representative organic molecules (Figure SI20), due to their different chemical properties and strong molar extinction coefficients. Calcein possess 4 carboxyl groups are negatively charged at pH 8.0 and can chelate Ca^{2+}, while CV possesses terminal amine groups, which are positively charged at pH 8.0.

The actual adsorption of the selected dye molecules on the surface of the calcite crystals with different dominant crystal faces has been visualized by confocal microscopy. Figure 5 shows typical microphotographs of the plate-like and rhombohedral calcite crystals labeled with calcein. It can be seen that, within the limits of the resolution of the technique, calcein only partially covers the {00.1} faces, while a more intense fluorescence is observed on the {10.4} planes. The respective confocal images showing the consecutive crystal sections of the same motif are shown in Figure SI21.

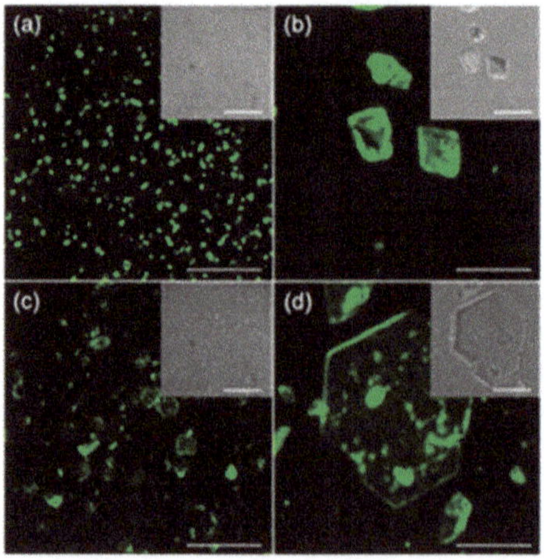

Figure 5. Confocal microscopy images of calcite crystals with adsorbed calcein. (**a**, **b**) {10.4} calcite, (**c**, **d**) {00.1} calcite. The insets show the corresponding transmission optical microscopy images. The scale bars are 50 μm (**a**,**c**) and 10 μm (**b**,**d**).

The adsorption properties of the calcite crystals of different morphologies have been determined by equilibrating them in solutions containing respective dye molecules for 24 h and by measuring the residual concentration after the crystals have been separated. No further adsorption was observed after 24 h. The kinetics of the adsorption have been determined as well (Figures SI22 and SI23), and it could be seen that the concentration of adsorbed CV is much higher in comparison to calcein and for both crystal morphologies. In order to additionally understand a mode of specific molecule/surface interactions, the experiments designed in a different way will be performed. The equilibrium concentration of dye adsorbed on the surfaces of different calcite crystals is presented in Figure 6, which shows stronger interactions of CV with both morphologies, while the adsorption of calcein on {10.4} faces is about twice as strong as the adsorption on {00.1} faces.

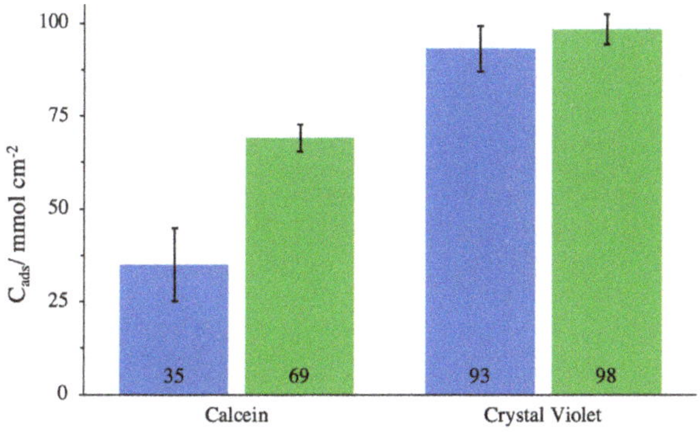

Figure 6. The concentration of respective dye molecules adsorbed after 24 h at the surface of {00.1} (blue) or {104} (green) calcite crystals.

The obtained results and different abilities for adsorption can be ascribed either to the different structure of the crystal planes, or to the chemical nature of the adsorbate molecule. {00.1} calcite exposes alternating planes of Ca^{2+} and CO_3^{2-} ions, with a preference for the latter [27], while the {10.4} face exposes both ions. The different chemical structure, and the net charge which exists at the conditions applied, are responsible for the observed different adsorption properties. Thus, positively charged CV is more likely to adsorb on calcite crystals, which typically possesses a net negative charge at wide range of solution composition and pH, as determined previously by the zeta potential measurements [15]. The interactions of calcein are significantly stronger with the {10.4} than with the {00.1} calcite surfaces, which is consistent with previous studies in which a Langmuir isotherm was proposed to describe the process [51]. On the other hand, it is known that the surface of {00.1} faces exposes the incorporated Li^+ in the form of lithium carbonate islands [25,26], so the surface complexation of calcein molecules, which interact by their carboxylate groups with Ca^{2+}, is limited or even suppressed. It follows that on the tabular {00.1} faces, the interaction with the calcein can predominantly happen on the steps and kinks exposed by the {00.1} faces. The interaction between the CV and the surface of calcite crystals is purely electrostatic, unlike calcein which requires the complexion of Ca^{2+}, and is therefore less selective with regards to the structure and coordination of surfaces. Indeed, as shown in the Figure 6, the amount of adsorbed CV was similar for both calcite morphologies.

4. Conclusions

A procedure for the preparation of uniform and thin tabular {001} calcite was proposed. Tabular calcite, which exhibited triangular and hexagonal faces, was precipitated from the Li^+-doped systems. In addition, the kinetics of formation of calcite crystals with a specific tabular morphology were demonstrated during the solution-mediated transformation process, in which predominantly vaterite initially appeared. With that, it was shown that higher Li^+ concentrations caused a slower conversion to calcite. The tabular {001} calcite, optimal for investigation of adsorption, was obtained at moderate Li^+ concentration, $c(Li^+) = 0.3$ mol·dm^{-3}: the crystals were relatively uniform in size and {00.1} predominated over the other faces.

The adsorption properties of the tabular {001} calcite was investigated by using two model organic dye molecules. The results of adsorption of calcein and crystal violet were compared to their adsorption on more common {10.4} calcite morphologies, in order to understand their interfacial dynamic and reactivity. It was found that the adsorption capability of the tabular {001} calcite is lower than the capability of pure {104} faced crystals and that the adsorption of the selected organic molecules depends on their functional groups and charge.

Since it is known that dissolved organic molecules play an important role in the regulation of crystal growth during the biomineralization process, which is a consequence of their adsorption and absorption on/in crystal face, this research contributes to understanding a mode and extent of interfacial interactions at different crystal planes.

Supplementary Materials: The following are available online at http://www.mdpi.com/2073-4352/9/1/16/s1, Figures SI1–SI6: FTIR spectra of the precipitates obtained in the system with different concentration of Li^+; Table SI1: Assignment of IR bands in FTIR spectra in all systems during the transformation; Figure SI7: The progress curves of different precipitation systems; Figure SI8: SEM images of calcite samples obtained in differently agitated systems; Figures SI9–SI14: PXRD patterns of the precipitates obtained in the system with different concentrations of Li^+; Table SI2: Assignment of peaks in PXRD patterns in all systems during the transformation; Figures SI15–SI16: Scanning electron micrographs of calcite samples obtained in the systems with different Li^+ addition; SI17: Vaterite samples prepared in the systems with different Li^+ addition; Figure SI18: Relative intensities of {00.1} with respect to the {10.4} calcite diffraction peaks and the aspect ratios, shown as a function of solution concentration of Li^+; Figure SI19: Amount of lithium in precipitate separated from the solution after 10 minutes of sonication and after 5 days of aging; Figure SI20: Molecular structure of calcein and crystal violet; Figure SI21: Consecutive confocal images of calcite crystals labeled with calceine; Figure SI22: The kinetics of adsorption of calcein and crystal violet on rhombohedral calcite crystals; Figure SI23: The kinetics of adsorption of calcein and crystal violet on tabular calcite crystals.

Author Contributions: Conceptualization, D.K. and G.F.; investigation, N.M., G.M., F.S. and S.F.; formal analysis, N.M., G.M., F.S. and S.F.; writing–original draft preparation, G.F. and D.K.; writing–review and editing, D.K. and G.F.

Funding: This work has been supported in part (N.M. and D.K.) by Croatian Science Foundation under the project (IP-2013-11-5055) and from the European Social Fund under Call number HR.3.2.01 for the project "Networks for professional training of young scientists in interdisciplinary research of innovative surfaces and material", MIPoMaT, (D.K.).

Conflicts of Interest: The authors declare no conflict of interest.

References

1. Magnabosco, G.; Giosia, M.D.; Polishchuk, I.; Weber, E.; Fermani, S.; Bottoni, A.; Zerbetto, F.; Pelicci, P.G.; Pokroy, B.; Rapino, S.; et al. Calcite Single Crystals as Hosts for Atomic-Scale Entrapment and Slow Release of Drugs. *Adv. Healthc. Mater.* **2015**, *4*, 1510–1516. [CrossRef] [PubMed]
2. Nudelman, F.; Sommerdijk, N.A. Biomineralization as an inspiration for materials chemistry. *Angew. Chem. Int. Ed.* **2012**, *51*, 6582–6596. [CrossRef] [PubMed]
3. Green, D.C.; Ihli, J.; Thornton, P.D.; Holden, M.A.; Marzec, B.; Kim, Y.Y.; Kulak, A.N.; Levenstein, M.A.; Tang, C.; Lynch, C.; et al. 3D visualization of additive occlusion and tunable full-spectrum fluorescence in calcite. *Nat. Commun.* **2016**, *7*, 13524. [CrossRef] [PubMed]
4. Weiner, S.; Addadi, L. Crystallization pathways in biomineralization. *Annu. Rev. Mater. Res.* **2011**, *41*, 21–40. [CrossRef]
5. Addadi, L.; Weiner, S. Interactions between acidic proteins and crystals: Stereochemical requirements in biomineralization. *Proc. Natl. Acad. Sci. USA* **1985**, *82*, 4110–4114. [CrossRef]
6. Magnabosco, G.; Polishchuk, I.; Erez, J.; Fermani, S.; Pokroy, B.; Falini, G. Insights on the interaction of calcein with calcium carbonate and its implications in biomineralization studies. *CrystEngComm* **2018**, *20*, 4221–4224. [CrossRef]

7. Ukrainczyk, M.; Stelling, J.; Vučak, M.; Neumann, T. Influence of etidronic acid and tartaric acid on the growth of different calcite morphologies. *J. Cryst. Growth* **2013**, *369*, 21–31. [CrossRef]
8. Wang, L.; Ruiz-Agudo, E.; Putnis, C.V.; Putnis, A. Direct observations of the modification of calcite growth morphology by Li+through selectively stabilizing an energetically unfavourable face. *CrystEngComm* **2011**, *13*, 3962–3966. [CrossRef]
9. Njegić-Džakula, B.; Falini, G.; Brečević, L.; Skoko, Ž.; Kralj, D. Effects of initial supersaturation on spontaneous precipitation of calcium carbonate in the presence of charged poly-l-amino acids. *J. Colloid Interface Sci.* **2010**, *343*, 553–563. [CrossRef]
10. Lowenstam, H.A.; Weiner, S. *On Biomineralization*; Oxford University Press: Oxford, UK, 1989; ISBN 0195364198.
11. Falini, G.; Fermani, S.; Goffredo, S. Coral biomineralization: A focus on intra-skeletal organic matrix and calcification. *Semin. Cell Dev. Biol.* **2015**, *46*, 17–26. [CrossRef]
12. Aizenberg, J.; Tkachenko, A.; Weiner, S.; Addadi, L.; Hendler, G. Calcitic microlenses as part of the photoreceptor system in brittlestars. *Nature* **2001**, *412*, 819–822. [CrossRef] [PubMed]
13. Addadi, L.; Raz, S.; Weiner, S. Taking advantage of disorder: Amorphous calcium carbonate and its roles in biomineralization. *Adv. Mater.* **2003**, *15*, 959–970. [CrossRef]
14. Ukrainczyk, M.; Kontrec, J.; Kralj, D. Precipitation of different calcite crystal morphologies in the presence of sodium stearate. *J. Colloid Interface Sci.* **2009**, *329*, 89–96. [CrossRef] [PubMed]
15. Njegić-Džakula, B.; Brečević, L.; Falini, G.; Kralj, D. Calcite Crystal Growth Kinetics in the Presence of Charged Synthetic Polypeptides. *Cryst. Growth Des.* **2009**, *9*, 2425–2434. [CrossRef]
16. Njegić-Džakula, B.; Brečević, L.; Falini, G.; Kralj, D. Kinetic Approach to Biomineralization: Interactions of Synthetic Polypeptides with Calcium Carbonate Polymorphs. *Croat. Chem. Acta* **2011**, *84*, 301–314. [CrossRef]
17. Magnabosco, G.; Polishchuk, I.; Pokroy, B.; Rosenberg, R.; Cölfen, H.; Falini, G. Synthesis of calcium carbonate in trace water environments. *Chem. Commun.* **2017**, *53*, 4811–4814. [CrossRef] [PubMed]
18. Davis, K.J.; Dove, P.M.; Wasylenki, L.E.; De Yoreo, J.J. Morphological consequences of differential Mg2+incorporation at structurally distinct steps on calcite. *Am. Mineral.* **2004**, *89*, 714–720. [CrossRef]
19. Rajam, S.; Mann, S. Selective stabilization of the (001) face of calcite in the presence of lithium. *J. Chem. Soc. Chem. Commun.* **1990**, *0*, 1789–1791. [CrossRef]
20. Song, R.Q.; Cölfen, H. Additive controlled crystallization. *CrystEngComm* **2011**, *13*, 1249–1276. [CrossRef]
21. Titiloye, J.O.; Parker, S.C.; Osguthorpe, D.J.; Mann, S. Predicting the influence of growth additives on the morphology of ionic crystals. *J. Chem. Soc. Chem. Commun.* **1991**, 1494–1496. [CrossRef]
22. Kitano, Y. The Behavior of Various Inorganic Ions in the Separation of Calcium Carbonate from a Bicarbonate Solution. *Bull. Chem. Soc. Jpn.* **1962**, *35*, 1973–1980. [CrossRef]
23. Bruno, M.; Massaro, F.R.; Prencipe, M.; Aquilano, D. Surface reconstructions and relaxation effects in a centre-symmetrical crystal: The {00.1} form of calcite ($CaCO_3$). *CrystEngComm* **2010**, *12*, 3626–3633. [CrossRef]
24. Pastero, L.; Costa, E.; Bruno, M.; Rubbo, M.; Sgualdino, G.; Aquilano, D. Morphology of calcite ($CaCO_3$) crystals growing from aqueous solutions in the presence of Li+ ions. Surface behavior of the {0001} form. *Cryst. Growth Des.* **2004**, *4*, 485–490. [CrossRef]
25. Pastero, L.; Aquilano, D. $CaCO_3$ (Calcite)/Li_2CO_3 (zabuyelite) anomalous mixed crystals. Sector zoning and growth mechanisms. *Cryst. Growth Des.* **2008**, *8*, 3451–3460. [CrossRef]
26. Massaro, F.R.; Pastero, L.; Costa, E.; Sgualdino, G.; Aquilano, D. Single and twinned Li_2CO_3 crystals (zabuyelite) epitaxially grown on {0001} and {10$\bar{1}$4} forms of $CaCO_3$ (calcite) crystals. *Cryst. Growth Des.* **2008**, *8*, 2041–2046. [CrossRef]
27. Aquilano, D.; Pastero, L. Anomalous mixed crystals: A peculiar case of adsorption/absorption. *Cryst. Res. Technol.* **2013**, *48*, 819–839. [CrossRef]
28. Pastero, L.; Aquilano, D.; Costa, E.; Rubbo, M. 2D epitaxy of lithium carbonate inducing growth mechanism transitions on {0 0 0 1}-K and {0 1 1¯ 8}-S forms of calcite crystals. *J. Cryst. Growth* **2005**, *275*, 1625–1630. [CrossRef]
29. Titiloye, J.O.; Parker, S.C.; Mann, S. Atomistic simulation of calcite surfaces and the influence of growth additives on their morphology. *J. Cryst. Growth* **1993**, *131*, 533–545. [CrossRef]
30. Bohr, J.; Wogelius, R.A.; Morris, P.M.; Stipp, S.L.S. Thickness and structure of the water film deposited from vapour on calcite surfaces. *Geochim. Cosmochim. Acta* **2010**, *74*, 5985–5999. [CrossRef]

31. Sand, K.K.; Yang, M.; Makovicky, E.; Cooke, D.J.; Hassenkam, T.; Bechgaard, K.; Stipp, S.L.S. Binding of Ethanol on Calcite: The Role of the OH Bond and Its Relevance to Biomineralization. *Langmuir* **2010**, *26*, 15239–15247. [CrossRef] [PubMed]
32. Bovet, N.; Yang, M.; Javadi, M.S.; Stipp, S.L.S. Interaction of alcohols with the calcite surface. *Phys. Chem. Chem. Phys.* **2015**, *17*, 3490–3496. [CrossRef] [PubMed]
33. Aschauer, U.; Spagnoli, D.; Bowen, P.; Parker, S.C. Growth modification of seeded calcite using carboxylic acids: Atomistic simulations. *J. Colloid Interface Sci.* **2010**, *346*, 226–231. [CrossRef] [PubMed]
34. Mureşan, L.; Sinha, P.; Maroni, P.; Borkovec, M. Adsorption and surface-induced precipitation of poly(acrylic acid) on calcite revealed with atomic force microscopy. *Colloids Surf. A Physicochem. Eng. Asp.* **2011**, *390*, 225–230. [CrossRef]
35. Hazen, R.M.; Filley, T.R.; Goodfriend, G.A. Selective adsorption of L- and D-amino acids on calcite: Implications for biochemical homochirality. *Proc. Natl. Acad. Sci. USA* **2001**, *98*, 5487–5490. [CrossRef] [PubMed]
36. Orme, C.A.; Noy, A.; Wierzbicki, A.; Mcbride, M.T.; Grantham, M.; Teng, H.H.; Dove, P.M.; Deyoreo, J.J. Formation of chiral morphologies through selective binding of amino acids to calcite surface steps. *Nature* **2001**, *411*, 775–779. [CrossRef] [PubMed]
37. Wierzbicki, A.; Sikes, C.S.; Madura, J.D.; Drake, B. Atomic force microscopy and molecular modeling of protein and peptide binding to calcite. *Calcif. Tissue Int.* **1994**, *54*, 133–141. [CrossRef]
38. Sonnenberg, L.; Luo, Y.; Schlaad, H.; Seitz, M.; Cölfen, H.; Gaub, H.E. Quantitative single molecule measurements on the interaction forces of poly(L-glutamic acid) with calcite crystals. *J. Am. Chem. Soc.* **2007**, *129*, 15364–15371. [CrossRef] [PubMed]
39. Yang, M.; Mark Rodger, P.; Harding, J.H.; Stipp, S.L.S. Molecular dynamics simulations of peptides on calcite surface. *Mol. Simul.* **2009**, *35*, 547–553. [CrossRef]
40. Freeman, C.L.; Harding, J.H.; Quigley, D.; Rodger, P.M. Simulations of ovocleidin-17 binding to calcite surfaces and its implications for eggshell formation. *J. Phys. Chem. C* **2011**, *115*, 8175–8183. [CrossRef]
41. Yang, M.; Stipp, S.L.S.; Harding, J. Biological control on calcite crystallization by polysaccharides. *Cryst. Growth Des.* **2008**, *8*, 4066–4074. [CrossRef]
42. Dalas, E. The effect of ultrasonic field on calcium carbonate scale formation. *J. Cryst. Growth* **2001**, *222*, 287–292. [CrossRef]
43. Nishida, I. Precipitation of calcium carbonate by ultrasonic irradiation. *Ultrason. Sonochem.* **2004**, *11*, 423–428. [CrossRef] [PubMed]
44. Berdonosov, S.S.; Znamenskaya, I.V.; Melikhov, I.V. Mechanism of the vaterite-to-calcite phase transition under sonication. *Inorg. Mater.* **2005**, *41*, 1308–1312. [CrossRef]
45. Njegić Džakula, B.; Kontrec, J.; Ukrainczyk, M.; Sviben, S.; Kralj, D. Polymorphic composition and morphology of calcium carbonate as a function of ultrasonic irradiation. *Cryst. Res. Technol.* **2014**, *49*, 244–256. [CrossRef]
46. Luo, J.; Kong, F.; Ma, X. Role of Aspartic Acid in the Synthesis of Spherical Vaterite by the $Ca(OH)_2$–CO_2 Reaction. *Cryst. Growth Des.* **2016**, *16*, 728–736. [CrossRef]
47. Arvidson, R.S.; Ertan, I.E.; Amonette, J.E.; Luttge, A. Variation in calcite dissolution rates: A fundamental problem? *Geochim. Cosmochim. Acta* **2003**, *67*, 1623–1634. [CrossRef]
48. De Giudici, G. Surface control vs. diffusion control during calcite dissolution: Dependence of step-edge velocity upon solution pH. *Am. Mineral.* **2002**, *87*, 1279–1285. [CrossRef]
49. Beck, R.; Andreassen, J.P. Spherulitic growth of calcium carbonate. *Cryst. Growth Des.* **2010**, *10*, 2934–2947. [CrossRef]
50. Andreassen, J.P. Formation mechanism and morphology in precipitation of vaterite—Nano-aggregation or crystal growth? *J. Cryst. Growth* **2005**, *274*, 256–264. [CrossRef]
51. Atun, G.; Acar, E.T. Competitive adsorption of basic dyes onto calcite in single and binary component systems. *Sep. Sci. Technol.* **2010**, *45*, 1471–1481. [CrossRef]

crystals

MDPI

Article

Calcium Carbonate Mineralization in a Surface-Tension-Confined Droplets Array

Zhong He [1], Zengzilu Xia [2], Mengying Zhang [3], Jinbo Wu [1,*] and Weijia Wen [1,4,*]

1 Materials Genome Institute, Shanghai University, Shanghai 200444, China; harryhe1994@163.com
2 Key Laboratory of Biorheological Science and Technology of the Ministry of Education, College of Bioengineering, Chongqing University, Chongqing 400044, China; zzlxia@cqu.edu.cn
3 College of Science, Shanghai University, Shanghai 200444, China; zhang.my@t.shu.edu.cn
4 Nano Science and Technology Program, Hong Kong University of Science and Technology, Clear Water Bay, Kowloon, Hong Kong, China
* Correspondence: jinbowu@t.shu.edu.cn (J.W.); phwen@ust.hk (W.W.)

Received: 2 April 2019; Accepted: 29 May 2019; Published: 30 May 2019

Abstract: Calcium carbonate biomimetic crystallization remains a topic of interest with respect to biomineralization areas in recent research. It is not easy to conduct high-throughput experiments with only a few macromolecule reagents using conventional experimental methods. However, the emergence of microdroplet array technology provides the possibility to solve these issues efficiently. In this article, surface-tension-confined droplet arrays were used to fabricate calcium carbonate. It was found that calcium carbonate crystallization can be conducted in surface-tension-confined droplets. Defects were found on the surface of some crystals, which were caused by liquid flow inside the droplet and the rapid drop in droplet height during the evaporation. The diameter and number of crystals were related to the droplet diameter. Polyacrylic acid (PAA), added as a modified organic molecule control, changed the $CaCO_3$ morphology from calcite to vaterite. The material products of the above experiments were compared with bulk-synthesized calcium carbonate by scanning electron microscopy (SEM), Raman spectroscopy and other characterization methods. Our work proves the possibility of performing biomimetic crystallization and biomineralization experiments on surface-tension-confined microdroplet arrays.

Keywords: droplet array; crystal growth; calcium carbonate; high-throughput; biomimetic crystallization; biomineralization; polyacrylic acid

1. Introduction

The inorganic-organic advanced hybrid materials formed by the biomineralization process have excellent physical and chemical properties, such as good wear resistance and extremely high fracture toughness and strength, which are unmatched by those of synthetic materials [1–3]. Calcium carbonate ($CaCO_3$) is one of the most abundant biominerals in nature [4–7]. The preparation of highly regulated $CaCO_3$ with fine structures under ambient conditions has attracted much attention [8–10]. Biomineralized calcium carbonate products have good biocompatibility and can be used not only as structural support for organisms [2] but also as biosensors [11], controlled released drug carriers [12,13], and so on. Nucleation and template effects in the context of $CaCO_3$ synthesis and crystalline phase changes have been studied for years [14,15]. During the biomineralization process, both the water-soluble fraction and the insoluble matrix of organic materials are considered to play essential roles [16]. Researchers have been able to control the morphology, nucleation, growth, and alignment of inorganic particles by using specific templates or macromolecules [17,18] The design and preparation of organic matrices (soluble and insoluble) has become an active area of biomineralization research [19,20].

Interactions between ionic surfactants and water-soluble polymers in aqueous solution have been studied with various techniques for decades [15,21].

However, conventional experiments to study the biomineralization of calcium carbonate are usually carried out in a beaker, which consumes a substantial amount of biomacromolecule reagents and makes it difficult to conduct high-throughput experiments. Some research groups have implemented advances in microfabrication techniques to create microstructure environments in which they can employ multiple strategies to control crystallization [22–24]. A reversibly sealed T-junction microfluidic device was fabricated by Yin et al. to investigate the influence of extrapallial (EP) fluid proteins in the polymorph control of crystal formation in mollusc shells [25]. Gong et al.'s research provided a new approach to biomimetic crystallization by using crystal hotels [26]. Zeng et al.'s work demonstrated a microfluidic approach towards the study of the formation and transformation of ACC (amorphous calcium carbonate) by using microfluidic technology [27]. However, these experiments were mainly carried out in microfluidic chips, which constrained the methods of characterization and hindered the use of high-throughput experiments.

In this work, calcium carbonate biomimetic crystallization experiments were conducted on a surface-tension-confined droplet array, which is a facile, controllable and high-throughput method to fabricate droplet arrays with controlled size, geometry and position on a patterned surface. Droplet arrays with different diameters from 50 μm to 700 μm were fabricated and used to carry out $CaCO_3$ crystallization experiments. The diameter and number of the $CaCO_3$ crystals were related to the droplet diameter and crystal surface defects were found on the crystals. In previous reports, some research groups have used ionic surfactant polyacrylic acid (PAA) to change the morphology of $CaCO_3$ crystals [21,28]. The crystallization of $CaCO_3$ particles from aqueous solutions in the absence and presence of PAA was studied in this work. It was found that different concentrations of PAA result in different morphologies of vaterite, such as dumbbells. The material products of the above experiments were compared with bulk-calculated calcium carbonate by scanning electron microscopy (SEM), Raman spectroscopy and other characterization methods.

2. Materials and Methods

2.1. Chemicals and Materials

AZ9260 photoresist and AZ400K developer were purchased from Suzhou Wenhao Microfluidic Technology Co., Ltd. (Suzhou, China). 1H,1H,2H,2H-Perfluorooctyltriethoxysilane (POTS) was obtained from Sigma–Aldrich (USA). Calcium chloride anhydrous and poly acrylic acid (PAA) with an average molecular weight of ca. 2000 were purchased from Aladdin (Shanghai, China). Sodium carbonate and other chemicals were obtained from Sinopharm Chemical Reagent Co., Ltd. (Shanghai, China). Deionized water (18.2 MΩ cm, S30CF, Master Touch, Shanghai, China) was used to prepare aqueous solutions. Silicon wafer substrates were purchased from Ningbo Sibranch International Trading Co., Ltd. (Ningbo, China).

2.2. Fabrication of Surface-Tension-Confined Droplet Arrays

A schematic diagram of STC (surface-tension-confined) technology is shown in Figure 1a. First, nonwettable substrates were fabricated. An air plasma cleaner (PDC-002, Harrick Plasma) was used to clean the glass substrates or silicon wafers, and then the POTS was evaporated onto the substrates at 120 °C for 1 h in an oven (FD115, Binder, Germany). Second, wettable patterns on the nonwettable surface were prepared. AZ9260 photoresist was spin-coated onto the substrate at 500 rpm for 6 s and subsequently at 2400 rpm for 60 s. After that, the coated substrate was baked on a hotplate at a constant temperature of 105 °C for 330 s. After baking, a designed photomask (Figure 1b) was placed on the substrate (several photomasks were designed) which was then exposed it to UV (Ultraviolet) light (MJB4, SUSS MicroTech). Next, the patterned substrate was developed in AZ400K aqueous solution (AZ400K: H_2O = 1:2) for 150 s, after which deionized water was used to remove the residual developer.

Subsequently, the developed substrate was treated with plasma for two more minutes to make the exposed area selectively wettable. The air plasma not only removed the contamination but also oxidized the exposed surface. Finally, the substrate was washed with acetone and ethyl alcohol to remove the photoresist coating on the silicon wafer. We have published a detailed description of this process [29].

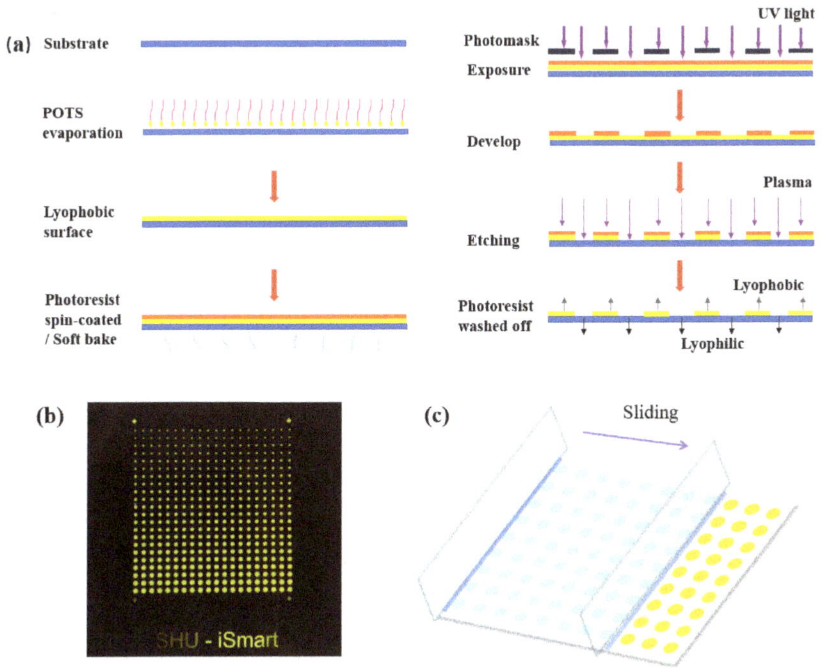

Figure 1. (**a**) Schematic illustration of the process followed to fabricate high-throughput surface-tension-confined microarrays. (**b**) Picture of lithographic mask with a spot gradient from 200 μm to 700 μm. (**c**) Schematic illustration of the sliding process, which forms the droplet array by sliding a liquid strip on the patterned substrate.

2.3. Fabrication of $CaCO_3$

$CaCO_3$ crystals were formed in small droplets that were generated on the STC chip. Equal volumes (100 μL) of equimolar aqueous solutions of calcium chloride (0.1 M) and sodium carbonate (0.1 M) prepared in deionized water in advance were first mixed on the chip. After setting up the sliding speed and slit height well, we used a piece of glass to slide the solutions from left to the right at a constant speed, as controlled by a custom-designed stepping motor (Figure 1c). Then, a long liquid strip was formed by capillary force. Next, we placed the STC chip into a petri dish in a constant temperature & humidity chamber (LHS-80HC-11, Bluepard, Shanghai, China) at 25 °C and 80% humidity. After 24 h of growth, the device was removed and dried on a 35 °C hot plate for 3 days. In this work, calcium carbonate was fabricated under two conditions, without PAA and with PAA mixed with $CaCl_2$ & Na_2CO_3. PAA was used at three concentrations: 0.1 mg/mL, 0.2 mg/mL and 0.4 mg/mL.

To compare the material products of the experiment described above with the bulk-synthesized calcium carbonate, a parallel comparison experiment was performed in a beaker. We mixed equal volumes (1 mL) of 0.1 M $CaCl_2$ and 0.1 M Na_2CO_3 solutions in a beaker in an ultrasonic water bath for 5 min. The remaining steps were performed as in the previous experiment and the conditions were controlled in the same way.

2.4. Characterization

The calcium carbonate crystal morphology was studied by field-emission scanning electron microscopy (FE-SEM, Regulus-8230, Hitachi, Tokyo, Japan) and an optical microscope (LV100ND, Nikon, Tokyo, Japan). Room-temperature Raman spectra were obtained using a standard micro-area Raman system. One droplet in ten was randomly selected for each size of droplet, and the crystals formed in the droplet were observed and counted using an optical microscope. The volumes of the droplets were calculated based on the equation of spherical cap. The process of droplet evaporation was observed under the microscope.

3. Results and Discussions

Calcium carbonate crystals were synthesized using a surface-tension-confined droplet array, and their morphology was changed by adding PAA. In addition, the results were analyzed and compared with that of calcium carbonate synthesized in the bulk phase.

3.1. Microdroplet Diameter Control

Calcium carbonate was synthesized in droplets with different diameters from 50 μm to 700 μm and observed using an optical microscope. In the surface-tension-confined microdroplets, the relationship between the contact angle and droplet diameter was reported in previous articles [30,31], from which the contact angle data are used in this part. The height and volume of the droplets can be calculated using Equations (1) and (2), respectively, which are based on the geometry of the droplet as a spherical cap. A cross-sectional view of a surface tension-confined droplet can be seen in Figure 2a, where d is the diameter of the droplet, h is the height of the droplet, v is the volumes of the droplet, and θ is the contact angle of the droplet. The relationship between the droplet diameter and liquid volume can be seen in Figure 2b. The volume of the droplet greatly affects the time required for complete evaporation. Figure 2c shows the time required for droplets of different diameters to evaporate completely. The volume of the droplets whose diameters are smaller than 100 μm is less than 200 picoliters. Thus, the solution evaporated less than 5 s-before the crystal growth and no crystals were observed in most of the droplets (more than 80%).

$$h = \frac{d}{2} * \tan\frac{\theta}{2} \tag{1}$$

$$V = \frac{\pi h\left(3d^2 + 4h^2\right)}{24} \tag{2}$$

(a)

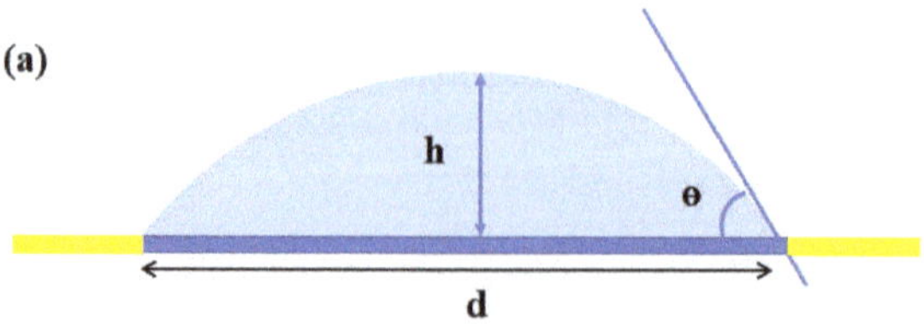

(a)

Figure 2. *Cont.*

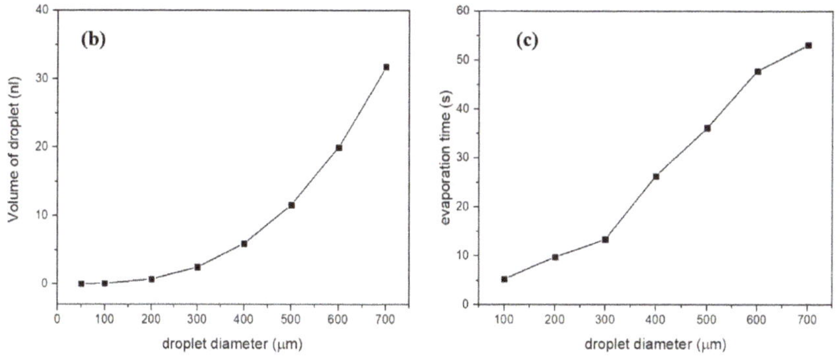

(b)

Figure 2. (**a**) Cross-sectional view of a surface-tension-confined droplet. (**b**) The relation between the droplet volume and droplet diameter. (**c**) The time required for evaporation of droplets with different diameters (25 °C, 30%RH).

As shown in Figure 3a, only one crystal of calcium carbonate with a size of less than 3 μm was observed in the 50 μm droplets. The $CaCO_3$ crystal was surrounded by amorphous precipitation and sodium chloride particles as a by-product of the reaction. Since the experiment was conducted in droplets on an open platform, all products remained on the surface except for volatile gases. The reaction by-product NaCl precipitates could exist only around the $CaCO_3$ crystal on the substrate after the solution in the droplets volatilized. As shown in Figure 3c, we observed more than 17 crystals of different sizes from 3 μm to 12 μm in a 400 μm droplet. Four crystals of $CaCO_3$ with diameters larger than 10 μm and approximately 10 crystals with diameters of approximately 3 μm were observed. The size of the crystal observed in the droplet was related to the droplet diameter. As shown in Figure 3d, the largest crystal size in a 300 μm diameter droplet is 7.24 μm, and in the image shown in Figure 3f we observed two crystals of over 20 μm in a 700 μm diameter droplet.

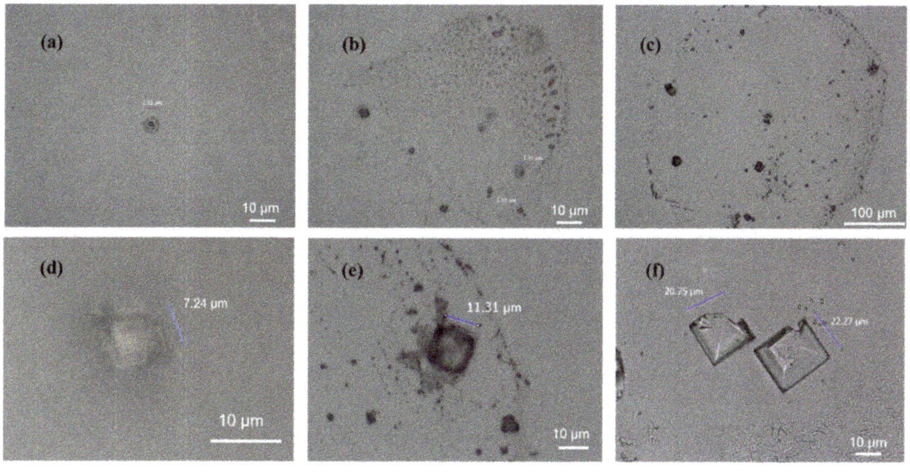

Figure 3. *Cont.*

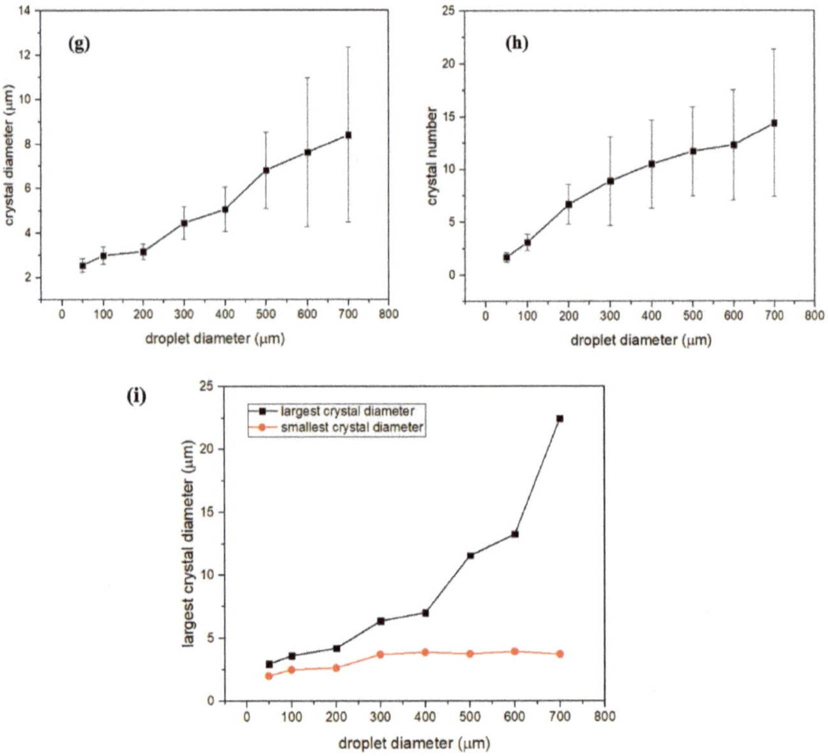

Figure 3. Optical micrograph of $CaCO_3$ formed in microdroplets at different diameters and statistical data graph: (**a**) 50 μm, (**b**) 100 μm, (**c**) 400 μm. (**d**), (**e**), (**f**) crystals formed in the droplet with 300 μm, 400 μm, 700 μm respectively. (**g**) The relation between average diameter of crystals and droplet diameter. (**h**) The relation between average amount of crystals and droplet diameter. (**i**) The relation largest & smallest crystal diameter of crystals and droplet diameter. Scale bars in (**a**), (**b**), (**d**), (**e**), (**f**) is 10 μm, Scale bars in (**c**) is 100 μm.

Statistically, it can clearly be seen that the size and number of $CaCO_3$ crystals are positively correlated with droplet diameter. As the diameter of the droplets increased from 50 μm to 700 μm, the average diameter of the crystals in the droplets grew from 3.17 μm to 8.42 μm (Figure 3g). As shown in Figure 3h, more crystals can be observed in the droplet as the droplet diameter became larger. As the droplet diameter increased from 50 μm to 700 μm, the maximum crystal diameter of the crystal formed in the droplets increased from 2.96 μm to 22.42 μm while the diameter of the smallest crystal was stabilized between 2–4 μm (Figure 3i). The size and number of crystals formed in droplets of 200 μm diameter are relatively stable. The size and number of crystals formed in the droplets were more dispersed as the diameter of the droplet increased. It can be explained that there is a competitive relationship between nucleation and crystal growth in droplets when the ions start to react. In the small droplets, since the liquid is an anisotropic environment and volatilizes faster, the relationship between crystal growth and nucleation is unbalanced.

3.2. *Crystals Formed without PAA*

SEM and representative Raman spectrum were used to further study the morphology and structure of $CaCO_3$ crystals formed in droplets. As shown in Figure 4a, several crystals of different sizes can be seen, and there are defects in the surface of some crystals. We used SEM to observe the crystals more clearly (Figure 4b). The majority of crystals were formed in cubic-shaped, corresponding to representative calcite spectra in the Raman spectrum [32] (Figure 4c). Among them, the peak at 520 cm^{-1} belonged to the silicon substrate. Compared with the synthesis of $CaCO_3$ crystals in the bulk phase (Figure 4d), the synthesis of $CaCO_3$ crystals in microdroplets has a greater chance of producing crystal surface defects. The effect of evaporation of droplets generated on the surface on the crystallization of materials has been studied in previous researchers' work, and we analyzed the causes of defects on the surface of calcium carbonate crystals based on their conclusions [33–35]. We supposed that the growth of crystals in microdroplets coexists with the volatilization of solution, which makes the evaporation rate of the microdroplets have a large effect on the crystal growth. Two kinds of droplet evaporation mode are shown in Figure 4e,f. In the experiment, droplets were first evaporated by a constant contact radius evaporation process and then evaporated in a constant contact angle or a mixture of two evaporation modes. The volatilization of the liquid caused the height of the droplet to decrease before the diameter shrink, resulting in insufficient ion accumulation of the upper crystal plane. We suspect that capillary flow and Marangoni flow play an important role in the evaporation progress of the sessile droplets as well [36]. Due to the existence of capillary flow and Marangoni flow during liquid volatilization, ions diffuse to the bottom of the microdroplets. These effects may result in insufficient ion accumulation in the upper layer of the crystal, which causes defects on the crystal surface.

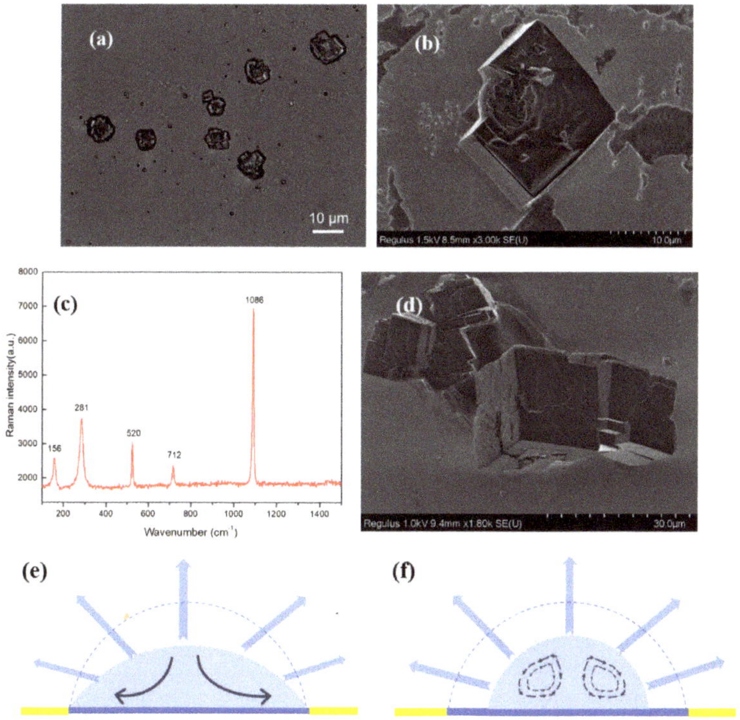

Figure 4. (**a**) Optical micrograph of $CaCO_3$ crystals formed in droplets; (**b**) Scanning electron microscopy (SEM) image of one crystal formed on a 400 μm diameter droplet. (**c**) Representative Raman spectrum of crystals with cubic-shaped, which shows its polymorph of calcite at 156, 281, 712, and 1086 cm^{-1}; (**d**) SEM image of crystals formed by bulk phase synthesis. (**e**) Constant contact radius evaporation and internal capillary flow. (**f**) Constant contact angle evaporation, internal Marangoni flow.

3.3. Crystals Formed with PAA

We further studied the crystallization of $CaCO_3$ particles from aqueous solutions in the presence of PAA. PAA (0.1 mg/mL) causes $CaCO_3$ crystals to aggregates to into a dumbbell shape (Figure 5a). The shape of the crystal becomes more rounded and grows into a spherical shape when the concentration of PAA rises (Figure 5b,c). The addition of PAA in bulk synthesis can also affect the growth of calcium carbonate. The increase of PAA concentration causes the proportion of spherical calcium carbonate to increase, and the spherical crystal is more rounded (Figure 5d–f). PAA changes both the nucleation frequency and the growth habit of crystals. In addition, the sphere-shaped crystals that formed in droplets show they Raman spectrum of vaterite [32] (Figure 5g). This polymer (PAA) concentration affects the crystallization and aggregation behavior of calcium carbonate. It has been reported in the literature that a bridging flocculation mechanism applies [37], which indicates that the morphology of the calcium carbonate microspheres is controlled by the bridging effect of polymers on nanosized particles. The structure of carboxylic acid in PAA is hydrophilic, while a large number of alkyl groups are hydrophobic, which makes PAA form a folded molecular structure in small droplets [38,39]. The carboxyl group induces calcium ion binding through electrostatic interactions to form a spherical template. The spherical template reduces the surface energy of vaterite $CaCO_3$, thereby stabilizing the $CaCO_3$ in the vaterite phase and preventing its conversion to thermodynamically stable calcite [14]. During the subsequent growth process, the steady-state phase of the vaterite nanoparticles is linked by the PAA molecular chain to form a vaterite carbonate microsphere with a particle size of 5–10 microns. Pan et al. observed a similar erosion behavior of calcite rhombs when only PAA was used [15], giving rise to rough rhombs with uneven surfaces due to strong interactions between PAA carboxyl acid groups and $CaCO_3$. Their experiments resulted in similar mean diameters of the calcite grains, even if the latter were transformed into hollow spheres with the addition of surfactants. As seen from the Figure 5h, the position of the crystallization in the droplet changes after the addition of the PAA. Small droplets are easily split into microdroplets before volatilization due to the changes of the solution surface activity caused by the surfactant PAA. The ions in the liquid are dispersed to the periphery of the droplets, so that the formed crystals are deposited around the droplets to form the morphology shown in Figure 5h.

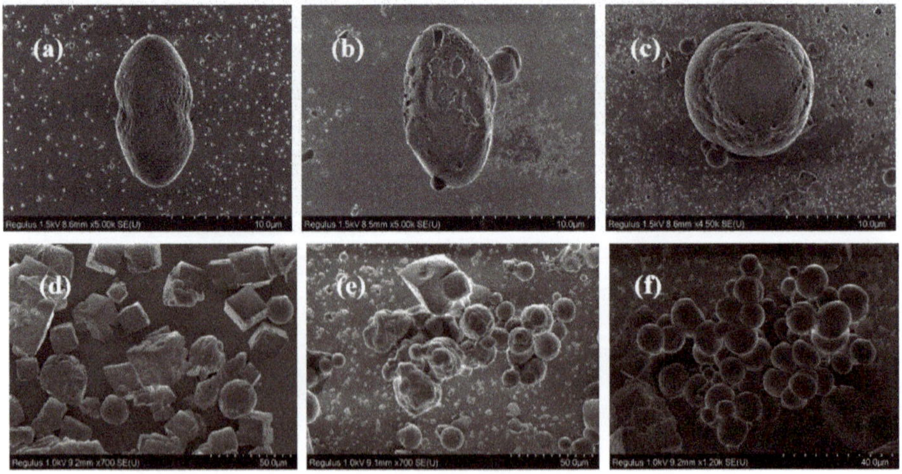

Figure 5. *Cont.*

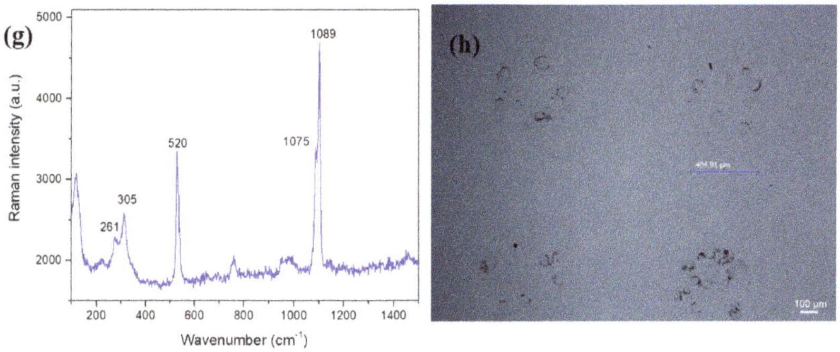

Figure 5. SEM images of crystals formed in a droplet with a polyacrylic acid (PAA) content of (**a**) 0.1 mg/mL, (**b**) 0.2 mg/mL and (**c**) 0.4 mg/mL; SEM images of crystals formed on bulk phase synthesis with PAA(**d**) 0.1 mg/mL, (**e**) 0.2 mg/mL and (**f**) 0.4 mg/mL; (**g**) Representative Raman spectrum of sphere-shaped crystals, which has the characteristic peaks of vaterite at 261, 305, 1075, and 1089 cm^{-1}; (**h**) Optical micrograph of $CaCO_3$ crystals formed in 400 μm diameter droplets.

4. Conclusions

In this work, calcium carbonate was synthesized by using a surface-tension-confined droplets array, and PAA was used to change the morphology of calcium carbonate. Our approach took advantage of STC technology, which is a facile, controllable and high-throughput method to fabricate droplet arrays with controlled size, geometry and position on a patterned surface. Calcite formed in the small droplets without PAA, and the crystal size and number were positively correlated with the droplet diameter. The size and number of the crystals formed in the droplet became more dispersed as the diameter of the droplet increased. The maximum crystal diameter of the crystal formed in the droplet increased from 2.96 μm to 22.42 μm while the diameter of the smallest crystal is stabilized between 2–4 μm as the droplet diameter increased. It has been found that calcite formed in small droplets tends to exhibit defects on the surface of the crystal compared to the case of bulk-synthesized calcium carbonate, which is related to the height of the droplets and the mode of liquid evaporation. Vaterite was formed and the position of crystallization in the droplet changed when the ionic surfactant PAA was added. Since a silicon wafer is used as the substrate, some preparation steps can be simpler when $CaCO_3$ is characterized by Raman spectroscopy and SEM—just cut the wafer to the appropriate shape. The 200 μm diameter droplet will be selected for $CaCO_3$ biomineralization experiments in future work because of the relatively stable crystal size and quantity formed. With this droplet diameter we can make a 100*100 array chip on a 4-inch silicon wafer, providing 10,000 volumes of 0.7 nanoliter droplets formed. This step will help us to achieve high-throughput $CaCO_3$ biomimetic crystallization and biomineralization experiments with less reagent consumption in subsequent work.

Author Contributions: Z.H. performed all the experiments, interpreted the results and prepared the manuscript. J.W. and M.Z. helped to designed and directed the experiments. Z.X. participated in the discussion and helped write the manuscript. W.W. supervised this work.

Funding: This research was funded by the National Natural Science Foundation (Grants No. 11674210) of China.

Acknowledgments: The authors wish to thank Y Gu, C Sun, H Du, J Liu for their characterization assistance and discussions.

Conflicts of Interest: The authors declare no conflict of interest.

References

1. Matsumura, S.; Kajiyama, S.; Nishimura, T.; Kato, T. Formation of Helically Structured Chitin/$CaCO_3$ Hybrids through an Approach Inspired by the Biomineralization Processes of Crustacean Cuticles. *Small* **2015**, *11*, 5127–5133. [CrossRef]
2. Jin, W.; Jiang, S.; Pan, H.; Tang, R. Amorphous Phase Mediated Crystallization: Fundamentals of Biomineralization. *Crystals* **2018**, *8*, 48. [CrossRef]
3. Mann, S. Molecular tectonics in biomineralization and biomimetic materials chemistry. *Nature* **1993**, *365*, 499–505. [CrossRef]
4. Pokroy, B.; Zolotoyabko, E.; Adir, N. Purification and Functional Analysis of a 40 kD Protein Extracted from the Strombus decorus persicus Mollusk Shells. *Biomacromolecules* **2006**, *7*, 550–556. [CrossRef] [PubMed]
5. Sun, W.; Jayaraman, S.; Chen, W.; Persson, K.A.; Ceder, G. Nucleation of metastable aragonite $CaCO_3$ in seawater. *Proc. Natl. Acad. Sci. USA* **2015**, *112*, 3199–3204. [CrossRef] [PubMed]
6. Zou, Z.; Habraken, W.J.E.M.; Matveeva, G.; Jensen, A.C.S.; Bertinetti, L.; Hood, M.A.; Sun, C.; Gilbert, P.U.P.A.; Polishchuk, I.; Pokroy, B.; et al. A hydrated crystalline calcium carbonate phase: Calcium carbonate hemihydrate. *Science* **2019**, *363*, 396–400. [CrossRef] [PubMed]
7. Cusack, M.; Freer, A. Biomineralization: Elemental and organic influence in carbonate systems. *Chem. Rev.* **2008**, *108*, 4433–4454. [CrossRef] [PubMed]
8. Shahlori, R.; McDougall, D.R.; Waterhouse, G.I.N.; Yao, F.; Mata, J.P.; Nelson, A.R.J.; McGillivray, D.J. Biomineralization of Calcium Phosphate and Calcium Carbonate within Iridescent Chitosan/Iota-Carrageenan Multilayered Films. *Langmuir* **2018**, *34*, 8994–9003. [CrossRef]
9. Kuang, W.; Liu, Z.; Yu, H.; Kang, G.; Jie, X.; Jin, Y.; Cao, Y. Investigation of internal concentration polarization reduction in forward osmosis membrane using nano-$CaCO_3$ particles as sacrificial component. *J. Membr. Sci.* **2016**, *497*, 485–493. [CrossRef]
10. Li, M.; Chen, Y.; Mao, L.B.; Jiang, Y.; Liu, M.F.; Huang, Q.; Yu, Z.; Wang, S.; Yu, S.H.; Lin, C.; et al. Seeded Mineralization Leads to Hierarchical $CaCO_3$ Thin Coatings on Fibers for Oil/Water Separation Applications. *Langmuir* **2018**, *34*, 2942–2951. [CrossRef]
11. Bujduveanu, M.-R.; Yao, W.; Le Goff, A.; Gorgy, K.; Shan, D.; Diao, G.-W.; Ungureanu, E.-M.; Cosnier, S. Multiwalled Carbon Nanotube-$CaCO_3$Nanoparticle Composites for the Construction of a Tyrosinase-Based Amperometric Dopamine Biosensor. *Electroanalysis* **2013**, *25*, 613–619. [CrossRef]
12. Dong, Z.; Feng, L.; Hao, Y.; Chen, M.; Gao, M.; Chao, Y.; Zhao, H.; Zhu, W.; Liu, J.; Liang, C.; et al. Synthesis of Hollow Biomineralized $CaCO_3$-Polydopamine Nanoparticles for Multimodal Imaging-Guided Cancer Photodynamic Therapy with Reduced Skin Photosensitivity. *J. Am. Chem. Soc.* **2018**, *140*, 2165–2178. [CrossRef]
13. Hanafy, N.A.N.; El-Kemary, M.; Leporatti, S. Reduction diameter of $CaCO_3$ crystals by using poly acrylic acid might improve cellular uptake of encapsulated curcumin in breast cancer. *J. Nanomed. Res.* **2018**, *7*, 235–239. [CrossRef]
14. Sommerdijk, N.A.J.M.; With, G. Biomimetic $CaCO_3$ Mineralization using Designer Molecules and Interfaces. *Chem. Rev.* **2008**, *108*, 4499–4550. [CrossRef] [PubMed]
15. Pan, Y.; Guo, Y.-P.; Zhao, X.; Wang, Z. Influence of Surfactant-polymer Complexes on Crystallization and Aggregation of $CaCO_3$. *Chem. Res. Chin. Univ.* **2012**, *28*, 737–742.
16. Liu, M.-F.; Lu, Z.; Zhang, Z.; Xiao, C.; Li, M.; Huang, Y.-X.; Liu, X.Y.; Jiang, Y. Correlations of crystal shape and lateral orientation in bioinspired $CaCO_3$ mineralization. *CrystEngComm* **2018**, *20*, 5241–5248. [CrossRef]
17. Xu, A.W.; Antonietti, M.; Cölfen, H.; Fang, Y.P. Uniform Hexagonal Plates of Vaterite $CaCO_3$ Mesocrystals Formed by Biomimetic Mineralization. *Adv. Funct. Mater.* **2006**, *16*, 903–908. [CrossRef]
18. Meier, A.; Kastner, A.; Harries, D.; Wierzbicka-Wieczorek, M.; Majzlan, J.; Büchel, G.; Kothe, E. Calcium carbonates: Induced biomineralization with controlled macromorphology. *Biogeosciences* **2017**, *14*, 4867–4878. [CrossRef]
19. He, L.; Xue, R.; Song, R. Formation of calcium carbonate films on chitosan substrates in the presence of polyacrylic acid. *J. Solid State Chem.* **2009**, *182*, 1082–1087. [CrossRef]
20. Shen, Y.; Nyström, G.; Mezzenga, R. Amyloid Fibrils form Hybrid Colloidal Gels and Aerogels with Dispersed $CaCO_3$ Nanoparticles. *Adv. Funct. Mater.* **2017**, *27*, 1700897. [CrossRef]

21. Ouhenia, S.; Chateigner, D.; Belkhir, M.A.; Guilmeau, E.; Krauss, C. Synthesis of calcium carbonate polymorphs in the presence of polyacrylic acid. *J. Cryst. Growth* **2008**, *310*, 2832–2841. [CrossRef]
22. Aizenberg, J.; Black, A.J.; Whitesides, G.M. Control of crystal nucleation by patterned self-assembled monolayers. *Nature* **1999**, *398*, 495–498. [CrossRef]
23. Li, L.; Sanchez, J.R.; Kohler, F.; Røyne, A.; Dysthe, D.K. Microfluidic Control of Nucleation and Growth of $CaCO_3$. *Cryst. Growth Des.* **2018**, *18*, 4528–4535. [CrossRef]
24. Komatsu, S.; Ikedo, Y.; Asoh, T.A.; Ishihara, R.; Kikuchi, A. Fabrication of Hybrid Capsules via $CaCO_3$ Crystallization on Degradable Coacervate Droplets. *Langmuir* **2018**, *34*, 3981–3986. [CrossRef] [PubMed]
25. Yin, H.; Ji, B.; Dobson, P.S.; Mosbahi, K.; Glidle, A.; Gadegaard, N.; Freer, A.; Cooper, J.M.; Cusack, M. Screening of Biomineralization using Microfluidics. *Anal. Chem.* **2009**, *81*, 473–478. [CrossRef] [PubMed]
26. Gong, X.; Wang, Y.W.; Ihli, J.; Kim, Y.Y.; Li, S.; Walshaw, R.; Chen, L.; Meldrum, F.C. The Crystal Hotel: A Microfluidic Approach to Biomimetic Crystallization. *Adv. Mater.* **2015**, *27*, 7395–7400. [CrossRef] [PubMed]
27. Zeng, Y.; Cao, J.; Wang, Z.; Guo, J.; Lu, J. Formation of Amorphous Calcium Carbonate and Its Transformation Mechanism to Crystalline $CaCO_3$ in Laminar Microfluidics. *Cryst. Growth Des.* **2018**, *18*, 1710–1721. [CrossRef]
28. Tang, Y.; Yang, W.; Yin, X.; Liu, Y.; Yin, P.; Wang, J. Investigation of $CaCO_3$ scale inhibition by PAA, ATMP and PAPEMP. *Desalination* **2008**, *228*, 55–60. [CrossRef]
29. Lin, Y.; Wu, Z.; Gao, Y.; Wu, J.; Wen, W. High-throughput controllable generation of droplet arrays with low consumption. *Appl. Surf. Sci.* **2018**, *442*, 189–194. [CrossRef]
30. Lin, Y.; Wu, Z.; Zhang, M.; Wu, J.; Wen, W. Lateral Size Scaling Effect during Discontinuous Dewetting. *Adv. Mater. Int.* **2018**, *5*, 1800729. [CrossRef]
31. Wu, H.; Chen, X.; Gao, X.; Zhang, M.; Wu, J.; Wen, W. High-Throughput Generation of Durable Droplet Arrays for Single-Cell Encapsulation, Culture, and Monitoring. *Anal. Chem.* **2018**, *90*, 4303–4309. [CrossRef]
32. Dandeu, A.; Humbert, B.; Carteret, C.; Muhr, H.; Plasari, E.; Bossoutrot, J.M. Raman Spectroscopy—A Powerful Tool for the Quantitative Determination of the Composition of Polymorph Mixtures: Application to $CaCO_3$ Polymorph Mixtures. *Chem. Eng. Technol.* **2006**, *29*, 221–225. [CrossRef]
33. Shahidzadeh, N.; Schut, M.F.; Desarnaud, J.; Prat, M.; Bonn, D. Salt stains from evaporating droplets. *Sci Rep.* **2015**, *5*, 10335. [CrossRef]
34. Shahidzadeh-Bonn, N.; Rafaı, S.; Bonn, D.; Wegdam, G. Salt Crystallization during Evaporation: Impact of Interfacial Properties. *Langmuir* **2008**, *24*, 8599–8605. [CrossRef] [PubMed]
35. Przybylek, M.; Cysewski, P.; Pawelec, M.; Ziolkowska, D.; Kobierski, M. On the origin of surface imposed anisotropic growth of salicylic and acetylsalicylic acids crystals during droplet evaporation. *J. Mol. Model.* **2015**, *21*, 49. [CrossRef]
36. Wu, Z.; Lin, Y.; Xing, J.; Zhang, M.; Wu, J. Surface-tension-confined assembly of a metal–organic framework in femtoliter droplet arrays. *RSC Adv.* **2018**, *8*, 3680–3686. [CrossRef]
37. Pan, Y.; Zhao, X.; Guo, Y.; Lv, X.; Ren, S.; Yuan, M.; Wang, Z. Controlled synthesis of hollow calcite microspheres modulated by polyacrylic acid and sodium dodecyl sulfonate. *Mater. Lett.* **2007**, *61*, 2810–2813. [CrossRef]
38. Schmidt, B.V.; Fechler, N.; Falkenhagen, J.; Lutz, J.F. Controlled folding of synthetic polymer chains through the formation of positionable covalent bridges. *Nat. Chem.* **2011**, *3*, 234–238. [CrossRef]
39. Guan, L.; Xu, H.; Huang, D. The investigation on states of water in different hydrophilic polymers by DSC and FTIR. *J. Polym. Res.* **2010**, *18*, 681–689. [CrossRef]

crystals

MDPI

Article

Induced Nucleation of Biomimetic Nanoapatites on Exfoliated Graphene Biomolecule Flakes by Vapor Diffusion in Microdroplets

Jaime Gómez-Morales [1,*], Luis Antonio González-Ramírez [1], Cristóbal Verdugo-Escamilla [1], Raquel Fernández Penas [1], Francesca Oltolina [2], Maria Prat [2] and Giuseppe Falini [3,*]

1 Laboratorio de Estudios Cristalográficos, IACT, CSIC-UGR. Avda. Las Palmeras 4, 18100 Armilla, Granada, Spain

2 Dipartimento di Scienze della Salute, Università del Piemonte Orientale, Via Solaroli, 17, 28100 Novara, Italy

3 Dipartimento di Chimica "Giacomo Ciamician", Alma Mater Studiorum Università di Bologna, via Selmi 2, 40126 Bologna, Italy

* Correspondence: jaime@lec.csic.es (J.G.-M.); giuseppe.falini@unibo.it (G.F.); Tel.: +34-958-230000 (J.G.-M.); +39-051-209-9484 (G.F.)

Received: 4 June 2019; Accepted: 1 July 2019; Published: 3 July 2019

Abstract: The nucleation of apatite nanoparticles on exfoliated graphene nanoflakes has been successfully carried out by the sitting drop vapor diffusion method, with the aim of producing cytocompatible hybrid nanocomposites of both components. The graphene flakes were prepared by the sonication-assisted, liquid-phase exfoliation technique, using the following biomolecules as dispersing surfactants: lysozyme, L-tryptophan, N-acetyl-D-glucosamine, and chitosan. Results from mineralogical, spectroscopic, and microscopic characterization (X-ray diffraction (XRD), Fourier transform infrared spectroscopy (FTIR), Raman, Variable pressure scanning electron microscopy (VPSEM), and transmission electron microscopy (TEM)) indicate that flakes were stacked in multilayers (>5 layers) and most likely intercalated and functionalized with the biomolecules, while the apatite nanoparticles were found forming a coating on the graphene surfaces. It is worthwhile to mention that when using chitosan-exfoliated graphene, the composites were more homogeneous than when using the other biomolecule graphene flakes, suggesting that this polysaccharide, extremely rich in –OH groups, must be arranged on the graphene surface with the –OH groups pointing toward the solution, forming a more regular pattern for apatite nucleation. The findings by XRD and morphological analysis point to the role of "functionalized graphene" as a template, which induces heterogeneous nucleation and favors the growth of apatite on the flakes' surfaces. The cytocompatibility tests of the resulting composites, evaluated by the 3-(4,5-Dimethylthiazol-2-yl)-2,5-diphenyltetrazolium bromide (MTT) colorimetric assay in a dose–dependent manner on GTL-16 cells, a human gastric carcinoma cell line, and on m17.ASC cells, a murine mesenchymal stem cell line with osteogenic potential, reveal that in all cases, full cytocompatibility was found.

Keywords: nanoapatites; graphene; crystallization; nanocomposites; lysozyme; L-tryptophan; N-acetyl-D-glucosamine; chitosan; MTT assay; GTL-16 cells

1. Introduction

In the last decade there has been increasing interest in the preparation of graphene/apatite hybrid nanocomposites and their derivatives, combining the bioactive properties of the nanocrystalline apatite with the mechanical strength of graphene for uses in load-bearing applications in bone tissue engineering [1]. Nanocrystalline apatites are the main inorganic components of bone and dentine. Compared to stoichiometric hydroxyapatite $Ca_5(OH)(PO_4)_3$, which is the most stable calcium phosphate at ambient temperature, nanocrystalline apatites are non-stoichiometric calcium- and OH- deficient, plate-shaped, and present a series of substituted ions within their crystal structure, including carbonate (4%–6%), Na (0.9%), Mg (0.5%), and other minor elements [2]. Biomimetic nanocrystalline apatites exhibit excellent biological properties, such as biocompatibility, osteoconductivity, and osteoinductivity [2]. In clinical practice, for applications in dental and orthopedic surgery, they are usually employed as a bioactive coating on a metallic support, combining the biological properties of the apatite with the mechanical strength of the support, thus promoting a fast fixation within the surrounding bony tissue. To this end, different calcium phosphate (CaP) coating methods, employing biocompatible titanium and titanium alloy supports, were developed in the past [3–7]. When the apatite is used as a synthetic bone graft, however, because of its intrinsic brittleness and low fracture toughness, the use of a reinforcing material to form a composite scaffold with better mechanical properties is necessary. The presence of this second component in the composite is aimed at improving the long-term functionality of the graft after clinical surgery, under load-bearing conditions. Among the used materials, it is worth mentioning different polymers, such as poly(L-lactic acid) (PLA) [8], poly(ε-caprolactone) (PCL) [9], ceramics [10], carbon nanotubes [11], and more recently, graphene or graphene derivatives [12]. Few of the materials tested displayed favorable biocompatibility with sufficient mechanical properties [12]. For example, PCL/apatite composites displayed excellent bioactivity and improved mechanical properties by increasing the apatite content [9]. However, carbon nanotubes are good reinforcing materials, but still present certain toxicity [12]. The use of graphene has been proposed as an alternative component to overcome these troubles.

Graphene is the model of a two-dimensional (2D) nanomaterial, which consists of a single sheet of carbon atoms covalently bonded in a hexagonal network, therefore presenting an extremely large specific surface area (theoretically 2700 m^2/g). The importance of this material has been recognized since it was isolated in 2004 [13]. This carbon allotrope exhibits exceptional properties, such as high mechanical strength, optimal thermal conductivity, and excellent electrical conductivity [14,15]. Graphene has shown promise as 2D scaffold for controlled growth and the osteogenic differentiation of human mesenchymal stem cells [16], or as a substrate to promote the adherence of human osteoblasts and mesenchymal stromal cells [17]. In addition, graphene and its derivatives, such as graphene oxide and reduced graphene oxide, have shown a good biocompatibility, which is a requirement for their biomedical applications [18].

Different preparation methods of hybrid graphene/apatite nanocomposites and their derivative nanomaterials have been proposed over the last 10 years of research, including in situ synthesis, biomimetic mineralization, hydrothermal synthesis, and chemical vapor deposition [1]. During this period, we reported for the first time a new methodology to precipitate biomimetic apatite nanoparticles [19] and other calcium phosphates [20] in microliter droplets by the vapor diffusion sitting drop (VDSD) micromethod. Vapor diffusion in milliliter scale vessels was employed contemporaneously by other research teams to obtain biomimetic calcium phosphates [21]. The VDSD method was previously employed for in vitro studies of $CaCO_3$ biocrystallization, in both the absence and presence of proteins [22,23] and biological fluids [24]. The vapor diffusion technique was also used to precipitate $CaCO_3$ calcite single crystals induced by the graphene biomolecule adduct [25]. The two main features of the VDSD technique are the control of the gas diffusion rate of NH_3 and CO_2 by simply changing the concentration of NH_4HCO_3 in a gas generation chamber, which acts as a reagent reservoir, and the confinement of the nucleation in isolated microdroplets located in a precipitation chamber. These microdroplets closely mimic the in vivo microenvironments where biominerals form. The disadvantage of the technique is the very small amount of precipitate for its subsequent characterization. Our team has recently used this method to induce the heterogeneous nucleation and growth of CaP films on mica sheets [26]. In the present work, we propose to extend this precipitation micromethod to induce the heterogeneous nucleation of nanocrystalline apatites on exfoliated graphene flakes, as the model of a bidimensional material scaffold. The final goal is to assess the usefulness of this methodology to obtain cytocompatible hybrid nanocomposites consisting of graphene and biomimetic apatite.

2. Materials and Methods

2.1. *Exfoliated Graphene Flakes*

Graphene flakes were prepared by the sonication-assisted liquid-phase exfoliation (LPE) technique [27]. Basically, the graphene flakes were prepared by exfoliating graphite suspensions in an ultrasonic bath in the presence of the following biomolecules, acting as dispersing surfactants (Figure 1, left): lysozyme (Lys), L-tryptophan (Try), N-acetyl-D-glucosamine (NAce), and chitosan (Chi) (high purity > 99%, from Sigma-Aldrich S.r.l., Milan, Italy). Typically, 20 mL glass vials containing 10 mL of water suspensions, prepared by mixing 100 mg of graphite powder (Sigma-Aldrich, purity 99.99%) and either 1.44 mg Lys, 2 mg Try, 2.2 mg NAce, or 40 mg Chi in ultrapure water, were sonicated in an ice bath for 5 hours. The suspensions were centrifuged at 3500 rpm for 5 min, and the supernatant was again sonicated for 1 hour and centrifuged for 5 min. After centrifugation, the supernatant was carefully removed with a micropipette. The as-prepared graphene dispersions were stored at 4 °C prior to the mineralization experiments. The dispersions were stable for more than one week.

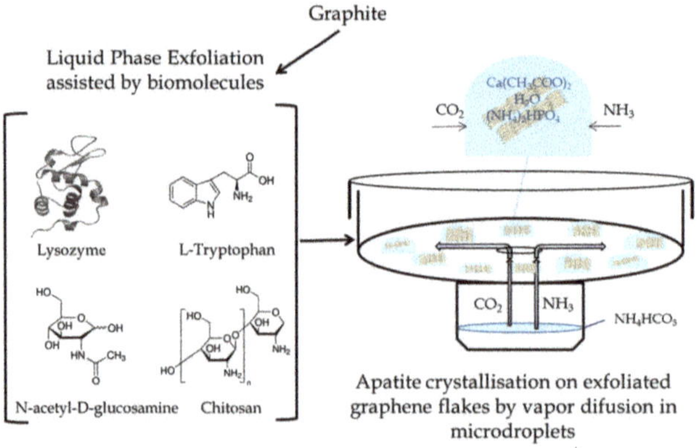

Figure 1. Schema showing the preparation of graphene flakes by sonication-assisted liquid-phase exfoliation of graphite in the presence of either lysozyme (Lys), L-tryptophan (Try), N-acetyl-D-glucosamine (NAce), or chitosan (Chi), and the experiment of the mineralization of the graphene flakes with apatites implemented in a "crystallization mushroom".

2.2. *Precipitation Method*

Precipitation experiments were carried by the VDSD method on several "crystallization mushrooms" (Triana Sci. & Tech, S.L., Armilla, Spain) at 20 °C and 1 atm total pressure (Figure 1, right). This microdevice, made of glass Pyrex, is composed of two cylindrical chambers connected through a hole of 6 mm diameter to allow the vapor diffusion [19]. Each mushroom hosted 10 droplets of 40 μL each in the upper chamber (crystallization zone), prepared by mixing 20 μL of each graphene suspension plus 2, 10, or 50 mM $Ca(CH_3COO)_2$ with 20 μL of 1.2, 6.0, or 30.0 mM $(NH_4)_2HPO_4$ to reach a final Ca/P ratio of 5:3. The lowest chamber (reservoir of reagent for generation of CO_2 and NH_3) of the different devices contained 3 mL of a 40 mM NH_4HCO_3 aqueous solution. The concentrations of $Ca(CH_3COO)_2$, $(NH_4)_2HPO_4$, and NH_4HCO_3 were optimized in a previous work to produce the nanosized apatite [19]. Control droplets without graphene were included in each mushroom. The glass cover and the upper chamber of the mushrooms were sealed with silicone grease, to isolate the experiment from the surrounding atmosphere. Experiments lasted five days. At the end of the experiments, the mushrooms were opened and the specimens were rinsed with deionized water and left to dry at ambient temperature for one day. Only the experiments performed with the highest $Ca(CH_3COO)_2$ and $(NH_4)_2HPO_4$ concentrations yielded enough precipitate for further characterization. The precipitated phase by this technique is a Ca- and OH-deficient carbonate apatite, whose average formula is written as $Ca_{(5\text{-}x)}OH_{(1\text{-}x)}(PO_4)_{(3\text{-}x)}(CO_3)_x$, with $0 \leq x \leq 1$ [20].

2.3. Characterization

Characterizations were carried out by X-ray diffraction (XRD), Fourier transform infrared spectroscopy (FTIR), Raman microspectroscopy, and electron microscopies (variable pressure scanning electron microscopy (VPSEM) and transmission electron microscopy (TEM)). XRD was performed using Cu Kα radiation (1.5418 Å) on a PANalytical X'Pert PRO diffractometer (Almelo, Netherlands) equipped with a PIXcel detector, operating at 45 kV and 40 mA. For the incident and diffracted beams, automatic-variable antiscatter slits with a constant irradiated length of 10 mm were used. The 2θ range was from 4° to 80°, with a step size of 0.026°. Raman microspectroscopy and FTIR characterizations were performed with a JASCO NRS-5100 Micro-Raman spectrometer (JASCO, Tokyo, Japan) (λ_{exc} = 532 nm), and a JASCO 6200 spectrometer (JASCO, Tokyo, Japan) provided of an attenuated total reflectance (ATR) accessory of diamond crystal, respectively. Scanning electron microscopy was performed with a variable-pressure Zeiss SUPRA40VP scanning electron microscope (VPSEM) (Carl Zeiss, Jena, Germany), coupled to a Renishaw inVia SCA-Raman spectrometer (λ_{exc} = 532 nm). Transmission electron microscopy (TEM) and selected area electron diffractions (SAED) were performed with a Carl Zeiss Libra 120 TEM microscope (Carl Zeiss, Jena, Germany) operating at 80 kV. The powder samples were ultrasonically dispersed in ethanol (absolute, ≥ 99.8%), and then a few droplets of the slurry were deposited on formvar-coated copper microgrids prior to observation.

2.4. The Cytocompatibility of Graphene–Apatite Nanocomposites

GTL-16 (a human gastric carcinoma cell line, obtained after cloning the MNK45 cell line, and at passage 70) [28] cells and m17.ASC (a spontaneously immortalized mouse mesenchymal stem cell clone from subcutaneous adipose tissue at passage 103) [29] cells (12,000 and 5000 cells/well in 96-well plates, respectively) were incubated for 24 hours. The different concentrations of the different graphene–apatite nanocomposites, ranging from 0.1 to 100 μg/ml, were added in 100 μL of fresh medium. After 72 hours incubation, cell viability was evaluated by the 3-(4,5-Dimethylthiazol-¬2-yl)-2,5-diphenyltetrazolium bromide (MTT; Sigma) colorimetric assay. Briefly, 20 μL of MTT solution (5 mg/ml in a phosphate buffered saline (PBS) solution, pH = 7.2) were added to each well. The plate was then incubated at 37 °C for 2 hours. After the removal of the solution, 125 μl of isopropanol and 0.2 M HCl were added to dissolve the formazan crystals. Then, 100 μl were removed carefully, and the optical density was measured in a multi-well reader (2030 Multilabel Reader Victor TM X4, PerkinElmer) at 570 nm. The viability of parallel cultures of untreated cells was taken as 100% viability, and values obtained from cells undergoing the different treatments were referred to this value. Experiments were performed four times, using three replicates for each sample.

3. Results and Discussion

3.1. Crystallographic and Spectroscopic Features of the Nanocomposites

The XRD patterns of the samples precipitated in the presence of graphene nanoflakes prepared by the LPE technique in the presence of Chi, NAce, Try, and Lys are reported in Figure 2. All samples show rather similar patterns. They display the main distinguishing reflections of the apatite phase (PDF 01-1008), with peaks at 25.87° 2θ corresponding to the (002) plane; peaks at 31.77°, 32.19°, and 32.90° 2θ, corresponding to the planes (211), (112), and (300), respectively; reflections at 33.9° and 39.81° (planes (202) and 310), respectively); and other minor peaks in the 2θ range from 40–55°. For the sake of comparison, the XRD pattern of a sample of mature, nanocrystalline-carbonated apatite prepared by the citrate-based thermal decomplexing method [30] is plotted at the bottom of the graph.

The reflections in the 2θ range of 31–34° are seen as a broad unresolved peak in all experiments, reflecting the nanocrystalline nature of the apatite particles.

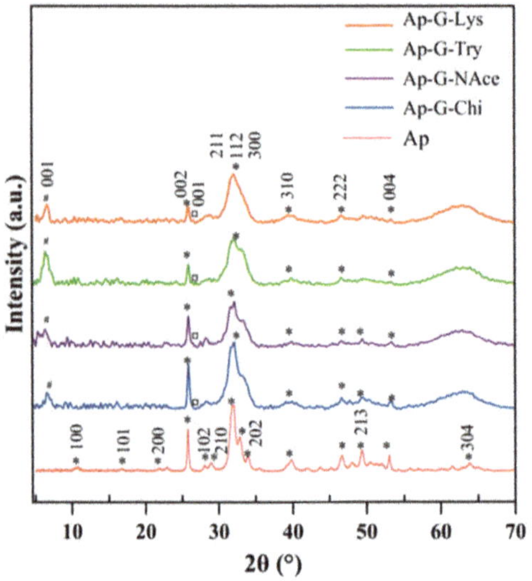

Figure 2. X-ray diffraction (XRD) patterns of nanocrystalline apatite nucleated by the vapor diffusion sitting drop (VDSD) method on graphene nanoflakes prepared by liquid-phase exfoliation (LPE) in the presence of Lys, Try, NAcel and Chi. (*) denotes apatite, (#) denotes the (001) plane of graphene, and (▫) denotes the (001) plane of residual graphite.

The absence of octacalcium phosphate (OCP) in the samples is witnessed by the lack of its main reflection at 4.74°, or plane (100) (OCP, powder diffraction file PDF 44-0778). The OCP phase was found to be a precursor of experiments of precipitation of apatite by VDSD and gel methods, carried out in absence of supports [19,31], and precipitated and stabilized in the experiments performed in the presence of mica muscovite sheets [15]. These differences reveal that in the presence of functionalized graphene flakes in the crystallization media, the heterogeneous nucleation and growth of the apatite phase could be directly induced on the surface of the nanoflakes, without the precipitation of the precursor OCP phase; this event significantly modifies the progression of the apatite precipitation in respect to previous experiments [15,19].

On the other hand, the reflections at around 6–7° 2θ most likely correspond to the (001) basal plane of the exfoliated graphene, similar to what was found for graphite oxide (GO) [30]. The shift of this reflection from around 10° 2θ to lower angles (higher *d*-spacings) in the case of graphite oxide exfoliated in the presence of some polymers was reported to be caused by the increase in both the relative humidity and the concentration of polymers that intercalated the GO lattice. These authors found that at the higher polymer concentration used the GO appeared to be completely exfoliated, based on the absence of any *d* (001) peak [32]. In our experiments, the presence of the (001) reflection at lower angles reflects the intercalation of the used biomolecules, in addition to H_2O, and the presence of a small peak at 26.6° 2θ would correspond to the (001) reflection of the residual, non-exfoliated graphite.

The characterization of the spectral features of the composites in the 1800–400 cm^{-1} range is shown in Figure 3a. The 4000–1800 cm^{-1} region (not shown) exhibits a broadband between 3600 cm^{-1} and 2800 cm^{-1}, corresponding to –OH stretching of adsorbed water, while no clear indication of apatitic –OH bands around 3580 cm^{-1} was found, as is usual for biomimetic apatites. The spectral region from 400 to 1800 cm^{-1} was analyzed in more detail, allowing the basic visualization of the spectral features of the apatitic part of the composite. Only a small band at around 1570 cm^{-1}, related to NH_2

bending vibration, and a band of negligible intensity at around 1650 cm^{-1}, related to carboxamide O=C–NHR [33] in the spectra Ap–G–NAce and Ap–G–Chi, reveal the presence of the biopolymers. These bands are absent in the FTIR spectrum of Ap. The main band at 1022–1026 cm^{-1} corresponds to the asymmetric stretching mode of PO_4^{3-} groups (ν_3PO4). The band at ~962 cm^{-1} is ascribed to the symmetric stretching (ν_1PO4), while less intense bands at ~607 and 565 cm^{-1} are due to the bending mode of PO_4^{3-} groups (ν_4PO4). The band at ~532 cm^{-1} in the ν_4PO_4 domain can be assigned to non-apatitic (surface) HPO_4^{2-} ions, which points to the biomimetic nature of the precipitated nanocrystalline apatite in these nanocomposites [34]. Finally, the small band at 474 cm^{-1} corresponds to the ν_2PO_4 mode. No clear indications of spectral features of graphene in the FTIR spectra could be observed.

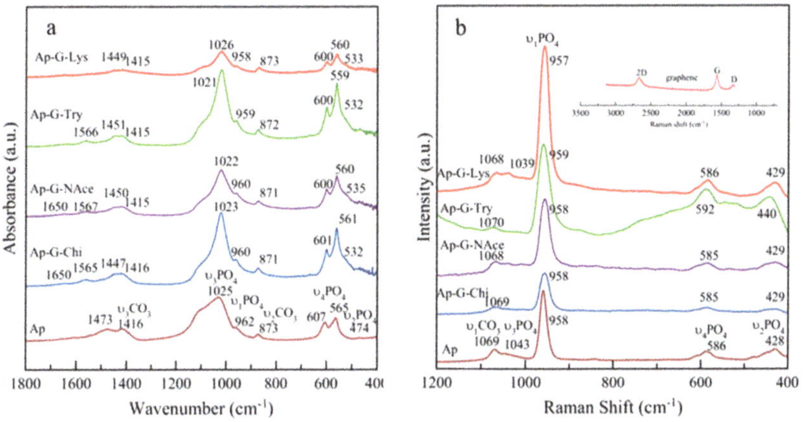

Figure 3. (**a**) FTIR spectra and (**b**) Raman spectra of nanocrystalline apatites nucleated by the VDSD method on graphene nanoflakes, prepared by the LPE method in the presence of Lys, Try, NAce, and Chi.

In all FTIR spectra, the presence of carbonate (CO_3^{2-}) bands was attested by vibrational signatures due to the ν_3CO_3 mode, with maxima around ~1416 cm^{-1} and 1473 cm^{-1}, and the ν_2CO_3 mode, with a peak around 873 cm^{-1}. The ν_2CO_3 region shows a broad peak distinctive of the different contributions characterizing different chemical environments of the CO_3 group within the apatite structure, i.e., A- and B-type (with carbonate ions replacing OH^- and PO_4^{3-} lattice ions, respectively) and labile carbonate species (belonging to the hydrated non-apatitic layer on the surface of nanocrystals) [35]. The existence of B-type substitutions was also witnessed by bands at 1515 and 1450 cm^{-1} [36].

Figure 3b shows the complementary characterization of the spectral features of the nanocomposites by Raman microspectroscopy in the 400–1200 cm^{-1} range, with a JASCO NRS-5100 Micro-Raman spectrometer (λ_{exc} = 532 nm). The spectral features of apatite arise at 957–959 cm^{-1} (ν_1PO_4), 428 cm^{-1} (ν_2PO_4), 586 cm^{-1} (ν_4PO_4), and 1039–1043 cm^{-1} (ν_3PO_4), with those of CO_3 (ν_1CO_3) arising at around 1069 cm^{-1}; no clear indications of the spectral features of graphene could be observed, likely due to its low intensity compared to bands of apatite. Raman spectrometer coupled to the VPSEM microscope was necessary to reveal the characteristic D, G, and 2D signals of the graphene flakes at around 1330, 1560, and 2670 cm^{-1}, respectively (Figure 3b, inset in red color). These bands displayed a 2D/G ratio lower than 1, and a full width at half maximum (FWHM) of the 2D band of 78 cm^{-1}, indicating they are stacked in multiple layers (FWHM > 66 cm^{-1}, > 5 layers) [37].

3.2. *Morphological Characteristics of Hybrid Apatite–Graphene Nanocomposites*

Figure 4a,c,e,g (left panel) shows the representative VPSEM micrographs of hybrid apatite–graphene nanocomposites, prepared with graphene nanoflakes and exfoliated in presence of Lys, Try, NAce, and Chi, respectively. Graphene flakes were visible at very low voltage (1–3 kV), while they become transparent to the electron beam at higher voltages. All samples were composed of a polydisperse population of flakes with an irregular shape, and nanocrystalline apatites with elongated shapes. The largest graphene flakes were produced in presence of Lys and Try (from 200–2000 nm), while the smallest ones were obtained in presence of NAce and Chi (200–800 nm). Only a few graphene flakes appeared greater than these sizes. In the last case, the composite was more homogeneous, with a more intimate apatite–graphene interaction and a closer spatial relationship. A more detailed size analysis is done in the TEM pictures (Figure 4b,d,f,h, on the left panel), where graphene flakes as large as 800–1000 nm can be observed in Figure 4b,d, prepared in presence of Lyz and Try, respectively.

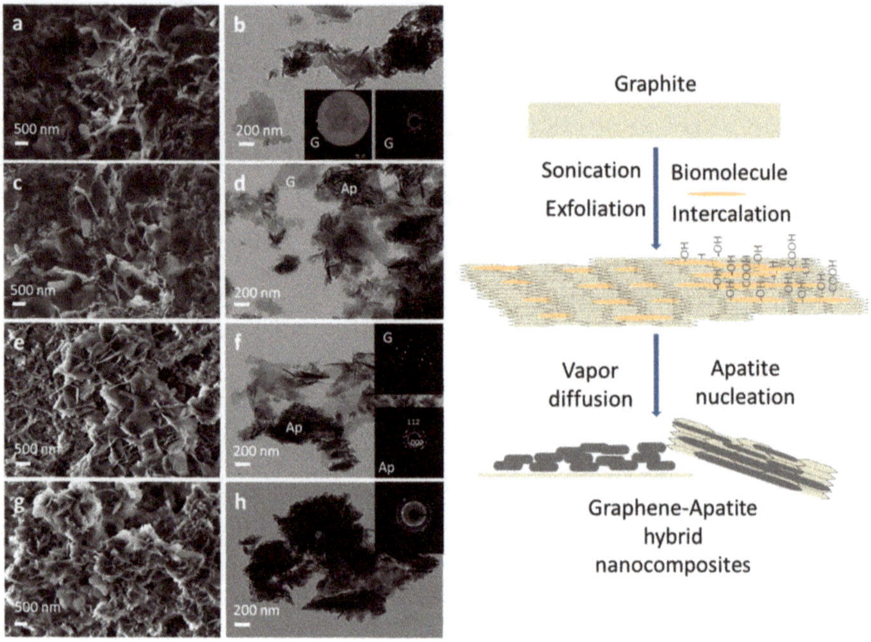

Figure 4. Left panel. (**a**,**c**,**e**,**g**) VPSEM and (**b**,**d**,**f**,**h**) TEM micrographs of nanocrystalline apatites nucleated by the VDSD method on graphene nanoflakes, prepared by liquid phase exfoliation in presence of Lys (**a**,**b**), Try (**c**,**d**), NAce (**e**,**f**), and Chi (**g**,**h**). Insets in the TEM micrographs are the selected area electron diffraction (SAED) patterns of the graphene flakes (G) and of the apatite (Ap). **Right panel.** Schematic illustration showing the formation process of graphene–apatite hybrid nanocomposites.

In these pictures, the nanocrystalline apatites displayed the following average lengths (L) and widths (W): $L = 105 \pm 4$ nm and $W = 13 \pm 2$ nm for Lyz-exfoliated graphene and L = 152 ± 49 nm and W = 16 ± 5 nm for Try-exfoliated graphene. In Figure 4f, apatite nanoparticles ($L = 90 \pm 12$ nm and $W = 8 \pm 3$ nm) coating the graphene flakes exfoliated in presence of NAce and some isolated flakes are observed. The SAED patterns of both components, graphene (G) and apatite (Ap), are shown as insets in this micrograph. Summing up, the TEM pictures of these three composites reveal a heterogeneous distribution of apatite-coated graphene and free graphene flakes. Finally, the full coating with apatite nanoparticles ($L = 69 \pm 8$ nm, $W = 7 \pm 3$ nm) of the graphene flakes exfoliated in presence of chitosan (Figure 4h), confirms the greater impact of these Chi–graphene flakes on the homogeneity of the composite. As a general rule, the apatite nanoparticles always appeared as coating the graphene flakes, but this finding was more clearly evidenced when the sheets were exfoliated in the presence of chitosan.

These observations point to the functionalization of the flakes with different biomolecules during sonication. The labeling of the flakes with the biomolecules favors the stability of the graphene suspensions for time periods longer than 1 week, as well as the interaction between the graphene and the apatite nanoparticles during the early stages of nucleation at the surface–solution interfaces (see Figure 4, right panel). The biomolecules holding hydrophilic groups, such as –COOH, –OH, $-NH_2$, =NH, or $-NH-C{=}O-CH_3$ must be placed, not only intercalating the graphene layers of the flakes, but also being adsorbed on their surfaces, with some ionizable functional groups pointing toward the solution and providing electrostatic repulsion, thus preventing graphene from re-aggregation, as suggested when LPE is assisted by ionic surfactants [38]. These functional groups can act as nucleators of the calcium phosphate. The other mechanism of graphene stabilization is by steric repulsion. In addition to the electrostatic considerations, this mechanism should be envisaged in the case of the macromolecules employed in this paper. The better homogeneity of the Ap–G–Chi composites must be related to the more regular array of the nucleation points on the functionalized surfaces. Indeed, Chi is a polysaccharide (polymer of N-acetyl-glucosamine and D-glucosamine [39]) extremely rich in –OH groups, which forms sheets by the lateral aggregation of chains. In these sheets, one side exposes $-NH_2$ groups (pKa ~ 6.5). At acidic pHs, Chi behaves as a polycation. However, at neutral to basic pHs, the $-NH_2$ groups are uncharged and must be the anchoring points with the more hydrophobic graphene surface. The other side of Chi exposes –OH groups, which can act as nucleators. The detailed mechanism of graphite intercalation and graphene labeling with the different biomolecules deserves a deeper experimental investigation, which is out of the scope of this work.

On the other hand, the findings from X-ray diffraction and the morphological analysis point to the role of the "functionalized graphene" as a template that diminishes the energy barrier for nucleation, thus favoring the heterogeneous nucleation and the growth of the apatite on its surface, similarly to results reported by Liu et al. for the synthesis of apatite-reduced graphite oxide nanocomposites [12]. The presence of the template in the supersaturated solution can also explain why the apatite particles displayed nanosized dimensions. Indeed, the decrease of Gibbs free energy in a supersaturated system can be produced either by growth of the millions of nuclei or by primary aggregation of the early formed particles. If the second mechanism is active, as was reported by Iafisco et al. [40] when producing apatite in a Ca-citrate/phosphate/H_2O system, the template might interact with the primary particles formed by heterogeneous nucleation, stabilizing them and thus minimizing their aggregation tendency and favoring the formation of nanoscale crystals, instead of larger crystals [19]. Therefore, the template effect can plausibly explain the formation of nanocrystalline apatites coating the graphene surfaces.

3.3. *Cytocompatibility of Hybrid Nanocomposites*

In view of the possibility of using these nanocomposites in future biomedical applications, their cytocompatibility was tested in an MTT assay on GTL-16 human carcinoma cells, after incubation at concentrations ranging from 0.1 to 100 µg/mL. No toxicity was observed on GTL-16 cells at any composite concentration, since in all cases full cell viability was observed (Figure 5). Furthermore, in view of using these nanocomposites for bone regenerative medicine, we tested their cytocompatibility also on the m17.ASC murine mesenchymal stem cell line, which has displayed osteogenic potential [29]. This cell line was somehow more sensitive to the contact with the graphene–apatite nanoparticles. Indeed, a certain level of toxicity was observed at the higher nanocomposite concentrations; however, in no case was viability lower than 80%, well above the 70% that is the cut-off indicated by ISO 10993-5:2009 [41].

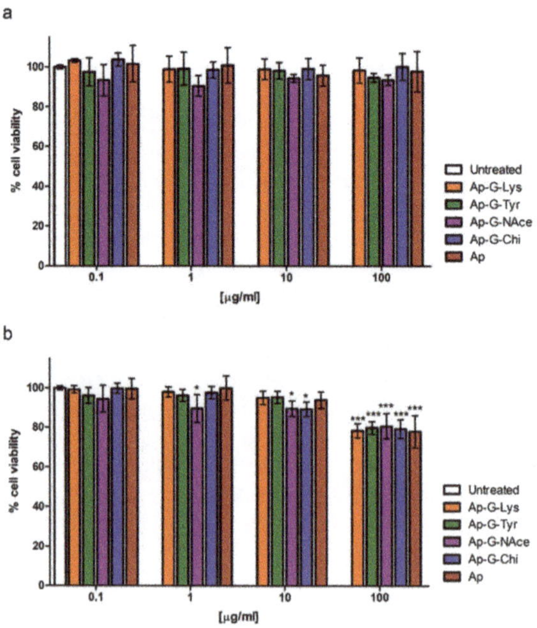

Figure 5. Viability of GTL-16 cells (**a**) and m17.ASC cells (**b**) incubated with graphene–apatite biohybrid composites (Ap-G) for 72 hours in 3-(4,5-Dimethylthiazol-2-yl)-2,5-diphenyltetrazolium bromide (MTT) essays. The graphene flakes were prepared by the liquid-phase exfoliation method in presence of Lys, Try, NAce, and Chi. Data represent means ± SD of four independent experiments performed in triplicate, and statistical analyses were carried on using one-way ANOVA, with a Bonferroni comparison test. For statistical analysis, all data were compared to untreated samples.

4. Conclusions

In this research, the nucleation of apatite nanoparticles on exfoliated graphene flakes has been successfully carried out by the sitting drop vapor diffusion technique. The graphene flakes were prepared by the sonication-assisted liquid-phase exfoliation technique of graphite in the presence of lysozyme, L-tryptophan, N-acetyl-D-glucosamine, and chitosan as dispersing surfactants. They were stacked in multiple layers (>5 layers), and most likely intercalated and functionalized with the biomolecules. The apatite nanoparticles were found forming a coating on the graphene surfaces. When using Lys-, Try-, and NAce-exfoliated graphene, the apatite–graphene composites displayed a heterogeneous distribution of apatite-coated graphene and free graphene flakes, while more homogeneous composites were obtained when using Chi-exfoliated graphene flakes. The cytocompatibility tests were performed in a dose-dependent manner on GTL-16 cells, a human gastric carcinoma cell line, and on m17.ASC cells, a murine mesenchymal stem cell line with osteogenic potential. These tests revealed that in all cases, these nanocomposites are fully cytocompatible.

Author Contributions: Conceptualization, J.G.-M. and G.F.; Investigation, J.G.-M., L.A.G.-R., R.F.P., F.O., M.P., and C.V.-E.; Methodology, J.G.-M. and G.F.; Supervision, J.G.-M.; Writing (original draft), J.G.-M., G.F.

Funding: Grant numbers MAT2014-60533-R and PGC2018-102047-B-I00 (MCIU/AEI/FEDER, UE).

Acknowledgments: The authors acknowledge the projects MAT2014-60533-R and PGC2018-102047-B-I00, funded by Agencia Estatal de Investigación of the Spanish Ministerio de Ciencia, Innovación y Universidades, co-funded by European Regional Development Fund (FEDER, UE), and a grant from the Università del Piemonte Orientale A. Avogadro, Italy to M.P. The Scientific Instrumentation Centre of the University of Granada is also acknowledged for technical assistance in SEM, TEM, FTIR and Raman spectroscopy. C.V.-E. acknowledges Spanish MINEICO for his PTA2015-11103-I contract.

Conflicts of Interest: The authors declare no conflict of interest. The funding sponsors had no role in the design of the study; in the collection, analyses, or interpretation of data; in the writing of the manuscript, and in the decision to publish the results.

References

1. Li, M.; Xiong, P.; Yan, F.; Li, S.; Ren, C.; Yin, Z.; Li, A.; Li, H.; Ji, X.; Zheng, Y.; et al. An overview of graphene-based hydroxyapatite composites for orthopedic applications. *Bioact. Mater.* **2018**, *3*, 1–18. [CrossRef] [PubMed]
2. Gómez–Morales, J.; Iafisco, M.; Delgado–López, J.M.; Sarda, S.; Druet, C. Progress on the preparation of nanocrystalline apatites and surface characterization: Overview of fundamental and applied aspects. *Prog. Cryst. Growth Charact. Mater.* **2013**, *59*, 1–46. [CrossRef]
3. De Groot, K.; Geesink, R.; Klein, C.P.A.T.; Serekian, P. Plasma sprayed coatings of hydroxylapatite. *J. Biomed. Mater. Res.* **1987**, *21*, 1375–1381. [CrossRef] [PubMed]
4. Pichugin, V.F.; Eshenko, E.V.; Surmenev, R.A.; Shesterikov, E.V.; Tverdokhlebov, S.I.; Ryabtseva, M.A.; Sokhoreva, V.V.; Khlusov, I.A. Application of high-frequency magnetron sputtering to deposit thin calcium-phosphate biocompatible coatings on a Titanium surface. *J. Surf. Invest. X.-ray, Synchr. Neutr. Tech.* **2007**, *1*, 679–682. [CrossRef]
5. Asri, R.I.M.; Harun, W.S.W.; Hassan, M.A.; Ghan, S.A.C.; Buyong, Z. A review of hydroxyapatite-based coating techniques: Sol–gel and electrochemical depositions on biocompatible metals. *J. Mechan. Behav. Biomed. Mater.* **2016**, *57*, 95–108. [CrossRef] [PubMed]
6. Iafisco, M.; Bosco, R.; Leeuwenburgh, R.S.C.G.; van den Beucken, J.J.J.P.; Jansen, J.A.; Prat, M.; Roveri, N. Electrostatic spray deposition of biomimetic nanocrystalline apatite coatings onto titanium. *Adv. Eng. Mater.* **2012**, *14*, B13–B20. [CrossRef]
7. Gómez-Morales, J.; Rodríguez-Clemente, R.; Armas, B.; Combescure, C.; Berjoan, R.; Cubo, J.; Martínez, E.; García-Carmona, J.; Garelik, S.; Murtra, J.; et al. Controlled nucleation and growth of thin hydroxyapatite layers on titanium implants by using induction heating technique. *Langmuir* **2004**, *20*, 5174–5178. [CrossRef]
8. Zhou, S.; Zheng, X.; Yu, X.; Wang, J.; Weng, J.; Li, X.; Feng, B.; Yin, M. Hydrogen Bonding Interaction of Poly(d,l-Lactide)/hydroxyapatite nanocomposites. *Chem. Mater.* **2007**, *19*, 247–253. [CrossRef]
9. Kim, J.-W.; Shin, K.-H.; Koh, Y.-H.; Hah, M.J.; Moon, J.; Kim, H.-E. Production of Poly(ε–Caprolactone)/Hydroxyapatite Composite Scaffolds with a Tailored Macro/Micro-Porous Structure, High Mechanical Properties, and Excellent Bioactivity. *Materials* **2017**, *10*, 1123. [CrossRef]
10. Li, J.; Fartash, B.; Hermansson, L. Hydroxyapatite-alumina composites and bone-bonding. *Biomaterials.* **1995**, *16*, 417–422. [CrossRef]
11. White, A.A.; Best, S.M. Hydroxyapatite–carbon nanotube composites for biomedical applications: A review. *Int. J. Appl. Ceram. Technol.* **2007**, *4*, 1–13. [CrossRef]
12. Liu, Y.; Huang, J.; Li, H. Synthesis of hydroxyapatite–reduced graphite oxide nanocomposites for biomedical applications: Oriented nucleation and epitaxial growth of hydroxyapatite. *J. Mater. Chem. B* **2013**, *1*, 1826–1834. [CrossRef]
13. Novoselov, K.S.; Geim, A.K.; Morozov, S.V.; Jiang, D.; Zhang, Y.; Dubonos, S.V.; Grigorieva, I.V.; Firsov, A.A. Electric field effect in atomically thin carbon films. *Science* **2004**, *306*, 666–669. [CrossRef] [PubMed]
14. Lee, C.; Wei, X.; Kysar, J.; Hone, V. Measurement of the elastic properties and intrinsic strength of monolayer graphene. *Science* **2008**, *321*, 385–388. [CrossRef] [PubMed]
15. Westervelt, R.M. Graphene nanoelectronics. *Science* **2008**, *320*, 324–325. [CrossRef] [PubMed]
16. Nayak, T.R.; Andersen, H.; Makam, V.S.; Khaw, C.; Bae, S.; Xu, X.; Ee, P.L.R.; Ahn, J.H.; Hong, B.H.; Pastorin, G.; et al. Graphene for controlled and accelerated osteogenic differentiation of human mesenchymal stem cells. *ACS Nano* **2011**, *5*, 4670–4678. [CrossRef]
17. Kalbacova, M.; Broz, A.; Kong, J.; Kalbac, M. Graphene substrates promote adherence of human osteoblasts and mesenchymal stromal cells. *Carbon* **2010**, *48*, 4323–4329. [CrossRef]

18. Yang, K.; Gong, H.; Shi, X.; Wan, J.; Zhang, Y.; Liu, Z. In vivo biodistribution and toxicology of functionalized nano-graphene oxide in mice after oral and intraperitoneal administration. *Biomaterials* **2013**, *34*, 2787–2795. [CrossRef]
19. Iafisco, M.; Gomez-Morales, J.; Hernandez-Hernandez, M.A.; García-Ruiz, J.M.; Roveri, N. Biomimetic Carbonate–Hydroxyapatite Nanocrystals Prepared by Vapor Diffusion. Adv. *Eng. Mater.* **2010**, *12*, 218–223.
20. Gomez-Morales, J.; Delgado-Lopez, J.M.; Iafisco, M.; Hernandez-Hernandez, M.A.; Prat, M. Amino acidic control of calcium phosphate precipitation by using the vapor diffusion method in microdroplets. *Cryst. Growth Des.* **2011**, *11*, 4802–4809. [CrossRef]
21. Nassif, N.; Martineau, F.; Syzgantseva, O.; Gobeaux, F.; Willinger, M.; Coradin, T.; Cassaignon, S.; Azais, T.; Giraud-Guille, M.M. In Vivo Inspired Conditions to Synthesize Biomimetic Hydroxyapatite. *Chem. Mater.* **2010**, *22*, 3653–3663. [CrossRef]
22. Gomez-Morales, J.; Hernandez-Hernandez, A.; Sazaki, G.; Garcia-Ruiz, J.M. Nucleation and Polymorphism of Calcium Carbonate by a Vapor Diffusion Sitting Drop Crystallization Technique. *Cryst. Growth Des.* **2010**, *10*, 963–969. [CrossRef]
23. Hernandez-Hernandez, A.; Rodriguez-Navarro, A.B.; Gomez-Morales, J.; Jimenez-Lopez, C.; Nys, Y.; Garcia-Ruiz, J.M. Influence of model globular proteins with different isoelectric points on the precipitation of calcium carbonate. *Cryst. Growth Des.* **2008**, *8*, 1495–1502. [CrossRef]
24. Hernandez-Hernandez, A.; Gómez-Morales, J.; Rodriguez-Navarro, A.B.; Gautron, J.; Nys, Y.; Garcia-Ruiz, J.M. Identification of some active proteins on the process of hen eggshell formation. *Cryst. Growth Des.* **2008**, *8*, 4330–4339. [CrossRef]
25. Calvaresi, M.; DiGiosia, M.; Ianiro, A.; Valle, F.; Fermani, S.; Polishchuk, I.; Pokroy, B.; Falini, G. Morphological changes of calcite single crystals induced by graphene–biomolecule adducts. *J. Cryst. Growth* **2017**, *457*, 356–361. [CrossRef]
26. Gómez-Morales, J.; Verdugo-Escamilla, C.; Gavira-Gallardo, J.A. Bioinspired calcium phosphate coated mica sheets by vapor diffusion and its effects on lysozyme assembly and crystallization. *Cryst. Growth Des.* **2016**, *16*, 5150–5158. [CrossRef]
27. Ciesielski, A.; Samori, P. Graphene via sonication assisted liquid-phase exfoliation. *Chem. Soc. Rev.* **2014**, *43*, 381–398. [CrossRef]
28. Giordano, S.; Ponzetto, C.; Di Renzo, M.F.; Cooper, C.S.; Comoglio, P.M. Tyrosine kinase receptor indistinguishable from the c-met protein. *Nature* **1989**, *339*, 155–156. [CrossRef]
29. Zamperone, A.; Pietronave, S.; Merlin, S.; Colangelo, D.; Ranaldo, G.; Medico, E.; di Scipio, F.; Berta, G.N.; Follenzi, A.; Prat, M. Isolation and characterization of a spontaneously immortalized multipotent mesenchymal cell line derived from mouse subcutaneous adipose tissue. *Stem Cells Dev.* **2013**, *22*, 2873–2884. [CrossRef]
30. Delgado-Lopez, J.M.; Iafisco, M.; Rodriguez-Ruiz, I.; Tampieri, A.; Prat, M.; Gómez–Morales, J. Crystallization of bioinspired citrate-functionalized nanoapatite with tailored carbonate content. *Acta Biomater.* **2012**, *8*, 3491–3499. [CrossRef]
31. Iafisco, M.; Marchetti, M.; Gómez-Morales, J.; Hernández-Hernández, M.A.; García-Ruiz, J.M.; Roveri, N. Silica gel template for calcium phosphates crystallization. *Cryst. Growth Des.* **2009**, *9*, 4912–4921. [CrossRef]
32. Blanton, T.N.; Majumdar, D. X-ray diffraction characterization of polymer intercalated graphite oxide. *Powder Diffr.* **2012**, *27*, 104–107. [CrossRef]
33. Kadir, M.F.Z.; Aspanut, Z.; Majid, S.R.; Arof, A.K. FTIR studies of plasticized poly(vinylalcohol)–chitosan blend doped with NH4NO3 polymer electrolyte membrane. *Spectrochim. Acta Mol. Biomol. Spectros.* **2011**, *78*, 1068–1074. [CrossRef] [PubMed]
34. Vandecandelaere, N.; Rey, C.; Drouet, C. Biomimetic apatite-based biomaterials: On the critical impact of synthesis and postsynthetic parameters. *J. Mater. Sci. Mater. Med.* **2012**, *23*, 2593–2606. [CrossRef] [PubMed]
35. Rey, C.; Combes, C.; Drouet, C.; Grossin, D. *Comprehensive Biomaterials* **2014**, 187–221.
36. Antonakos, A.; Liarokapis, E.; Leventouri, T. Micro-Raman and FTIR studies ofsynthetic and natural apatites. *Biomaterials* **2007**, *28*, 3043–3054. [CrossRef] [PubMed]
37. Hao, Y.; Wang, Y.; Wang, L.; Ni, Z.; Wang, Z.; Wang, R.; Koo, C.K.; Shen, Z.; Thong, J.T.L. Probing layer number and stacking order of few-layer graphene by Raman spectroscopy. *Small* **2010**, *6*, 195–200. [CrossRef]
38. Guardia, L.; Fernandez-Merino, M.J.; Paredes, J.I.; Solis-Fernandez, P.; Villar Rodil, S.; Martinez-Alonso, A.; Tascon, J.M.D. High-throughput production of pristine graphene in an aqueous dispersion assisted by non-ionic surfactants. *Carbon* **2011**, *49*, 1653–1662. [CrossRef]

39. Nilsen-Nygaard, J.; Strand, S.P.; Vårum, K.M.; Draget, K.I.; Nordgård, C.T. Chitosan: Gels and Interfacial Properties. *Polymers* **2015**, *7*, 552–579. [CrossRef]
40. Iafisco, M.; Ramírez-Rodríguez, G.B.; Sakhno, Y.; Tampieri, A.; Martra, G.; Gómez-Morales, J.; Delgado-López, J.M. The growth mechanism of apatite nanocrystals assisted by citrate: Relevance to bone biomineralization. *CrystEngComm* **2015**, *17*, 507–515. [CrossRef]
41. ISO 10993-5:2009. Biological Evaluation of Medical Devices—Part 5: Tests for In Vitro Cytotoxicity. Available online: https://www.iso.org/standard/36406.html (accessed on 22 December 2018).

crystals

MDPI

Article

A Simple Technique to Improve Microcrystals Using Gel Exclusion of Nucleation Inducing Elements

Adafih Blackburn [1,2,†], Shahla H. Partowmah [1,3,†], Haley M. Brennan [1,4,†], Kimberly E. Mestizo [1,5], Cristina D. Stivala [1,6], Julia Petreczky [1,7], Aleida Perez [1], Amanda Horn [1], Sean McSweeney [8,9] and Alexei S. Soares [8,*]

1 Office of Educational Programs, Brookhaven National Laboratory, Upton, NY 11973, USA; ajblackburn98@gmail.com (A.B.); shahla.partow@gmail.com (S.H.P.); hmbrennan@email.wm.edu (H.M.B.); Kimberly.mestizo@gmail.com (K.E.M.); cstivala19@ross.org (C.D.S.); julia.petreczky@gmail.com (J.P.); pereza@bnl.gov (A.P.); ahorn@bnl.gov (A.H.)
2 Department of Pharmacology, Stony Brook University, Stony Brook, NY 11794, USA
3 School of Business, State University of New York at Old Westbury, Old Westbury, NY 11568, USA
4 Department of Biology, College of William and Mary, Williamsburg, VA 23187, USA
5 General Douglas MacArthur High School, Levittown, NY 11756, USA
6 Ross School, East Hampton, NY 11937, USA
7 Shoreham-Wading River High School, Shoreham-Wading River Central School District, Shoreham, NY 11786, USA
8 Photon Sciences Directorate, Brookhaven National Laboratory, Upton, NY 11973, USA; smcsweeney@bnl.gov
9 Biosciences Department, Brookhaven National Laboratory, Upton, NY 11973, USA
* Correspondence: soares@bnl.gov; Tel.: +1-(631)344-7306
† These authors contributed equally to this work.

Received: 14 November 2018; Accepted: 10 December 2018; Published: 12 December 2018

Abstract: A technique is described for generating large well diffracting crystals from conditions that yield microcrystals. Crystallization using this technique is both rapid (crystals appear in <1 h) and robust (48 out of 48 co-crystallized with a fragment library, compared with 26 out of 48 using conventional hanging drop). Agarose gel is used to exclude nucleation inducing elements from the remaining crystallization cocktail. The chemicals in the crystallization cocktail are partitioned into high concentration components (presumed to induce aggregation by reducing water activity) and low concentration nucleation agents (presumed to induce nucleation through direct interaction). The nucleation agents are then combined with 2% agarose gel and deposited on the crystallization shelf of a conventional vapor diffusion plate. The remaining components are mixed with the protein and placed in contact with the agarose drop. This technique yielded well diffracting crystals of lysozyme, cubic insulin, proteinase k, and ferritin (ferritin crystals diffracted to 1.43 Å). The crystals grew rapidly, reaching large size in less than one hour (maximum size was achieved in 1–12 h). This technique is not suitable for poorly expressing proteins because small protein volumes diffuse out of the agarose gel too quickly. However, it is a useful technique for situations where crystals must grow rapidly (such as educational applications and preparation of beamline test specimens) and in situations where crystals must grow robustly (such as co-crystallization with a fragment library).

Keywords: drug discovery; education; crystallization; crystallography; nucleation; micro-crystals; agarose; ferritin; lysozyme; proteinase k; insulin

1. Introduction

It is convenient to have a robust crystallization protocol that yields high quality protein crystals even when the mother liquor is perturbed, for example by addition of chemicals from a fragment library [1,2]. Robust and rapid crystallization is also useful for applications such as to trouble-shooting diffraction equipment or in-classroom demonstrations. Here, we present a general technique for reliably and rapidly growing protein crystals by identifying elements that induce nucleation [3] and excluding these chemicals from the remaining crystallization solution by sequestering them in agarose gel [4]. This technique combines a vapor-diffusion approach to crystal growth [5] with a free-diffusion approach to nucleation [6]. A conventional vapor diffusion plate is used [7]. First, a pellet of gelled nucleation inducing chemicals is deposited on the crystallization shelf. Second, the solution of protein and precipitant is placed in contact with the pellet composed of gelled nucleation inducing chemicals. The reservoir contains the complete crystallization cocktail. If re-sealable crystallization plates are used (we used EasyXtal® 15-well crystallization plate from Qiagen), the gelled nucleation-inducing chemicals can be pre-deposited on the crystallization shelf so that crystal growth is initiated in one step by pipetting the stored solution of protein and precipitant into contact with the agarose. An added benefit is that hazardous nucleation-inducing chemicals are safer to handle when they are pre-applied as a gel to the crystallization shelf in a controlled laboratory environment (in contrast, precipitants frequently present few hazards if handled outside of the laboratory).

The protein experiences a concentration ratio for the gel sequestered nucleant that begins at 0 and terminates at 1 (where unity is the concentration in the reservoir). In contrast, the concentration of the precipitant varies by a smaller amount, from $\frac{1}{2}$ to 1 (again as a ratio of the concentration in the reservoir). A similar methodology was introduced when agar was first proposed as a crystallization facilitator [8], with the difference that all components were embedded in the agar matrix [9], and similar techniques continue to be used to simulate the advantages of growing crystals in microgravity [10]. In contrast, our goal is to introduce large concentration changes for nucleants, and more gradual concentration changes for precipitants. A comparable change in concentration can be induced in other ways, for example using button dialysis [11] and free diffusion methods [12]. The 4-corner method increases the likelihood of crystallization success by sampling four discreet concentrations of each precipitant [13]. One advantage of inducing large changes in concentration is that crystallization often becomes more robust. For example, co-crystallization in the presence of chemical libraries (such as fragment libraries) often perturbs conditions such that either no crystals are formed, or a shower of micro crystals are formed [14]. In contrast, inducing crystallization using a large precipitant gradient is often highly crystallogenic, and hence likely to robustly produce well diffracting crystals even when perturbed by chemical biology experiments. Robust conditions for co-crystallization are particularly valuable for drug discovery projects, where co-crystallization with chemicals of interest is often preferred over soaking because soaking can either take too much time [15] or can induce artificial artifacts [16].

2. Materials and Methods

We developed nucleant segregation strategies to crystallize four model proteins. Briefly, the crystallization cocktail is partitioned into two parts, and the nucleation component is segregated (using 2% agarose) from the protein and remaining chemicals (Figure 1). Our first model protein was lysozyme, chosen because of the affordability and predictability that make it an unsurprising test sample in many development projects. For many of the same reasons our second model protein was proteinase K. Our third model protein was cubic insulin, to test the compatibility of nucleant segregation techniques with established protocols for button dialysis. Our final target for crystallization by exclusion of nucleating agents was ferritin. Because of its importance in general understanding of biological function, as well as its relevance to disease processes, ferritin is intensively studied by structural methods. However, ferritin is difficult to crystallize, which has slowed progress in structure-based understanding of its function and role in disease. Nucleant segregation offers a novel strategy for growing ferritin crystals. Our process induces the formation of small ferritin protein crystals in under one hour. Large and well diffracting ferritin crystals are obtained overnight. Using conventional techniques, comparable quality crystals require extended growth times (crystals may form more rapidly using techniques such as external magnetic fields; [17]).

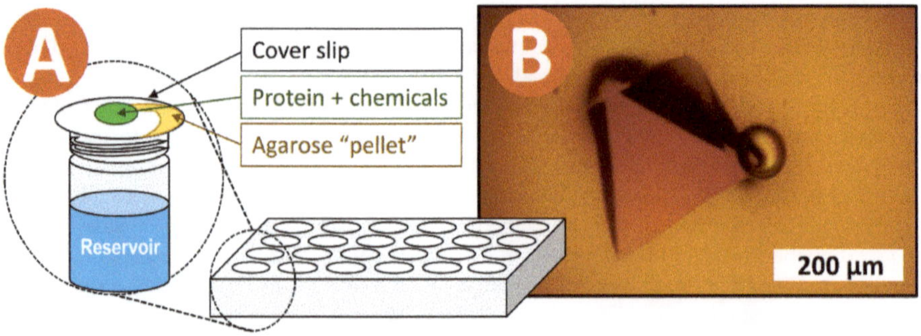

Figure 1. Crystallization using gel exclusion. (**A**) The gel exclusion setup that was used to crystallize test proteins (the shape of the agarose pellet shown was particularly important for ferritin). (**B**) Example of a ferritin crystal grown by gel exclusion (these crystals diffracted to ~1.5 Å resolution).

A faster crystallization protocol would be advantageous, for example for high throughput drug discovery projects (which require rapid crystal growth to screen large numbers of drug candidates). Fast and robust crystal growth can also be advantageous to basic science projects (for example, protein crystallization can be perturbed to generate insights into dynamic behavior).

2.1. *Lysozyme*

Agarose (2% *w*/*v*) was heated to 100 °C and combined with 0.4 M benzamidine hydrochloride, 4% sodium chloride, 100 mmol/L sodium acetate (pH 4.6), 10% glycerol, and 5% ethylene glycol. In the case of lysozyme, the entire crystallization cocktail was included in the nucleation pellet, as explained in Section 2.5. A 20 µL pellet of gelled nucleation solution was deposited on each cover slip and allowed to cool (Figure 1A). Lysozyme powder (Sigma L6876, Sigma Aldrich, Saint Louis, MO, USA) was dissolved (30 mg/mL in 100 mmol/L sodium acetate pH 4.6) and placed adjacent to the agarose pellet. Control crystals were grown in a similar way, with the difference that no agarose was added, such that the nucleating solution was added in the liquid state.

2.2. *Proteinase K*

Agarose (2% *w*/*v*) at 100 °C was combined with 0.8 M sodium nitrate. A 20 µL pellet of this gel was deposited on each cover slip and allowed to cool. Proteinase K powder (Sigma P2308) was dissolved (30 mg/mL 0.1 M BisTris pH 6.5 and 0.08 M $CaCl_2$) and placed adjacent to the agarose pellet. Similar to the growth of Lysozyme control crystals, proteinase K control crystals were grown with no agarose and the addition of nucleating solution was in the liquid state.

2.3. *Cubic Insulin*

Gelled nucleating solution (2% *w*/*v* agarose) at 100 °C was combined with 0.25 mol/L sodium phosphate pH 9.2 and deposited on individual cover slips as 20 µL pellets to cool. Human insulin powder (Sigma 91077C) was dissolved (30 mg/mL in 0.05 mol/L sodium phosphate pH 11) and placed adjacent to the pellet. Control crystals for insulin were not grown using previous techniques. Button dialysis [18] proved more effective with the button containing 20 µL of 30 mg/mL protein in 0.05 mol/L sodium phosphate pH 11, and the reservoir (around the button) containing excess of 0.25 mol/L sodium phosphate pH 9.2.

2.4. Ferritin

Mammalian ferritin assembles into a roughly spherical protein that is primarily responsible for oxidizing soluble ferrous iron into insoluble ferric iron ($Fe^{2+} \rightarrow Fe^{3+}$ is the ferroxidase activity). Ferritin then stores the insoluble iron in a 270 nm^3 cavity that can accommodate up to 4500 iron atoms [19]. Ferritin consists of 24 heterogeneous subunits, each of which can be either a light subunit (FTL – ferritin light chain) or a heavy subunit (FTH1 – ferritin heavy chain) [20]. FTL and FTH1 subunits are structurally similar, but FTH1 has increased ferroxidase activity, so that the greater the proportion of FTH1, the greater the ability of the ferritin assembly to rapidly incorporate iron [21]. Ferritin is intensively studied for two primary health related reasons. First, ferritin accumulates iron in response to infection (denying iron to invaders) [22]. Second, ferritin maintains a constant level of soluble iron (iron homeostasis), which is important in anemia and iron overload defects [23]. In addition to its direct applications to human health, ferritin's bio-mineralization and metal accumulation properties are frequently used as tools in biotechnology applications [24].

Cadmium is required for formation of ferritin crystals. In conventional crystallization, the cadmium is introduced along with the remaining crystallization chemicals. Since cadmium is very effective in inducing nucleation of ferritin crystals, this process requires a very fine balance for the cadmium concentration. If there is too much cadmium, a shower of microcrystals results. If there is too little, the protein either remains in solution, or grows exceedingly slowly. One consequence of this, is that for most protocols, much of the protein remains un-crystallized, since the crystallization drop runs out of cadmium much before it runs out of ferritin protein.

We first used conventional high throughput screening to identify crystallization conditions for ferritin (cadmium in different concentrations was included in all of our screens). We used established micro-crystallization protocols [25] using an acoustic screening robot [26] to identify three conditions that yielded ferritin crystals. We then used gel exclusion to segregate the cadmium from the remaining crystallization cocktail to identify the optimal crystallization condition for our method (3.5 mol/L $(NH_4)_2SO_4$, 100 mmol/L tris(hydroxymethyl)aminomethane (TRIS) pH 7.4, 1 mmol/L ethylenediaminetetraacetic acid (EDTA), 50 mmol/L $CdSO_4$. The cadmium was sequestered in a hydrogel by combining with 2% agarose and heating to 100 °C for 5 min.

We initially used a conventional "hanging drop" crystallization strategy, which consisted of a plate with 24 separate crystallization chambers. The protein + crystallization chemicals + cadmium-agarose pellet were positioned on the cover-slip. The cadmium-agarose pellet was of approximately equal size compared to the protein-chemical drop, and was placed on the cover-slips and left to equilibrate with the reservoir. The protein-chemical drop consisted of 50% ferritin and 50% crystallization solution. Once we identified an optimal crystallization condition, we began experimenting with altering the shape of the agar pellet, to further optimize crystallization. We used screw-on cover slips (because conventional cover slips contain limited real estate for designing different agar pellet shapes (screw-on cover slips also have the advantage of a plastic boundary that can be used in the shaping process; EasyXtal® 15-Well from Qiagen). We observed that an extended shape, which minimized the contact area between the cadmium-pellet and the crystallization solution, was optimal (Figure 1B). For comparison, control crystals were grown without using agarose.

2.5. Lysozyme Co-Crystallization with Fragment Library

Each chemical from a 48-fragment library was deposited on a cover slip and allowed to dry (10 µL of 100 mmol/L fragment dissolved in water). Gelled nucleating solution (2% *w*/*v* agarose) was heated to 100 °C and combined with 0.4 mol/L benzamidine hydrochloride, 4% sodium chloride, 100 mmol/L sodium acetate (pH 4.6), 10% glycerol, and 5% ethylene glycol. Then, 20 µL pellets of the gel was deposited on each cover slip and allowed to cool while in contact with each chemical. Lysozyme powder (Sigma L6876) was dissolved (30 mg/mL in 100 mmol/L sodium acetate pH 4.6) and placed adjacent to the agarose pellet. Control crystals were grown with no agarose added, having nucleating solution added in the liquid state.

The cover slips containing lysozyme, proteinase, and cubic insulin were placed on a VDX™ hanging drop plate (Hampton HR3-172, Hampton Research, Aliso Viejo, CA, USA), where the reservoir contained the complete respective crystallization cocktail. Crystallization plates and buttons were placed on a Leica MZ16 microscope (Leica Camera AG, Wizlar, Germany) equipped with an Olympus DP72 camera (Olympus, Shinjuku, Tokyo, Japan) operated with the *cellSens standard* (version 1.6, Olympus, Shinjuku, Tokyo, Japan) software. Pictures of the growing crystals were taken automatically using the process manager at intervals of 150 s (gel exclusion crystals) and 1500 s (control crystals). Growth rate data were obtained for three gel exclusion crystals and for three control crystals of each protein, and an overall crystal length "L" was determined for each crystal at each time point as described in Section 2.2. Cubic insulin crystals moved through the field of view during the growth process, so a single continuous growth curve could not be produced. Three student researchers measured the dimensions of each crystal by inspection using Image J software (version 1.51w, NIH, Bethesda, MD, USA) (the measurements from the three students were then averaged). The crystal volume was determined using the average x, y, and z dimensions of each crystal, and the cube root of the crystal volume was used as a measure of the observed length "L" of the crystal at each time point (Equation (1), Section 3).

3. Results

We observed that all crystals that we tested grew faster using gel exclusion compared to controls grown using conventional techniques. For both gel exclusion and controls, the crystal size was initially observed to increase according to a one term asymptotic function but was bounded by an upper limit that is less than the asymptotic value. Hence, the observed crystal size was empirically fitted to a two-part equation, where the initial growth follows a simple two-parameter asymptotic Function (1a) until reaching a maximum size limit, after which crystal size is constant (1b):

$$\begin{bmatrix} L = L_C\left(\frac{t}{t+\tau}\right) \; if \; L < L_{max} & (1a) \\ L = L_{max} \; if \; L > L_{max} & (1b) \end{bmatrix} \quad (1)$$

Here L is the predicted length of each side of the crystal, L_c is an adjustable constant that corresponds to the asymptotic value (length units), t is the time over which crystallization has occurred, τ is an adjustable constant that corresponds to the time needed for the crystal to reach $\frac{1}{2}$ L_c (time units), and L_{max} is an adjustable constant which indicates the upper limit for crystal size (length units, the point at which all soluble protein has been subsumed into the crystal such that no further growth is possible). The three adjustable constants were iteratively refined to minimize the discrepancy between the observed crystal length and the predicted crystal length (Table 1).

Compared to controls, lysozyme crystals grown using gel exclusion grew 4.6 times faster. Also, when combined with a 48-fragment library, 100% of the gel exclusion lysozyme co-crystals grew, compared with 38% of the control lysozyme crystals. Proteinase K crystals were also observed to grow faster using gel exclusion, compared to controls (10.7 times faster).

Cubic insulin is conventionally grown using button dialysis, which involves a laborious setup of the apparatus. In contrast, our gel exclusion crystallization setup was simple, and resulted in faster growing crystals (17.1 times faster).

Growing ferritin crystals requires a prolonged period of time to grow using conventional methods, particularly for iron containing ferritin. Using gel exclusion, we were able to grow large well diffracting ferritin crystals overnight. These crystals diffracted to higher resolution than any iron containing ferritin crystal in the protein data bank (PDB), 1.43 Å, compared to 2.22 Å; our best control crystal had a resolution of 2.54 Å. The highest resolution for iron containing equine ferritin in the PDB is 2.2 Å, which is comparable to our best control crystals at 2.5 Å. Data to 2.0 Å have been recorded from iron containing ferritin crystals grown under high magnetic fields but were not deposited in the PDB [27].

Table 1. Refined curve fitting parameters for crystal growth curves.

Lysozyme:	Gel Exclusion	Hanging Drop
Time step (s)	150	1500
Tau (h)	0.57 ± 0.02	2.66 ± 0.48
Lmax (μm)	512 ± 49	427 ± 71
Lc (μm)	605 ± 58	505 ± 84
Residual (%)	2.04% ± 0.36%	2.55% ± 0.56%
Proteinase K:	**Gel Exclusion**	**Hanging Drop**
Time step (s)	120	600
Tau (h)	1.01 ± 0.24	16.16 ± 2.60
Lmax (μm)	363 ± 81	361 ± 31
Lc (μm)	430 ± 21	534 ± 46
Residual (%)	5.66% ± 0.82%	2.80% ± 0.06%
Cubic insulin:	**Gel Exclusion**	**Hanging Drop**
Time step (s)	150	1500
Tau (h)	3.30 ± 0.00	56.33 ± 5.18
Lmax (μm)	222 ± 16	86 ± 10
Lc (μm)	555 ± 40	214 ± 25
Residual (%)	3.14% ± 0.63%	4.92% ± 1.97%

3.1. Lysozyme

Lysozyme crystals grew much faster using gel exclusion compared to controls. Crystals' growth using gel exclusion and controls had similar crystal habits, though the gel exclusion crystals tended to exhibit a less elongated crystal habit (Figure 2). The growth constants τ, L_c, and L_{max} (Equation (1)) for each of the three crystals grown using gel exclusion, and for each of the three crystals grown using conventional methods, are shown in Table 1. For visual illustration, we also fitted growth curves to the overall data obtained by averaging the lengths of all three crystals grown using gel exclusion and compared to the overall growth curve from the control crystals (Figure 1 inset).

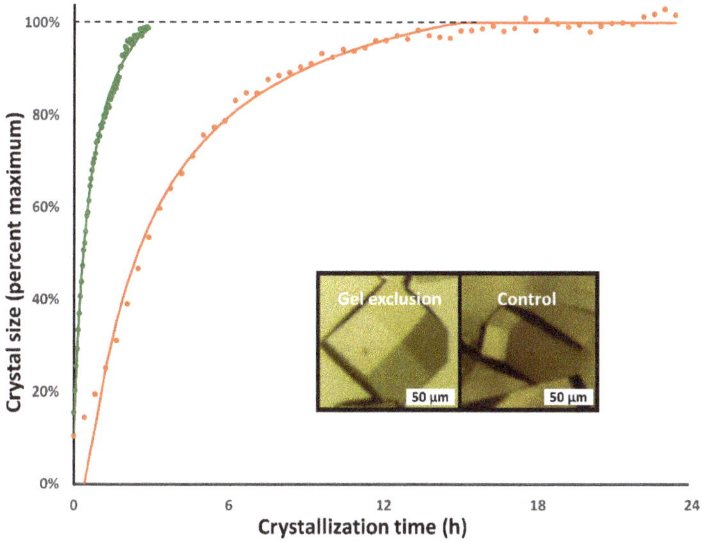

Figure 2. Growth rate for lysozyme crystals grown using gel exclusion (green) and conventional hanging drop (orange). Data are fitted by a single polynomial asymptotic curve with an upper bound that is lower than the asymptotic value. Crystal habits were similar (inset).

3.2. Proteinase K

Proteinase K crystals also grew much faster using gel exclusion compared to controls. Crystal growth using gel exclusion and controls led to virtually identical crystal habits (Figure 3), and the fitted parameters are shown in Table 1. As in Section 2.1, overall growth rates were also determined by averaging the sizes of the three crystals grown using gel exclusion and the three controls (Figure 3 inset).

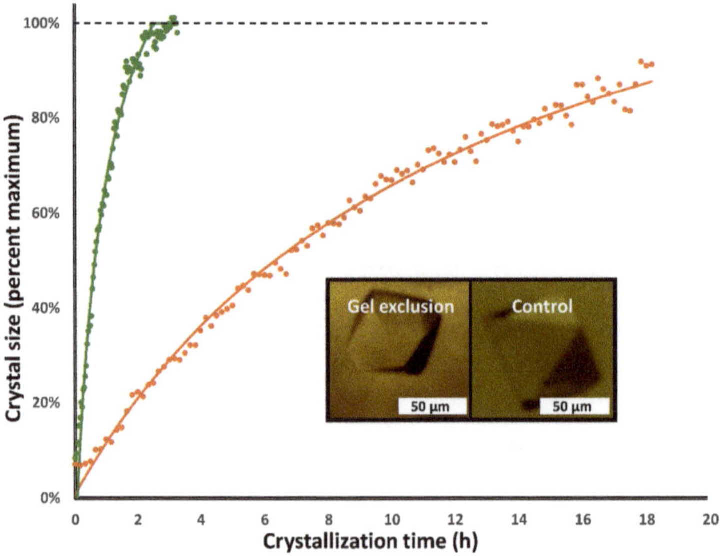

Figure 3. Growth rate for proteinase K crystals grown using gel exclusion (green) and conventional hanging drop (orange). Data are fitted by a single polynomial asymptotic curve with an upper bound that is lower than the asymptotic value. Crystal habits were virtually identical (inset).

3.3. Cubic Insulin

Gel exclusion proved to be a simple alternative to button dialysis, with many of the same benefits, but with a much simpler experimental setup. Cubic insulin crystals grown using both methods had identical habits. As previously, crystal growth was greatly accelerated using gel exclusion (average values in Table 1; continuous growth curves could not be generated due to crystal movement).

3.4. Ferritin

Using gel exclusion to sequester cadmium to fine-tune crystallization of ferritin yielded well diffracting crystals that grew in a few hours. As mentioned in Section 2.4, we noticed that there was a correlation between the shape of the cadmium pellet and the quality of the crystals that grew in that crystallization drop. After exploring many different shapes for the cadmium pellet, we observed that the optimum shape was a "comma" shape. The crystals that resulted from this improved procedure had a better, more symmetric crystal habit, and yielded improved diffraction quality.

Once we chose an optimal crystallization strategy (Section 2.4), we obtained an X-ray diffraction data set (Figure 4). Diffraction data were measured at the National Synchrotron Light Source II (NSLS II) beamline 17-ID-1 (AMX) and processed with *XDS* (version January 2018, MPI, Heidelberg, Germany) [28]. Data were further processed using *CTRUNCATE* in the *CCP4i* suite (version 0.8.9, CCP4, Oxon, UK) [29]. Structures were obtained by molecular substitution from published models and refined using *REFMAC* (v0.8.9, CCP4, Oxon, UK) [30] and *ArpWarp* [31] (version 7, EMBL, Hamburg, Germany) (starting models 1IER [32]). Structures were visually inspected using *coot* (version 0.8.9,

CCP4, Oxon, UK) [33]. The quality of these crystals was a significant improvement compared to the best crystals grown using conventional crystallization strategies (Table 2; PDB accession code 6MSX). We conclude that our proposed three-state crystallization strategy (with three distinct chemical environments in the hanging drop, the agar pellet, and the reservoir) allowed us to fine-tune the crystallogenesis of ferritin, yielding faster growing and higher quality crystals.

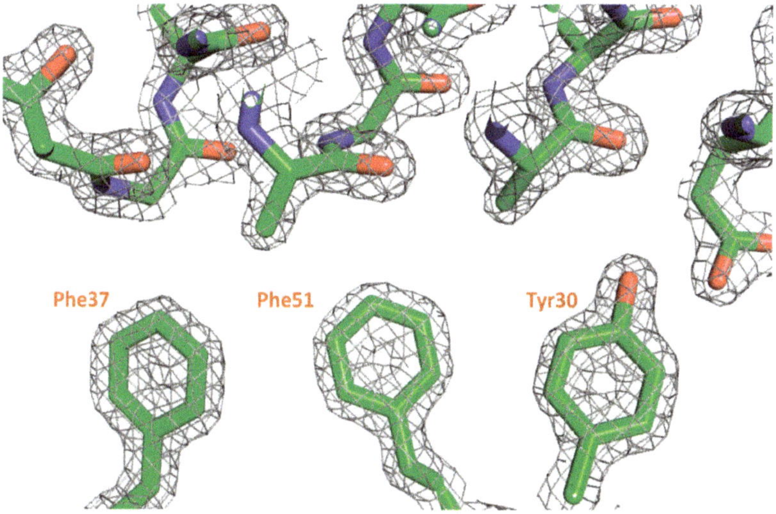

Figure 4. High quality crystals grown in 12 h yield accurate electron density for ferritin. Electron density is contoured at 3σ (2fo–fc). Three residues are shown, the remainder of the ferritin is represented as a helix ribbon (6MSX).

Table 2. Data reduction and structure refinement statistics.

Protein	Ferritin	Lysozyme	Lysozyme	Lysozyme
Fragment	n/a	3-amino-Phenol	Picolinic acid	Eosin Y/ Trp62
Resolution (Å)	1.43 (1.47)	1.35 (1.39)	1.23 (1.26)	1.56 (1.60)
R_{merge} (%)	11.6 (407)	7.0 (74.7)	6.7 (53.0)	9.0 (69.4)
$CC_{1/2}$	100.0 (49.9)	99.5 (84.6)	99.4 (82.4)	99.5 (77.7)
$<I/\sigma(I)>$	31.81 (1.17)	21.9 (3.2)	17.4(2.3)	17.8 (3.1)
Multiplicity	138.2 (48.9)	12.8 (11.3)	8.1 (5.5)	12.8 (11.2)
Unique reflections	47,743	27,063 (1906)	32,856 (1997)	17,660 (1235)
Completeness (%)	100.0 (100.0)	99.7 (96.0)	97.3 (83.1)	99.7 (96.3)
Rwork (%)	15.71	15.75	17.62	17.44
Rfree (%)	17.17	18.59	19.91	21.16
R.M.S. bond (Å)	0.03	0.03	0.03	0.02
rms angles (°)	2.65	2.53	2.51	1.83
Mean B (Å^2)	30	19	19	26
Mean ligand B (Å^2)	n/a	17	30	38
Figure	4	6a	6b	6c
PDB code	6MSX	6MX9	n/a	n/a

3.5. Lysozyme Co-Crystallization with Fragment Library

We originally designed the gel exclusion method as a convenience to easily generate crystals for applications that require rapid crystal growth, such as educational courses and benchmarking the performance of diffraction instrumentation. However, during testing we observed that gel exclusion was a robust crystallization technique that reliably yielded crystals even when mistakes were made during the crystallization setup. To test the usefulness of this observation, we co-crystallized lysozyme with 48 chemical fragments using gel exclusion and compared this to crystals generated using conventional hanging drop. Using gel exclusion, we observed lysozyme crystals in all 48 fragment screens (Figure 5, top), compared to 26 in the 48-control screen (Figure 5, bottom).

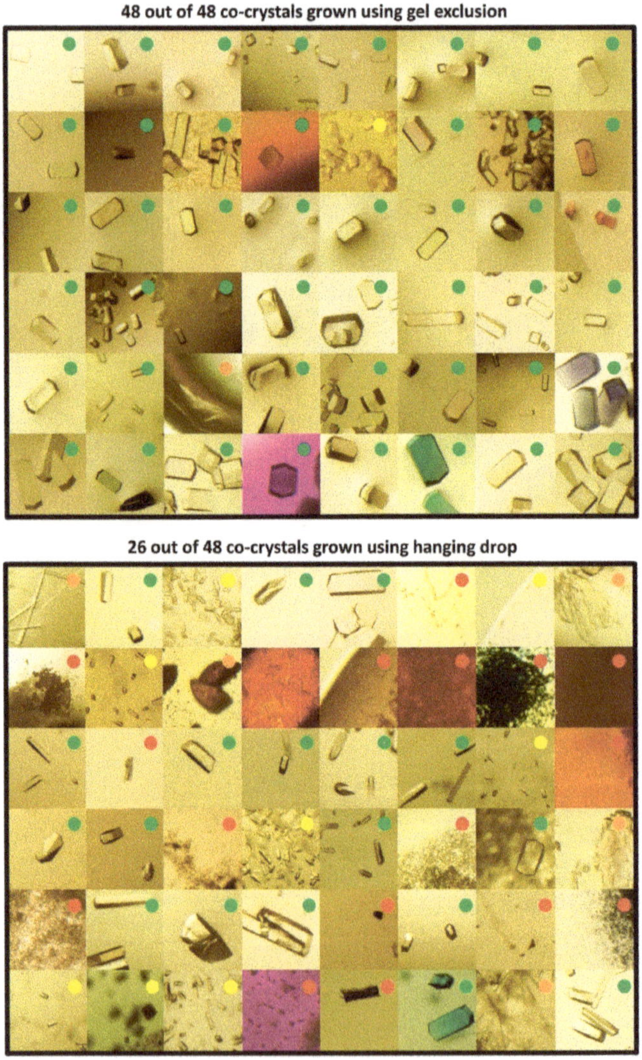

Figure 5. Co-crystallization of lysozyme with 48 fragments (gel exclusion on left, control on right). Green dots indicate large crystals, yellow dots indicate small crystals, orange dots indicate possible crystals, and red dots indicate no crystals). Gel exclusion co-crystallization yielded 48 crystals out of 48 screens, compared to 26 using conventional hanging drop (detailed description of fragment library in Supplementary materials section 1).

We obtained diffraction data from crystals drawn from all 48 gel exclusion screens. Diffraction data were measured and processed, and structures were solved, as described in Section 2.5 (starting models 4N8Z; [34]). Inspection of the structures revealed that three of the fragments tested induced observable changes in the electron density, 3-aminophenol, picolinic acid, and eosin Y (Figure 6A–C respectively). One observed ligand (3-aminophenol) was deposited in the PDB (6MX9).

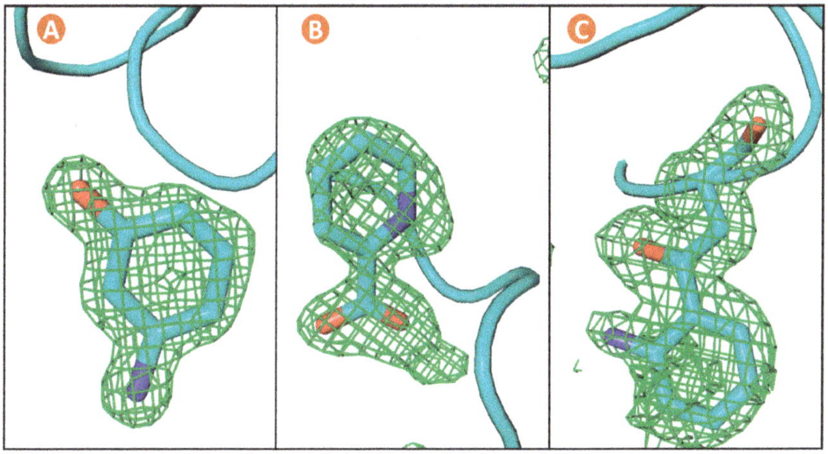

Figure 6. Electron density indicated that 3 out of 48 fragments perturbed the lysozyme starting model. Two of these were observed bound to the protein, (**A**) 3-aminophenol and (**B**) picolinic acid. (**C**) The third induced oxidation of tryptophan 62 (eosin Y) (**C**). Figure (**C**) depicts a tryptophan residue with an open ring and a changed confirmation (observed in the presence of eosin Y) [35]. Omit difference electron density is contoured at 3σ (**A**) or 2σ (**B**,**C**). Since picolinic acid resembles a crystallization component, (**B**) was re-crystallized without benzamidine.

4. Discussion

As discussed earlier, agar exclusion is a useful strategy for applications that require fast crystallization, such as educational applications and generating test crystals to test diffraction equipment. There are also safety advantages to sequestering toxic chemicals inside of an agar pellet (such as cadmium sulfate used to nucleate ferritin crystals), and further advantages in work flow compared to awkward crystallization setups such as button diffusion. In the case of ferritin, crystals not only grew very quickly, but yielded higher resolution data than conventionally grown crystals. High resolution data can improve the accuracy of ferritin models deduced from the diffraction (for example, we used the diffraction data to estimate the iron content inside the ferritin cavity using an iterative density modification and phase refinement procedure that is sensitive to data and model quality [36]). The integrated total number of iron atoms inside of the ferritin cavity was deduced to be 835 ± 22, starting from 20 models with very different initial average cavity density. This number is similar to a previously reported estimate of 800 Fe/molecule for native equine ferritin (measured using atomic absorption spectrophotometry [37,38]). All of these advantages are incremental conveniences that improve experimental timing, work flow, or quality—but do not potentiate new science that would otherwise not be possible.

The 100% co-crystallization rate for lysozyme with 48 fragments suggests that agar exclusion may be a robust approach to fragment screening. The fact that none of the three fragments that induced observable perturbations to the protein structure yielded crystals in the conventional hanging drop screens begs the question of whether fragments that interact with the protein surface are more likely to interfere in the crystallization process, compared to fragments that do not interact with the protein surface. This would be particularly important because it would imply that the miss rate for fragment screening is exaggerated because useful fragments are especially likely to interfere in co-crystallization. We can use the hypergeometric distribution to calculate that there is a 9% probability of observing 0 perturbations in the electron density from the set of 26 co-crystals that did grow in the control set, and observing 3 perturbations in the electron density from the set of 22 co-crystals that did not grow in the control set, if failure to co-crystallize and perturbation of the electron density are independent variables (see supplemental section 2). Hence, we conclude that lysozyme co-crystals are less likely to form in the presence of chemicals that perturb the structure of lysozyme, compared to chemicals that do not perturb the structure of lysozyme.

5. Conclusions

We found simple and effective partitions of the crystallization cocktail for four test proteins, with some of the chemicals sequestered in an agar pellet and some present in the protein drop. We observed that this gel exclusion approach yielded large well diffracting crystals far more rapidly than conventional crystallization. In the case of ferritin, crystals grown using gel exclusion were of higher quality than comparable crystals grown using hanging drop, and also of higher quality than iron containing crystals in the protein data bank. In the case of lysozyme, we observed that co-crystallization with a fragment library was more robust with agar exclusion compared to conventional hanging drop techniques. Notably, none of the three fragments that were observed to perturb the lysozyme structure resulted in crystals when co-crystallized using conventional hanging drop methods. We conclude that, in the case of proteins that are available in high quantities, agar exclusion is a promising technique for rapidly growing robust protein crystals.

Supplementary Materials: The following are available online at http://www.mdpi.com/2073-4352/8/12/464/s1.

Author Contributions: Conceptualization, A.S.S.; investigation, A.S.S., A.B., S.H.P., H.M.B., K.E.M., C.D.S. and J.P.; writing—original draft preparation, A.S.S.; writing—review and editing, A.S.S., A.B., S.H.P., H.M.B., K.E.M., C.D.S., J.P., A.P., A.H. and S.M. supervision, A.S.S., S.H.P., A.P., A.H. and S.M.

Funding: Major ongoing financial support for acoustic droplet ejection applications was through the Life Science Biomedical Technology Research resource, supported by the National Institute of Health, National Institute of General Medical Sciences (NIGMS) through a Biomedical Technology Research Resource P41 grant (P41GM111244), and by the DOE Office of Biological and Environmental Research (KP1605010).

Acknowledgments: Data for this study were measured at beamline 17-ID-1 (AMX) at the National Synchrotron Light Source II (NSLS II). Personnel for this study were recruited largely through the Science Undergraduate Laboratory Internships Program (SULI) and the High School Research Program (HSRP), supported through the U.S. Department of Energy, Office of Science, Office of Workforce Development for Teachers and Scientists (WDTS), DOE contract No. DE-SC0012704. SULI students contributed during the 2017 summer + fall, and during the 2018 summer. HSRP students contributed during the 2018 summer. Initial high throughput screening was performed during the 2017 Summer Science Exploration "Structural Biology: Solving Protein Structures at the Synchrotron" (supported by A. Horn). We thank the participants B. Scapellati, A. Morrison, R. Barnes, A. DeLury, S. Smiley, H. Dutt, J. Hart, J. Caesar, A. McCrea, R. Tiska, M. De Somma, M. De Somma, M. McKenna, S. Wallach, L. Davis, A. Dolich, T. Giannuzzi, R. Nagia, and Alyssa Arbuiso.

Conflicts of Interest: The authors declare no conflict of interest.

References

1. Lin, Y. What's happened over the last five years with high-throughput protein crystallization screening? *Expert Opin. Drug Discov.* **2018**, *13*, 691–695. [CrossRef]
2. Chayen, N.E.; Helliwell, J.R.; Snell, E.H. *Macromolecular Crystallization and Crystal Perfection*; Oxford University Press: Oxford, UK, 2010; Volume 24.
3. Santesson, S.; Cedergren-Zeppezauer, E.S.; Johansson, T.; Laurell, T.; Nilsson, J.; Nilsson, S. Screening of nucleation conditions using levitated drops for protein crystallization. *Anal. Chem.* **2003**, *75*, 1733–1740. [CrossRef] [PubMed]
4. Ericson, D.L.; Yin, X.; Scalia, A.; Samara, Y.N.; Stearns, R.; Vlahos, H.; Ellson, R.; Sweet, R.M.; Soares, A.S. Acoustic Methods to Monitor Protein Crystallization and to Detect Protein Crystals in Suspensions of Agarose and Lipidic Cubic Phase. *J. Lab. Autom.* **2016**, *21*, 107–114. [CrossRef] [PubMed]

5. Luft, J.R.; DeTitta, G.T. Chaperone salts, polyethylene glycol and rates of equilibration in vapor-diffusion crystallization. *Acta Crystallogr. Sect. D* **1995**, *51*, 780–785. [CrossRef] [PubMed]
6. Stojanoff, V.; Jakoncic, J.; Oren, D.A.; Nagarajan, V.; Navarro Poulsen, J.C.; Adams-Cioaba, M.A.; Bergfors, T.; Sommer, M.O. From screen to structure with a harvestable microfluidic device. *Acta Crystallogr. Sect. F* **2011**, *67*, 971–975. [CrossRef] [PubMed]
7. McPherson, A.; Gavira, J.A. Introduction to protein crystallization. *Acta Crystallogr. Sect. F* **2014**, *70*, 2–20. [CrossRef] [PubMed]
8. Robert, M.C.; Lefaucheux, F. Crystal growth in gels: Principle and applications. *J. Cryst. Growth* **1988**, *90*, 358–367. [CrossRef]
9. Provost, K.; Robert, M.C. Application of gel growth to hanging drop technique. *J. Cryst. Growth* **1991**, *110*, 258–264. [CrossRef]
10. Snell, E.H.; Helliwell, J.R. Macromolecular crystallization in microgravity. *Rep. Prog. Phys.* **2005**, *68*, 799. [CrossRef]
11. McPherson, A. Review Current approaches to macromolecular crystallization. In *EJB Reviews 1990*; Springer: Berlin/Heidelberg, Germany, 1990; pp. 49–71.
12. Salemme, F.R. A free interface diffusion technique for the crystallization of proteins for X-ray crystallography. *Arch. Biochem. Biophys.* **1972**, *151*, 533–539. [CrossRef]
13. Gorrec, F.; Löwe, J. Automated Protocols for Macromolecular Crystallization at the MRC Laboratory of Molecular Biology. *J. Vis. Exp.* **2018**. [CrossRef] [PubMed]
14. Hassell, A.M.; An, G.; Bledsoe, R.K.; Bynum, J.M.; Carter, H.L.; Deng, S.J.; Gampe, R.T.; Grisard, T.E.; Madauss, K.P.; Nolte, R.T.; et al. Crystallization of protein–ligand complexes. *Acta Crystallogr. Sect. D Biol. Crystallogr.* **2007**, *63*, 72–79. [CrossRef] [PubMed]
15. Cole, K.; Roessler, C.G.; Mule, E.A.; Benson-Xu, E.J.; Mullen, J.D.; Le, B.A.; Tieman, A.M.; Birone, C.; Brown, M.; Hernandez, J.; et al. A linear relationship between crystal size and fragment binding time observed crystallographically: Implications for fragment library screening using acoustic droplet ejection. *PLoS ONE* **2014**, *9*, e101036. [CrossRef] [PubMed]
16. Ehrmann, F.R.; Stojko, J.; Metz, A.; Debaene, F.; Barandun, L.J.; Heine, A.; Diederich, F.; Cianférani, S.; Reuter, K.; Klebe, G. Soaking suggests "alternative facts": Only co-crystallization discloses major ligand-induced interface rearrangements of a homodimeric tRNA-binding protein indicating a novel mode-of-inhibition. *PLoS ONE* **2017**, *12*, e0175723. [CrossRef] [PubMed]
17. Moreno, A.; Quiroz-García, B.; Yokaichiya, F.; Stojanoff, V.; Rudolph, P. Protein crystal growth in gels and stationary magnetic fields. *Cryst. Res. Technol. J. Exp. Ind. Crystallogr.* **2007**, *42*, 231–236. [CrossRef]
18. Thomas, D.H.; Rob, A.; Rice, D.W. A novel dialysis procedure for the crystallization of proteins. *Protein Eng. Des. Sel.* **1989**, *2*, 489–491. [CrossRef]
19. Pippard, M.J. Iron deficiency anemia, anemia of chronic disorders and iron overload. In *Blood and Bone Marrow Pathology*, 2nd ed.; Elsevier: Amsterdam, The Netherlands, 2011; pp. 173–195.
20. Honarmand Ebrahimi, K.; Hagedoorn, P.L.; Hagen, W.R. Unity in the biochemistry of the iron-storage proteins ferritin and bacterioferritin. *Chem. Rev.* **2014**, *115*, 295–326. [CrossRef] [PubMed]
21. Arosio, P.; Elia, L.; Poli, M. Ferritin, cellular iron storage and regulation. *IUBMB Life* **2017**, *69*, 414–422. [CrossRef]
22. Weiss, G. Iron and immunity: A double-edged sword. *Eur. J. Clin. Investig.* **2002**, *32*, 70–78. [CrossRef]
23. Arosio, P.; Levi, S. Ferritin, iron homeostasis, and oxidative damage 1, 2. *Free Radic. Biol. Med.* **2002**, *33*, 457–463. [CrossRef]
24. Jutz, G.; van Rijn, P.; Santos Miranda, B.; Böker, A. Ferritin: A versatile building block for bionanotechnology. *Chem. Rev.* **2015**, *115*, 1653–1701. [CrossRef] [PubMed]
25. Zipper, L.E.; Aristide, X.; Bishop, D.P.; Joshi, I.; Kharzeev, J.; Patel, K.B.; Santiago, B.M.; Joshi, K.; Dorsinvil, K.; Sweet, R.M.; et al. A simple technique to reduce evaporation of crystallization droplets by using plate lids with apertures for adding liquids. *Acta Crystallogr. Sect. F* **2014**, *70*, 1707–1713. [CrossRef] [PubMed]
26. Teplitsky, E.; Joshi, K.; Ericson, D.L.; Scalia, A.; Mullen, J.D.; Sweet, R.M.; Soares, A.S. High throughput screening using acoustic droplet ejection to combine protein crystals and chemical libraries on crystallization plates at high density. *J. Struct. Biol.* **2015**, *191*, 49–58. [CrossRef] [PubMed]

27. Gil-Alvaradejo, G.; Ruiz-Arellano, R.R.; Owen, C.; Rodríguez-Romero, A.; Rudino-Pinera, E.; Antwi, M.K.; Stojanoff, V.; Moreno, A. Novel protein crystal growth electrochemical cell for applications in X-ray diffraction and atomic force microscopy. *Cryst. Growth Des.* **2011**, *11*, 3917–3922. [CrossRef]
28. Kabsch, W. Xds. *Acta Crystallogr. Sect. D Biol. Crystallogr.* **2010**, *66*, 125–132. [CrossRef] [PubMed]
29. Winn, M.D.; Ballard, C.C.; Cowtan, K.D.; Dodson, E.J.; Emsley, P.; Evans, P.R.; Keegan, R.M.; Krissinel, E.B.; Leslie, A.G.W.; McCoy, A.; et al. Overview of the CCP4 suite and current developments. *Acta Crystallogr. Sect. D* **2011**, *67*, 235–242. [CrossRef]
30. Winn, M.D.; Murshudov, G.N.; Papiz, M.Z. Macromolecular TLS refinement in REFMAC at moderate resolutions. *Methods Enzymol.* **2003**, *374*, 300–321.
31. Lamzin, V.S.; Wilson, K.S. Automated refinement of protein models. *Acta Crystallogr. Sect. D Biol. Crystallogr.* **1993**, *49*, 129–147. [CrossRef]
32. Granier, T.; Gallois, B.; Dautant, A.; Langlois d'Estaintot, B.; Precigoux, G. Comparison of the Structures of the Cubic and Tetragonal Forms of Horse-Spleen Apoferritin. *Acta Crystallogr. Sect. D* **1997**, *53*, 580–587. [CrossRef]
33. Emsley, P.; Cowtan, K. Coot: Model-building tools for molecular graphics. *Acta Crystallogr. Sect. D Biol. Crystallogr.* **2004**, *60*, 2126–2132. [CrossRef]
34. Yin, X.; Scalia, A.; Leroy, L.; Cuttitta, C.M.; Polizzo, G.M.; Ericson, D.L.; Roessler, C.G.; Campos, O.; Ma, M.Y.; Agarwal, R.; et al. Hitting the target: Fragment screening with acoustic in situ co-crystallization of proteins plus fragment libraries on pin-mounted data-collection micromeshes. *Acta Crystallogr. Sect. D* **2014**, *70*, 1177–1189. [CrossRef] [PubMed]
35. Folzer, E.; Diepold, K.; Bomans, K.; Finkler, C.; Schmidt, R.; Bulau, P.; Huwyler, J.; Mahler, H.C.; Koulov, A.V. Selective Oxidation of Methionine and Tryptophan Residues in a Therapeutic IgG1 Molecule. *J. Pharm. Sci.* **2015**, *104*, 2824–2831. [CrossRef] [PubMed]
36. Soares, A.S.; Caspar, D.L. X-ray diffraction measurement of cosolvent accessible volume in rhombohedral insulin crystals. *J. Struct. Biol.* **2017**, *200*, 213–218. [CrossRef] [PubMed]
37. Bolann, B.J.; Ulvik, R.J. On the limited ability of superoxide to release iron from ferritin. *Eur. J. Biochem.* **1990**, *193*, 899–904. [CrossRef] [PubMed]
38. Smith, J.M.; Helliwell, J.R.; Papiz, M.Z. Absorption of X-radiation by single crystals of proteins containing labile metal components: The determination of the number of iron atoms within the central core of ferritin. *Inorg. Chim. Acta* **1985**, *106*, 193–196. [CrossRef]

crystals

MDPI

Article

Recent Insights into Protein Crystal Nucleation

Christo N. Nanev

Rostislaw Kaischew Institute of Physical Chemistry, Bulgarian Academy of Sciences, 1113 Sofia, Bulgaria; nanev@ipc.bas.bg; Tel.: +359-2-856-64-58; Fax: +359-2-971-26-88

Received: 27 April 2018; Accepted: 12 May 2018; Published: 17 May 2018

Abstract: Homogeneous nucleation of protein crystals in solution is tackled from both thermodynamic and energetic perspectives. The entropic contribution to the destructive action of water molecules which tend to tear up the crystals and to their bond energy is considered. It is argued that, in contrast to the crystals' bond energy, the magnitude of destructive energy depends on the imposed supersaturation. The rationale behind the consideration presented is that the critical nucleus size is determined by the balance between destructive and bond energies. By summing up all intra-crystal bonds, the breaking of which is needed to disintegrate a crystal into its constituting molecules, and using a crystallographic computer program, the bond energy of the closest-packed crystals is calculated (hexagonal closest-packed crystals are given as an example). This approach is compared to the classical mean work of separation (MWS) method of Stranski and Kaischew. While the latter is applied merely for the so-called Kossel-crystal and vapor grown crystals, the approach presented can be used to establish the supersaturation dependence of the protein crystal nucleus size of arbitrary lattice structures.

Keywords: protein crystal nucleation; thermodynamic and energetic approach; protein 'affinity' to water; solubility; balance between crystal bond energy and destructive surface energies; supersaturation dependence of the crystal nucleus size

1. Introduction

Crystallization is the most efficient and economical way of obtaining chemically pure compounds. That is why it is widely used in the pharmaceutical, fertilizer and sugar industries. Single crystal X-ray diffraction, being the most universal, powerful and accurate tool for biological macromolecule structure analysis and protein–substrate interactions, requires relatively large and well-diffracting crystals. However, the major stumbling stone in X-ray diffraction analysis of protein crystals is the lack of a recipe (or definite indications) for growing crystals of newly expressed proteins. The issue is finding conditions that make the homogeneously scattered protein molecules in a solution form stable crystal nuclei; once nucleated, the crystals continue growing spontaneously. Evidently, nucleation is the crucial step that determines the difference between success and failure in protein crystallization trials. Regardless of the numerous auxiliary crystallization tools employed, such as automation and miniaturization of crystallization trials by means of robots, Dynamic Light Scattering, crystallization screening kits, etc., it is researchers' creativeness and acumen that remain indispensable. Even with state-of-the-art tools, it is exceptionally challenging to probe the nucleation processes in real time. Only most recently, this problem has been addressed successfully by Van Driessche et al. [1]. Nevertheless, some of the most intimate moments of molecule-by-molecule assembly to form crystal nuclei are still elusive and require further elucidation.

Following the fundamental notion of a kink position (*Halbkristalllage* in German) introduced by Kossel [2] and Stranski [3], the Bulgarian scientists Stranski and Kaischew [4–6] were the first to apply a molecular kinetic (and energetic) approach to crystal nucleation. They proposed the so-called mean work of separation (MWS) method. To calculate the MWS value, all of the different kinds of bonds (between the first, second and third neighboring crystal-building blocks) are counted separately, and each number is multiplied by the corresponding bonding energy. Then, the products are summed up, and the result is divided by the total number of blocks in the corresponding crystal element (crystal face, cluster edge). It has been argued that for vapor phase crystallization MWS value is equal to the chemical potential, taken with a negative sign, plus a substance and temperature dependent constant [7]. To calculate the work required (energy barrier) for nucleus formation, crystal bond energies are used as well. They are calculated by summing up the number of all intra-crystal bonds, the breaking of which is needed to totally disintegrate a crystal into all its constituting molecules. The same approach is used in the present study. Applying the MWS method to the so-called Kossel-crystal (a crystal build by small cubes held together by equal forces in a cubic primitive crystal lattice) allows the equilibrium crystal shapes to be determined.

A big advantage of the MWS method is its simplicity. It uses the relative bond energies between separate crystal building units instead of the absolute bond energy values, which frequently are unknown. However, the MWS method has two major drawbacks: it is applicable only to the Kossel-crystal model (existing extremely rarely in nature) and considers solely enthalpic effects. Perhaps it is for these very reasons that the MWS method is used rarely nowadays, even though it enables semi-quantitative studies of crystal nucleation and growth.

To find the critical nucleus size, Garcia-Ruiz established the balance between the cohesive energy (ΔG_v), which maintains the integrity of a crystalline cluster, and the sum of destructive energies (ΔG_s), which tend to tear up the crystal, i.e., $-\Delta G_v + \Delta G_s = 0$ [8]. Using a cubic primitive crystal lattice formed by spheres, the author equilibrated the number of bonds shared by the crystal building units with the number of dangling bonds at the crystal surface, pointing toward the solution. This corresponds to the classical nucleation theory where ΔG_v is proportional to the crystal volume while ΔG_s is proportional to its surface, and the critical nucleus size is determined from the compensation of the large surface energy, which is inherent for the undercritical molecule clusters, by the faster volume energy increase resulting from the rising crystal size. Garcia-Ruiz's intuitive approach accounts for the water molecules acting on the apexes and edges that exist on the polyhedral crystal nuclei but are absent on the droplets. His model shows that at the crystal vertices, water molecules pull protein molecules towards the solution from three (perpendicular) directions, at the crystal edges from two directions, and at the crystal face from one direction only.

Recently, Garcia-Ruiz's brilliant idea was been further elaborated [9]. When the system is undersaturated, i.e., crystallization is impossible, the protein 'affinity' to water molecules prevails over the crystallization propensity. Therefore, to evoke crystallization, it is necessary to impose supersaturation and the higher the latter, the more thermodynamically stable the crystal is, with respect to the solution. Thus, it is feasible to assume that the imposed supersaturation decreases the protein-to-water affinity, i.e., supersaturation diminishes the destructive energy (ψ_d) per bond. This means that the tendency to tear up the crystal depends on the degree of supersaturation in contrast to the cohesive energy per bond in the crystal lattice (ψ_b) which is supersaturation independent. This means that any supersaturation increase will lead to an increase in ψ_b/ψ_d ratio. On this basis, the critical nucleus size dependence on supersaturation has been determined (from the balance between the sum of all intra-crystal bonds and the sum of surface destructive energies) for the Kossel-crystal model [9].

The objective of the present work is to shed additional light on the thermodynamic and molecular aspects of the homogeneous nucleation of non-Kossel crystals, such as nucleation of protein crystals in solution. To determine the supersaturation dependent critical nucleus size, the crystal bond energy and the sum of supersaturation dependent surface destructive energies are equilibrated (referred to as EBDE) and compared to the MWS method. In view of the nature of lattice binding forces among huge biomolecules, only the first nearest neighbor interactions are considered. For simplicity purposes, equal bond energy interactions throughout the whole crystal are assumed. The consequences of the highly anisotropic and multivalent interactions of proteins during crystal nucleation were considered elsewhere [10]. However, specifics of protein and small molecule crystal nucleation are kept in mind [11].

2. Equilibration of Crystal Bond Energy with Surface Destructive Energies (EBDE)

2.1. Thermodynamic Basis of EBDE and Definition of Protein-to-Water Affinity

In [9], ψ_d was loosely defined as accounting for entropy, including the sum of all nucleation disfavoring entropic contributions. All entropy increasing components (favoring crystal nucleation) were included in ψ_b. A more rigorous definition of protein-to-water 'affinity' which uses both nucleation process enthalpy and entropy is provided below.

Thermodynamics stipulates that enthalpy and entropy govern phase transition. In contrast to crystal nucleation from vapors, protein crystal nucleation evokes a simultaneous entropy change in both solutions and crystals [12]. Rearrangement and/or release of some associated water or trapping of even more water molecules occurs when protein molecules get together to form a new solid phase. Therefore, entropy accounts for the change in the number of molecules in both protein crystals and solutes. Many water molecules are released in the solution for each protein molecule bound into the crystal lattice. In the disordered bulk solvent, these water molecules have six degrees of freedom and increase the entropy of the whole crystallizing system, thus leading to a decrease in the Gibbs free energy of phase transition. Sometimes, the entropy gain resulting from the said water transfer is the main component in the driving energy for protein crystal nucleation. However, it is not only the release of water molecules from the contacting patches (during crystalline bonds formation) that affects crystallization thermodynamics. Being immobilized in the crystal lattice, protein molecules lose entropy due to the highly constrained translational and rotational degrees of freedom. In turn, protein molecule ordering (crystal nuclei formation) is stimulated by an entropy gain due to the newly acquired vibrational degrees of freedom that arise upon molecule attachment to the crystal. Together with the high crystallization enthalpy, entropy changes ensure the negative values of Gibbs free energy required for spontaneous crystallization are attained (for more details see [12,13]). The conformational entropy of residues involved in crystal contact has been recently reconsidered [14]. Simultaneously, study of hydrogen bonds, van der Waals contacts and electrostatic interactions has shown that hydrogen-mediated van der Waals interactions are the dominant force that maintains protein crystal lattice integrity. Unfortunately, however, non-Kossel crystal lattice analysis remains a theoretical challenge [13].

The relation between ψ_d and supersaturation ($\Delta\mu$) is derived via the protein-to-water 'affinity' (α) noting that the lesser the value of α, the lesser the crystal solubility (c_e). As a first approximation, proportionality with a constant κ is assumed:

$$\alpha = \kappa c_e. \tag{1}$$

Also noted is the equal probability of crystal nuclei formation and dissolution. This means that the nucleation process can be regarded as a reversible chemical reaction characterized by the equilibrium constant for crystallization (K_{eq}). Combining the standard Gibbs free energy (ΔG^o) for crystals,

$$\Delta G^o = \Delta H^o - T\Delta S^o, \tag{2}$$

where ΔH^o is the standard enthalpy change, and ΔS^o is the change in entropy of the nucleation process. T is the absolute temperature. The Gibbs free energy isotherm equation is

$$\Delta G^o = -RT\ln K_{eq},$$

and by denoting the universal gas constant as R, the following is obtained:

$$\ln K_{eq} = -(\Delta H^o/RT) + (\Delta S^o/R). \tag{3}$$

Considering the approximation suggested by Sleutel et al. [15] and assuming an ideal solution (i.e., the activity coefficient is a unit), the crystallization equilibrium constant (K_{eq}) can be represented as $K_{eq} \approx (c_e/c^o)^{-1}$, which gives

$$\ln(c_e/c^o) = (\Delta H^o/RT) - (\Delta S^o/R). \tag{4}$$

where $c^o = 1$ mol L^{-1} is the solution's concentration in the standard state. Thus, solubility, c_e, is expressed in terms of the change in the entropy and enthalpy of the nucleation process:

$$c_e = c^o \exp[(\Delta H^o/RT) - (\Delta S^o/R)], \text{ and} \tag{5}$$

$$\alpha = \kappa c^o \exp[(\Delta H^o/RT) - (\Delta S^o/R)]. \tag{6}$$

The protein crystal solubility, c_e (and hence, protein-to-water affinity, α), depends on the solution's composition, mainly the precipitant type and concentration, pH, and temperature. (Isothermal protein crystal nucleation is considered here.) The lesser the solubility (affinity), the larger the supersaturation at the same solution concentration, c:

$$\Delta\mu = k_B T \ln(c/c_e) = k_B T \ln\{c \exp[(\Delta S^o/R) - (\Delta H^o/RT)]/c^o)\} = k_B T \ln(\kappa c/\alpha), \tag{7}$$

where k_B is the Boltzmann constant.

It has been known for a long time that the disordered solvent phase fills the void between molecules in the protein crystal lattice [16]. The role of water included in protein crystals acting like an 'additional glue' to hold protein molecules together has been considered elsewhere [17] and is not within the scope of this paper.

2.2. *Energetics of Protein Crystal Nucleation*

From energetic point of view, it is evident why 3D nuclei are preferred over 2D nuclei built by the same number of molecules. A (forth) molecule arriving from solution bulk prefers to start forming a second layer (B) rather than attaching itself to the periphery of an already existing layer (A) (Figure 1). The reason is that by sitting in the hole between the three molecules of layer A, this molecule bonds to the others using energy $3\psi_b$. If it is to remain in layer A, it will acquire two bonds only, so it jumps and forms layer B (Figure 2). For the same reason, it is highly probable that a fifth molecule will attach itself to the opposite side of the molecule triplet (Figure 1). That is how the cluster of bond energy $9\psi_b$, arises, instead of the planar configuration of energy $7\psi_b$.

Figure 1. Three close-packed molecules form layer A.

Figure 2. A molecule jumps to start formation of a second layer (B).

Typical 3D protein crystals are polyhedral, which is why crystal nuclei are assumed to have polyhedral shapes. The thermodynamic definition of protein-to-water affinity presented above assumes that crystals are stable entities, even though molecules at crystal vertexes and edges are connected more loosely than those at the crystal face. In view of the rather high supersaturations needed for protein crystal nucleation, this assumption is quite reasonable (and can be regarded as a justification for EBDE application).

Keeping in mind Garcia-Ruiz's postulate about water molecules pulling protein molecules at crystal vertexes towards the solution more strongly (from three different directions) than those at crystal faces, we notice that small molecule clusters, comprised of up to six molecules, have all of their molecules sitting at crystal vertexes. From the EBDE perspective, such small molecule clusters can be critical nuclei only provided supersaturation is very high. Table 1 shows the ψ_b/ψ_d ratio for the number (n) of molecules in a cluster; note that the supersaturation is decreasing from left to right. A 3D cluster formed by four molecules does not predetermine the lattice type—hexagonal closest packed (HCP) or face centered cubic (FCC)—of the growing crystal, while a 3D cluster, comprised of five molecules (three in layer A and one on each side, i.e., in two layers B), results in an HCP crystal lattice. A FCC structure requires larger clusters of at least six molecules. However, it is not only the energetic aspect that determines the crystal lattice type. Wukovitz and Yeates [18] have pointed out that an appropriate symmetry is mandatory for the crystallization of biological macromolecules.

Table 1. The ψ_b/ψ_d ratio for the number (n) of molecules in a cluster.

n	2	3	4(in 3D)	5(HCP)	6(FCC)
ψ_b/ψ_d	6	3	2	≈1.7	≈1.6

Using the MWS method, Kaischew [19] showed that homogeneously formed crystal nuclei are larger than heterogeneous nuclei. The reason for this is that the volume of a heterogeneous nucleus decreases to a degree that depends on the substrate's nucleation activity (see Figure 3). Energetically preferred 3D HCP (Figure 4) and 3D FCC crystals arise by adding three molecules in the (corresponding) holes at the top, and another three molecules at the bottom of an initial layer A of spheres arranged in the shape of a hexagon; all of the tetrahedral holes are covered in the HCP structure, and all of the octahedral holes are covered in the FCC crystals. The result is a three-fold rise in bond energy, up to $36\psi_b$, compared with $12\psi_b$ in the planar configuration, and the ψ_b/ψ_d ratio is calculated by EBDE (see Table 2, in which the supersaturation is decreasing from left to right).

The EBDE method benefits from crystallographic computer programs rendering the sums of bond energies ($\Delta G_v{}^3$) of arbitrary sized 3D crystals and the number (N_s) of surface atoms exposed to water destructive action. The advantage of crystallographic computer programs is that they are applicable to diverse crystal structures. Given as an example here are HCP crystals with <0100> edge lengths determined by the numbers (L) of the molecules in them. The ATOMS (Version 5.0.4, 1999) computer program is used to calculate $\Delta G_v{}^3$ and N_s for the crystals of a truncated bipyramidal shape. These crystals are geometrical homologues to the crystal (shown in Figure 5), with $L = 3$ (the three blue balls), in which the number of molecules (N_t) in the topmost and correspondingly, in the bottommost,

{0001} faces depend on L (so that for the differently sized crystals, $N_t = 3$ for $L = 2$ (Figure 4); $N_t = 7$ for $L = 3$ (Figure 5); $N_t = 12$ for $L = 4$; $N_t = 19$ for $L = 5$; $N_t = 27$ for $L = 6$, etc.).

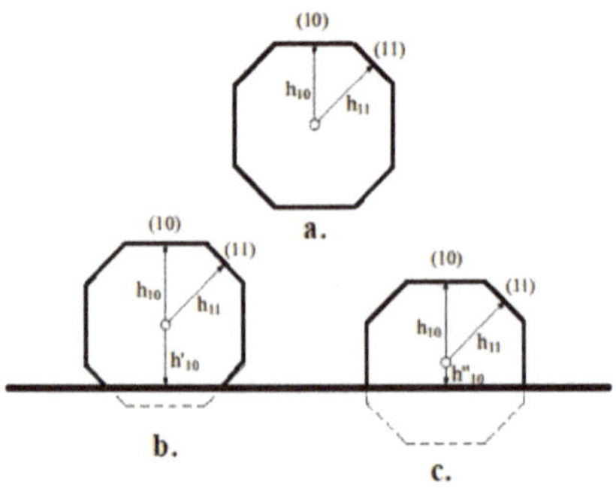

Figure 3. (**a–c**) 2D crystal nuclei: (**a**) homogeneous nucleus; (**b**,**c**) heterogeneously formed nuclei. Wulff's points are shown by small circles, and h_i denotes the corresponding distances (proportional to the respective surface free energies) used for the Currie–Wulff's constructions; the solid line represents the substratum surface.

Figure 4. Formation of a hexagonal closest packed (HCP) crystal.

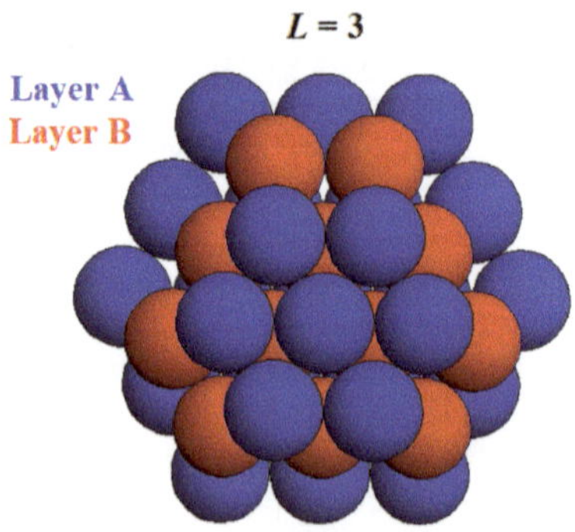

Figure 5. Top view of a five-layered HCP (truncated dihexagonal dipyramid) crystal, for which the two bottom layers are not seen. The crystal is formed on a closest packed monomolecular layer A of spheres (the blue balls) arranged into the shape of a complete hexagon with $L = 3$.

The plot of $\Delta G_v^3/\Delta G_s$ vs. L is shown in Figure 6. The reason for this plotting is that the energy balance ($-\Delta G_v + \Delta G_s = 0$) means $\Delta G_v/\Delta G_s = 1$, so the reciprocal of the ordinate value in Figure 6 (giving the corresponding ψ_b/ψ_d values) shows the decrease in supersaturation with an increase in L (Table 2). Here Garcia-Ruiz's postulate (that the molecules at the crystal vertexes are pulled towards the solution from three different directions, the molecules in the crystal edges are pulled from two directions, and the molecules in the crystal faces are pulled from one direction only) is also taken into account since it should apply to any crystal structure—water molecules are 'negligent' of the crystal lattice type. It is worth noting that the larger nucleus stands in equilibrium under the actual supersaturation until the larger ψ_b/ψ_d value in Table 2 (corresponding to higher supersaturation needed for formation the smaller crystal nucleus) is reached.

Table 2. The ψ_b/ψ_d ratio for the number (L) of molecules in the crystal edge.

L	2	3	4	5	6	7	8	9	10
ψ_b/ψ_d	1	0.43	0.25	0.17	0.13	0.10	0.087	0.07	0.06

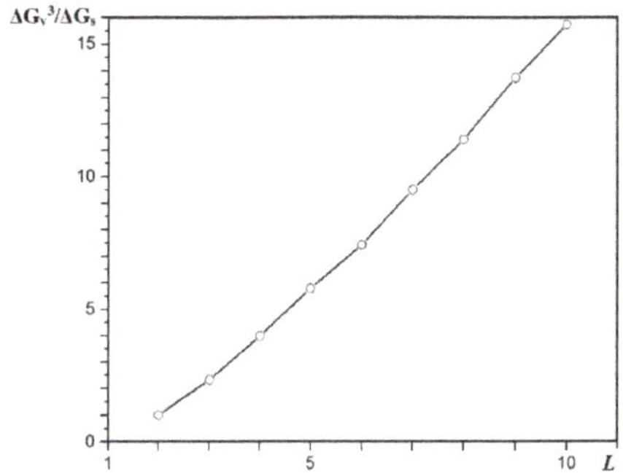

Figure 6. Plot of $\Delta G_v^3/N_s$ vs. L.

However, only crystals of modest size are of interest because though feasible from an energetic point of view, larger crystal nuclei are less likely to appear. There are kinetic reasons for this—bringing together a vast number of molecules via molecule-by-molecule assembly into a crystal nucleus involves very large fluctuations which in turn requires very long waiting times. For the same reason, the simplest crystal shapes are considered—any additional nuclei faces require a substantial increase in crystal size.

The discrete character of cluster-size alteration has been considered by Stoyanov, Milchev and Kaischew [20–23]. They established a step-wise, instead of a continuous, relationship between nucleus size and supersaturation, and a supersaturation interval instead of a fixed level of supersaturation corresponding to each critical nucleus.

3. MWS Method Application to Closest-Packed 3D Crystals; Comparison of EBDE and MWS

Crystallographic computer programs provide crystal lattice images and numerical data, in contrast to the classical MWS method which uses analytical expressions. Simple models of modestly sized 3D crystals with complete shapes were used to calculate the MWSs of faces forming the habitus of non-Kossel crystal nuclei, the reason being that incomplete crystalline clusters would have sites for subsequent attachment and formation of minimum surface free energy clusters. Considered herein are the closest-packed (HCP and FCC) crystals. To prepare such models, closest-packed monomolecular layers were stacked consecutively onto both sides of a basic A-layer. Three-layered (Figure 7), five-layered (Figure 5), etc. homologous HCP crystal nuclei with L = 3, 4, etc. were created. (It is worth noting that heaping 7 and more close-packed layers need larger foundations, i.e. larger

L values. As already mentioned however, a nucleus size limit is inevitable.) In contrast, to calculate the MWS of a {0001} crystal face, a whole upper mono-layer was stripped-off from the rest of the crystal. Then the layer disintegrated completely into its constituent molecules. The results made it obvious that MWS value calculations introduced uncertainty, caused by the alternatively stacked hexagonal and ditrigonal layers in the HCP crystals. This is why two different MWS values (but not one like by the Kossel-crystal) were calculated for the closest-packed surface layers. Oddly enough, depending on the type of surface layer to be disintegrated, the crystal stood in equilibrium with two different supersaturations.

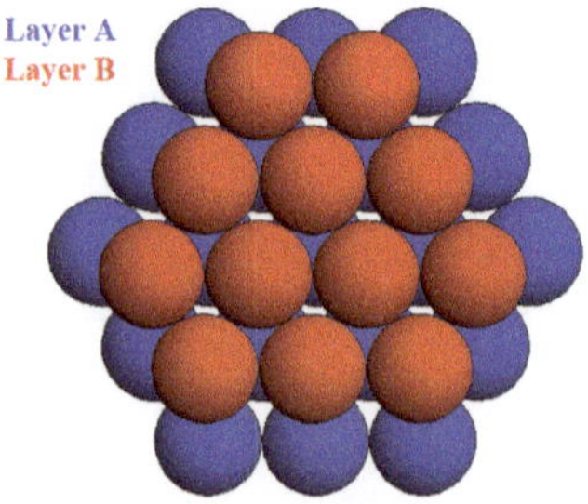

Figure 7. Top view of a three-layered HCP truncated dihexagonal dipyramid crystal (the bottom layer is unseen). The crystal is built by stacking the closest packed layers of spheres (the red balls) onto the single A layer (the blue balls, arranged into the shape of a complete hexagon with $L = 3$).

Well-known crystallography equations were applied for the MWS value calculations. Since each molecule in the closest-packed surface layer (regardless whether hexagonal or ditrigonal) is related to three molecules beneath it, the work (energy) needed for striping-off one such layer is always three times the number (z) of molecules in the layer. By denoting the number of molecules in the edge of a hexagonal layer by λ, we obtain

$$z = 3\lambda(\lambda - 1) + 1, \tag{8}$$

which gives z = 7, 19, 37, 61, 91, 127 ... for λ = 2, 3, 4, 5, 6, 7 ... respectively.

In addition, for the number (Z') of molecules in the ditrigonal layer, which is situated onto a hexagonal layer:

$$Z' = 3(\lambda - 1)^2, \tag{9}$$

which gives Z' = 3, 12, 27, 48, . . . for λ = 2, 3, 4, 5 ... respectively.

A complete disintegration of the hexagonal and ditrigonal layers into their constituent molecules requires different amounts of work ($\Delta G_v{}^2$) which, in turn, implies two different MWS values. Though disintegration can be done in various ways, the result must be one and the same. Here, the crystallographic formula concerning the number of bonds in the hexagonal crystal layer is used:

$$\Delta G_v{}^2 = (3\lambda - 3)(3\lambda - 2). \tag{10}$$

Thus, the MWS value (W) for HCP crystals with hexagonal {0001} surface layers is

$$W = 3\psi_b + (3\lambda - 3)(3\lambda - 2)\psi_b/[3\lambda(\lambda - 1) + 1] = 3\psi_b + [9\lambda^2 - 15\lambda + 6]\psi_b/[3\lambda^2 - 3\lambda + 1]. \tag{11}$$

Neglecting (for $L \rightarrow \infty$) all numbers smaller than the quadratic terms in Equation (11) gives $W \rightarrow 6\psi_b$, i.e., the bonding energy of the molecule in a kink position.

The formula for $\Delta G_v{}^2$ of ditrigonal layers is

$$\Delta G_v{}^2 = (3\lambda - 5)(3\lambda - 3), \tag{12}$$

and the MWS value (W) for HCP crystals with ditrigonal {0001} surface lattice planes is

$$W = 3\psi_b + (3\lambda - 5)(3\lambda - 3)\psi_b/3(\lambda - 1)^2 = 3\psi_b + [9\lambda^2 - 24\lambda + 15]\psi_b/[3\lambda^2 - 6\lambda - 3]. \tag{13}$$

Here, again, all numbers smaller than the quadratic terms in Equation (13) can be neglected for $L \rightarrow \infty$, and again, $W \rightarrow 6\psi_b$.

Relevant to the application of the MWS method to FCC crystals is the cubo-octahedron crystal shown in Figure 8 (which is composed by five layers, BCABC, and has eight equal octahedral faces of triangular shape, all of them containing six molecules). The EBDE method shows that this crystal can be a critical nucleus under supersaturation, which is determined from the ratio $\psi_b/\psi_d \approx 0.42$. It is almost equal to the ψ_b/ψ_d ratio for the HCP crystal shown in Figure 5 ($\psi_b/\psi_d \approx 0.43$, see Table 2). Note that the FCC crystal is comprised of 55 molecules, while the HCP crystal has two additional molecules.

Figure 8. Top view of a five-layered face centered cubic (FCC) lattice crystal. It is formed on the closest packed single layer of A spheres (the blue balls) arranged in the shape of a complete hexagon with $L = 3$.

By applying the MWS method, we see that the stripping-off of the whole upper triangular layer {111} (containing green balls, Figure 8) requires work (energy) amounting to $18\psi_b$, while each layer here has a bond energy of $9\psi_b$. This gives a MWS value of $4.5\psi_b$. The cubic faces {100} are built from nine molecules of bond energy $12\psi_b$, and the bond energy of the front cubic face in Figure 8 to the layer beneath it is $32\psi_b$. This gives a MSW value $\approx 4.9\psi_b$. The different MWSs of the {111} and {100} faces indicate that, according to the MSW method, this crystal does not have an equilibrium shape.

The difficulties encountered with the highly symmetrical HPC and FCC crystals indicate that the MWS method has low applicability to less-symmetrical crystals. This notwithstanding, historically, the classical MWS method has played an important role in the crystal growth theory, regardless of being demonstrated merely on the Kossel-crystal model [24]. Markedly, present-day computer programs are capable of elucidating at least one of the basic notions used in the classical MWS method, namely the meaning of an 'infinitely' large crystal. The latter is used as a benchmark in Stranski-Kaischew's molecular kinetic theory. Figure 9 shows how the 'infinitely' large crystal size is approached asymptotically. Here, the N_b/N_a ratio (where N_b is the total number of bonds in the crystal and N_a is the number of molecules in the truncated HCP bipyramid) is plotted versus the crystal edge length, L. It is well-known that the N_b/N_a for a 12-coordinated molecule is six, and the bond energy of a molecule in the kink position is $6\psi_b$.

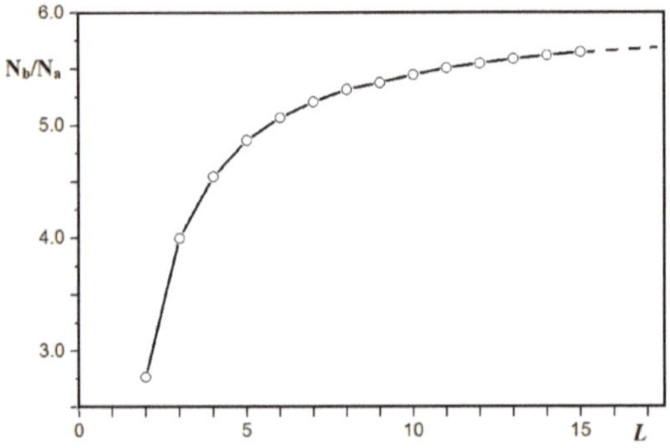

Figure 9. Plot of N_b/N_a ratio (where N_b is the total number of bonds in the crystal and N_a is the number of molecules in the truncated HCP bipyramid), which shows the influence of crystal vertexes and edges fading away with crystal enlargement. With $L \to \infty$ (L is the crystal edge length), the N_b/N_a ratio gradually approaches 6. Geometrical homologues to the HCP crystals like those in Figure 5 and in Figure 7 are considered.

4. Difference between Equilibrium and Growth Crystal Shapes

The nucleation stage predetermines the polymorphic crystal form (i.e., crystal lattice type) but not the habitus of a growing crystal (i.e., the type of faces on it). Crystal habitus depends on the growth rates of faces in normal directions. Historically, Stranski and Kaischew [25] were the first to recognize that the adsorption energy of a molecule at the middle of a crystal face can be used to characterize its growth rate. Later, Hartman and Bennema [26] proved that although dependent on the growth mechanism (via two-dimensional nucleation or spiral growth) and system conditions, face growth rate always increases with the increase in attachment energy per molecule on this crystal face. For instance, {110} faces grow faster than {100} and {111} faces because the steady-state growth shapes are determined by the number of their dangling bonds. The latter are five for {110} faces, four for {100} faces and three for {111} faces. In fact, ferritin crystals with the largest {111} faces have been observed most frequently—an indication that this is the slowest growing type of face in the normal direction and hence, the most morphologically important crystal face [27]. Using Atomic Force Microscopy, Yau and Vekilov [28] observed that near-critical-size apoferritin crystallites grow by attachment of molecules to the side {110} faces, thus forming a large {111} face. Growth of an apoferritin microcrystal consisting of three {111} layers of about 60 molecules in each was also detected. These nearly-critical crystallites were of quasi-planar (raft-like) shapes which is rather unusual for the FCC crystals.

5. Conclusions

An advantage of the EBDE method is that coupled with adequate computer programs, it can predict the nucleation of crystals of diverse lattice structures. As exemplified here with HCP crystals, modern computer programs (such as ATOMS) enable calculations considered unimaginable at the time of Stranski and Kaischew, over 80 years ago. Just as an example, the ordinate for $L = 10$ in Figure 9 is calculated for a HPC bipyramid crystal having $N_a = 2859$ atoms and $N_b = 15588$ bonds.

Founding: This work was co-financed by the National Science Fund of the Bulgarian Ministry of Science and Education, contract DCOST 01/22.

Acknowledgments: The author would like to acknowledge networking support by the COST Action CM1402 Crystallize. Also, the help of Kostadin Petrov is acknowledged. Crystallographic considerations and calculations, as well as Figures 5–9, were provided by him.

Conflicts of Interest: The author declares no conflict of interest. The founding sponsor had no role in the design of the study; in the collection, analyses, or interpretation of data; in the writing of the manuscript, and in the decision to publish the results.

References

1. Van Driessche, A.E.S.; Van Gerven, N.; Bomans, P.H.H.; Joosten, R.R.M.; Friedrich, H.; Gil-Carton, D.; Sommerdijk, N.A.J.M.; Sleutel, M. Molecular nucleation mechanisms and control strategies for crystal polymorph selection. *Nature* **2018**, *556*, 89–94. [CrossRef] [PubMed]
2. Kossel, W. Zur Theorie des Kristallwachstums. In *Nachrichten von der Gesellschaft der Wissenschaften zu Göttingen, Mathematisch-Physikalische Klasse*; Weidmannsche Buchhandlung: Berlin, Germany, 1927; pp. 135–143.
3. Stranski, I.N. Zur Theorie des Kristallwachstums. *Z. Phys. Chim. A* **1928**, *136*, 259–278. [CrossRef]
4. Stranski, I.N.; Kaischew, R.A. Über den Mechanismus des Gleichgewichtes kleiner Kriställchen, Part I. *Z. Phys. Chem. B* **1934**, *26*, 100–113.
5. Stranski, I.N.; Kaischew, R.A. Über den Mechanismus des Gleichgewichtes kleiner Kriställchen, Part II. *Z. Phys. Chem. B* **1934**, *26*, 114–116.
6. Stranski, I.N.; Kaischew, R.A. Über den Mechanismus des Gleichgewichtes kleiner Kriställchen, Part III. *Z. Phys. Chem. B* **1934**, *26*, 312–316.
7. Kaischew, R.A. Zur Theorie des Kristallwachstums. *Z. Phys.* **1936**, *102*, 684–690. [CrossRef]
8. Garcia-Ruiz, J.M. Nucleation of protein crystals. *J. Struct. Biol.* **2003**, *142*, 22–31. [CrossRef]
9. Nanev, C.N. On some aspects of crystallization process energetics, logistic new phase nucleation kinetics, crystal size distribution and Ostwald ripening. *J. Appl. Cryst.* **2017**, *50*, 1021–1027. [CrossRef]
10. Nanev, C.N. Kinetics and Intimate Mechanism of Protein Crystal Nucleation. *Prog. Cryst. Growth Character. Mater.* **2013**, *59*, 133–169. [CrossRef]
11. Nanev, C.N. Phenomenological Consideration of Protein Crystal Nucleation; the Physics and Biochemistry behind the Phenomenon. *Crystals* **2017**, *7*, 193. [CrossRef]
12. Vekilov, P.G.; Feeling-Taylor, A.; Yau, S.-T.; Petsev, D. Solvent entropy contribution to the free energy of protein crystallization. *Acta Cryst. D* **2002**, *58*, 1611–1616. [CrossRef]
13. Vekilov, P.G.; Chernov, A.A. The Physics of Protein Crystallization. *Solid State Phys.* **2003**, *57*, 1–147.
14. Dimova, M.; Devedjiev, Y.D. Protein crystal lattices are dynamic assemblies: The role of conformational entropy in the protein condensed phase. *IUCrJ* **2018**, *5*, 130–140. [CrossRef]
15. Sleutel, M.; Maes, D.; Van Driessche, A.E.S. Kinetics and Thermodynamics of Multistep Nucleation and Self-Assembly in Nanoscale Materials. In *Advances in Chemical Physics*; Chapter 9 What can Mesoscopic Level IN SITU Observations Teach us About Kinetics and Thermodynamics of Protein Crystallization? Nicolis, G., Maes, D., Eds.; Wiley-Blackwell: Malden, MA, USA, 2012; Volume 151, pp. 223–276.
16. McPherson, A. *Preparation and Analysis of Protein Crystals*; John Wiley & Sons: New York, NY, USA, 1982.
17. Nanev, C.N. Brittleness of protein crystals. *Cryst. Res. Technol.* **2012**, *47*, 922–927. [CrossRef]
18. Wukovitz, S.W.; Yeates, T.O. Why protein crystals favor some space-groups over others. *Nat. Struct. Biol.* **1995**, *2*, 1062–1067. [CrossRef] [PubMed]
19. Kaischew, R.A. Equilibrium shape and work for formation of crystal nuclei on substrates. *Common Bulg. Acad. Sci. Phys. Ser.* **1950**, *1*, 100–136.
20. Stoyanov, S. On the atomistic theory of nucleation rate. *Thin Solid Films* **1973**, *18*, 91–98. [CrossRef]
21. Milchev, A. Electrochemical phase formation on a foreign substrate—Basic theoretical concepts and some experimental results. *Contemp. Phys.* **1991**, *32*, 321–332. [CrossRef]
22. Milchev, A.; Stoyanov, S.; Kaischev, R.A. Atomistic theory of electrolytic nucleation: I. *Thin Solid Films* **1974**, *22*, 255–265. [CrossRef]
23. Milchev, A.; Stoyanov, S.; Kaischev, R.A. Atomistic theory of electrolytic nucleation: II. *Thin Solid Films* **1974**, *22*, 267–274. [CrossRef]
24. Tassev, V.L.; Bliss, D.F. Stranski, Krastanov, and Kaischew, and their influence on the founding of crystal growth theory. *J. Cryst. Growth* **2008**, *310*, 4209–4216. [CrossRef]
25. Stranski, I.N.; Kaischew, R.A. Gleichgewichtsformen homöopolarer Kristalle. *Z. Kristallogr.* **1931**, *78*, 373–385. [CrossRef]
26. Hartman, P.; Bennema, P. The attachment energy as a habit controlling factor: I. Theoretical considerations. *J. Cryst. Growth* **1980**, *49*, 145–156. [CrossRef]

27. Nanev, C.N. Bond selection during protein crystallization: Crystal shapes. *Cryst. Res. Technol.* **2015**, *50*, 451–457. [CrossRef]
28. Yau, S.-T.; Vekilov, P.G. Quasi-planar nucleus structure in apoferritin crystallization. *Nature* **2000**, *406*, 494–497. [PubMed]

crystals

MDPI

Review

Peculiarities of Protein Crystal Nucleation and Growth

Christo N. Nanev

Rostislaw Kaischew Institute of Physical Chemistry, Bulgarian Academy of Sciences, 1113 Sofia, Bulgaria; nanev@ipc.bas.bg; Tel.: +359-2-856-6458

Received: 18 October 2018; Accepted: 5 November 2018; Published: 8 November 2018

Abstract: This paper reviews investigations on protein crystallization. It aims to present a comprehensive rather than complete account of recent studies and efforts to elucidate the most intimate mechanisms of protein crystal nucleation. It is emphasized that both physical and biochemical factors are at play during this process. Recently-discovered molecular scale pathways for protein crystal nucleation are considered first. The bond selection during protein crystal lattice formation, which is a typical biochemically-conditioned peculiarity of the crystallization process, is revisited. Novel approaches allow us to quantitatively describe some protein crystallization cases. Additional light is shed on the protein crystal nucleation in pores and crevices by employing the so-called EBDE method (equilibration between crystal bond and destructive energies). Also, protein crystal nucleation in solution flow is considered.

Keywords: protein crystallization; biochemical aspects of the protein crystal nucleation; classical and two-step crystal nucleation mechanisms; bond selection during protein crystallization; equilibration between crystal bond and destructive energies; protein crystal nucleation in pores; crystallization in solution flow

---

1. Introduction

Crystallization is widespread in nature, in everyday live, meteorology (ice and snow), and even in biology, e.g., biomineralization of bone, teeth, and shells. Crystals are present in healthy (insulin) as well as in diseased human organisms (e.g., kidney and gallbladder stones, uric acid crystals in gout). Protein crystals also bear significant scientific, medical, and industrial relevance. Crystalline drug formulations are most appropriate to maintain protein stability during storage, transport, and upon administration.

Reports on protein crystallization date back some 180 years. Friedrich Ludwig Hünefeld observed crystallization of the hemoglobin from earthworm blood by accident [1]. Before the introduction of X-ray crystallography in biology 1934 [2], biochemists and physiologists used protein crystallization for purification and characterization purposes only. To date, relatively large and well-diffracting crystals are needed for X-ray (and neutron) diffraction studies, which are the most powerful methods for structure-function studies of biomolecules. The knowledge of 3D protein molecule structures and protein-substrate interactions is vital when it comes to understanding the mechanisms of life and the human genome, and developing novel, protein-based pharmaceuticals for structure-guided drug design and controlled drug delivery. Not surprisingly, already in 1962, the Nobel Prize in Chemistry was awarded jointly to Max Perutz and John Kendrew "for their studies of the structures of globular proteins", (hemoglobin and myoglobin). Other Nobel Prizes for X-ray structure determinations of bio-molecules and complexes have followed.

Unfortunately, growing crystals suitable for X-ray crystallography remains, even nowadays, the major stumbling block in the whole study. Despite intensive endeavors, there is no recipe for growing crystals of newly-expressed proteins. Instead, a tedious trial-and-error approach is applied. The numerous state-of-the-art crystallization tools employed, such as robots, automation and miniaturization of crystallization trials, Dynamic Light Scattering, crystallization screening kits, etc., do not exclude the need for researchers' creativity and acumen. Different approaches to evoke protein crystallization have been attempted with and without success.

Because of the lack of seed crystals, spontaneous crystallization is used with newly-expressed proteins. Inevitably, it starts with the formation of the smallest sized stable crystalline particles possible under the conditions present, coined as 'nuclei'. Therefore, the step that determines the difference between success and failure of the crystallization effort is to compel the protein molecules scattered homogeneously in the solution to form stable crystal nuclei; once nucleated, the crystals continue their growth spontaneously. So, a detailed understanding of crystal nucleation in general, and of the protein crystal nucleation, are obligatory. Such knowledge is needed because, being the first crystallization stage, nucleation predetermines important features of the subsequent crystal growth, such as polymorph selection, number of nucleated crystals, crystal size distribution, and frequently, crystal quality.

Protein crystal polymorphism is the ability of a same chemical composition substance to exist in more than one crystal structure. Whilst having identical chemical properties, polymorphs can differ markedly in their dissolvability and bioavailability (the fraction of an administered dose of unchanged drug that reaches the systemic circulation). Thus, depending on the crystal polymorphic form the same molecule may have, or may not have a therapeutic effect [3]. Furthermore, in the worst case, a change of the polymorphic form may render a drug toxic. Hence, of crucial importance is to identify all relevant polymorphs which are decisive for the therapeutic function of drug formulation.

The molecular-kinetic mechanism of protein crystal nucleation is extremely complex. This process involves a subtle interplay between physical and biochemical factors which enables a highly-precise self-assembly of biological macromolecules into stable clusters. For instance, a physical requirement for successful protein crystallization is that the protein-protein attraction strength should be moderate—the attraction should be large enough to promote crystallization while not being so large as to provoke amorphous precipitation. In other words, the pair-wise protein attraction must be carefully fine-tuned, which is achieved by selecting proper crystallization conditions. However, although some physical laws established previously for the crystallization of small inorganic molecules rule protein crystal nucleation as well, it is the large size of the protein molecules and their highly inhomogeneous and patchy surfaces that make protein crystal nucleation so peculiar. Physical and biochemical aspects of protein crystal nucleation can be distinguished in an appropriately-designed experimental setting, e.g., see [4].

Substantial difference, on both the molecular and macroscopic levels, is established by the crystallization of protein and small (inorganic) molecules. On a molecular-scale, independently of the spatial orientation of the meeting species, every hit between small molecules in supersaturated media has the potential to contribute for the formation of a crystal bond. The reason for this is that small molecules possess spherical interaction fields and a constant interaction potential. In contrast, the surface of protein molecules is highly patchy and heterogeneous, and only a limited number of discrete patches on it become attractive molecule portions under crystallization conditions. Due to the strict selection of the crystalline bonding patches, a successful collision between protein molecules, resulting in the formation of a crystalline connection, requires not only sufficiently close approach of the species, but also, their proper spatial orientation. Macroscopically, the difference between small molecules and proteins is manifested through the notorious reluctance of proteins to crystallize. Furthermore, although requiring unusually high supersaturation, protein crystal nucleation and growth occur much more slowly than that with small-molecule substances.

The so-called bond selection mechanism (BSM) was devised to explain the reduced rate of the protein crystal nucleation [5–8]. It accounts for the biochemical constraint associated with the strict selection of crystalline bonds, which also enforces specific bond orientations. Principally similar to BSM is the increasingly popular 'sticky patch' model, which was derived from colloid chemistry and soft matter physics, e.g., see [9,10]. The severe steric restriction to the protein crystal bond formation (arising due to the small size of the contacting patches) is mitigated to some degree by 'sticky' collisions, i.e., where two biomolecules in a water environment remain trapped close to each other after their first encounter. In contrast to small molecules, large biomolecules perform rotational diffusion, which involves multiple collisions (about nine collisions after their first encounter) [11]. During rotational diffusion, the biomolecules get a good chance of reorienting toward proper spatial positioning of the crystallization patches on the two meeting protein molecules; successful encounters become much more probable.

Bond selection is also a factor in protein crystal growth [12]; for a molecule to bond at the kink site, it must be in an adequate orientation. Typically, the protein crystals grow under unusually high supersaturations, i.e., around 100%, and even more. Despite this high supersaturation, their growth proceeds more slowly than that observed with small molecule crystals; a typical value of the step kinetic coefficient for inorganic crystals grown from solution is 2 to 3 orders of magnitude higher than that of protein crystals [13,14]. This fact can be explained because of a low probability of proper spatial orientation of an incoming protein molecule for its incorporation into the kink site. Perhaps, the higher protein concentration that is used at crystallization conditions is needed to mitigate this decelerating impact (through a higher attachment attempt frequency).

The aim of this review is to consider some biochemically-conditioned peculiarities of the protein crystal nucleation (and growth). Recently-discovered molecular scale pathways for protein crystal nucleation are discussed first. Novel proof for the BSM are presented. The so-called EBDE method (equilibration between crystal bond and destructive energies) is re-substantiated, and protein crystal nucleation in pores is revisited on this basis. Another aim of the present paper is to report a novel consideration of the crystallization in solution flow.

2. Classical or Two-Step Crystal Nucleation Mechanisms: What Is Currently Known?

Despite nucleation significance and nearly a century-long period of intensive study, crystallization onset is still debated. Quite often, classical nucleation theory (CNT) is employed to explain protein crystal nucleation (e.g., [15]). However, while providing an adequate explanation of the fluctuation-based nucleation mechanism and the origin of the nucleation barrier, there is an inadequacy between some measurements of crystal nucleation rates and CNT predictions. (It is worth noting that nucleation rates calculations according CNT suffer from uncertainty in determining the energy of the interface arising between the new phase and the mother phase. While such energies are not measurable for those nanoparticles, interface energy variation of only 10% can alter the nucleation rate by many orders of magnitude. The reason is that the nucleation rate depends exponentially on the nucleation energy barrier, which in turn is determined by the interface free energy in power three.) CNT also failed to account for observations in biomineralization processes that are responsible for the formation of invertebrate mineralized skeletal elements, e.g., the mollusk shell nacre layer (aragonite polymorph) and the sea urchin spicule (calcite polymorph) [16]. To explain this inadequacy, the so-called two-step nucleation mechanism (TSNM) has been proposed [17].

Though frequently contested, CNT has been confirmed by atomic force microscopy [18]. Molecular-scale images of sub-critical and super-critical crystals landing on apoferritin crystals under supersaturated conditions reveal a classical nucleation pathway. Yau and Vekilov observe that pre-nucleation clusters of apoferritin are also crystalline, and have the same molecular arrangement as those in bulk crystals. So, the nucleation pathway complies with the classical crystal nucleation pathway. Sleutel et al. [19] have also observed that glucose isomerase 2D crystal nucleation proceeds following the classical pattern, and proved the existence of a critical crystal size. This notwithstanding, a question arises of whether CNT can give a reliable physical rendition of protein crystal nucleation process?

TSNM denies the simultaneous densification and ordering during a single nucleation event. The theoretical basis for TSNM has been laid by ten Wolde and Frenkel [20]. Performing numerical simulations, they predict that the thermodynamically-favored nucleation pathway for colloidal and protein-like substances entails the initial formation of a liquid cluster to be subsequently transformed into a crystalline nucleus. However, this prediction is valid only for conditions close to the critical point of a phase diagram (the intersection of the liquid-liquid binodal and spinodal). Nucleation pathways that follow the CNT would result in different phase diagram regions. In reality, due to protein aggregation or gelation, the critical point is hardly accessible in protein crystallization experiments. This means that most protein crystallization experiments do not occur under conditions prescribed by ten Wolde and Frenkel (and the TSNM of proteins has not been experimentally demonstrated with proteins crystallizing under such conditions).

TSNM does not contest the basic CNT concept of a fluctuation-based nucleation mechanism. In its initial formulation, TSNM assumes nucleation initiation via a high-density liquid phase appearing in the bulk solution, with crystal nuclei being formed inside this dense liquid phase during the second TSNM step [21]. The intermediate phase preserves some similarity to the mother phase, since it is only densified. Therefore, the phase-transition energy barrier is lowered below the one needed for a direct crystal nucleus formation occurring via the CNT mechanism. The fact that TSNM breaks up a single large activation barrier into two smaller ones (the second barrier being for the ordering step) makes it intuitively more attractive [22]. For whatever reason, the TSNM idea has gained increasing popularity over the years, and despite the limited extent of clear experimental proof for the case of protein crystal nucleation, the TSNM idea is broadly accepted as a scientific fact, superseding CNT.

It is mainly mesoscale observations that corroborate TSNM. Sauter et al. [23] have reported a two-step pathway of protein nucleation for the protein/precipitant system β-lactoglobulin/$CdCl_2$. Vivares et al. [24] have observed TSNM formation of glucose isomerase crystals in a concentrated liquid phase. Using Dynamic Light Scattering, Schubert et al. [25] have observed the occurrence of liquid dense protein clusters with growing-over-time nanocrystals formed inside, as verified by transmission electron microscopy (TEM).

However, due to the molecular scale of the processes involved, insights into the earliest crystal nucleation stages have remained only very partially understood until recently. The remarkable advancement in instrumental techniques has enabled molecular-scale observation of protein crystal nucleation. For instance, no intermediate condensed liquid droplets, but only amorphous solid particles consisting of lysozyme molecules, are observed [26], without the formation of crystalline phases inside such amorphous particles. This speaks about the challenges faced by TSNM initial formulation.

Liquid-cell TEM reveals diverse nucleation pathways. Van Driessche et al. [27] used vitrified samples plunge frozen at various time intervals. In doing so, the authors imaged the nanoscale structure of pre-nucleation clusters along the way to the crystalline nuclei. Looking into the earliest stages of glucose isomerase crystal nucleation [28], Van Driessche et al. have shown distinct nucleation pathways for two glucose isomerase polymorphic crystal forms. For the rhombic polymorph, the authors have not observed any amorphous precursors of crystal nuclei being formed prior the appearance of crystalline nanoparticles (the latter being as small as 100 nm). Therefore, Van Driessche et al. [27] concluded homogeneous crystal nucleation from solution following the rules of the CNT. With the needle-like polymorphic form of glucose isomerase, they observed crystalline nanorods that measured 12 by 2 molecules, and appeared almost immediately after setting crystallization conditions. The smallest glucose isomerase rods captured at very early time points (just 20 s after adding a crystallizing precipitate) have the same crystalline order as the bulk crystals, which is in accordance with CNT. Afterwards, however, the rods undergo grouping and oriented attachment into larger structures, referred to as fibers. These fibers (having widths of 40 nm and lengths varying between 100 nm up to multiple microns) have a significant deviation from the crystallographic packing, as predicted by the two-step model. On the basis of these observations, Van Driessche et al. [27] have hypothesized that the formation of a crystalline nucleus takes place by the lateral merging of fibers mediated by oriented attachment. In conclusion, the authors found indications that elements of both CNT and TSNM may act during the nucleation of glucose isomerase crystals. A two-step crystal nucleation has also been seen using time-resolved potentiometry and turbidimetry combined with Dynamic Light Scattering, small-angle X-ray scattering, and in situ imaging by cryo-TEM (on frozen-hydrated cryo-samples) [29]. The authors have studied small molecule (Portland cement) crystallization.

Most recently, Sleutel and Van Driessche have introduced some balanced views on the subject [30]. Assessing up-to-date results from cryo-TEM in situ imaging with its strengths and weaknesses, the authors offer new insights into the earliest stages of the protein crystal nucleation. For proteins, neither the suggested ubiquity of TSNM can be confirmed unambiguously, nor CNT rejected completely; the reality might be diverse. The experimental proof thus far can only affirm the absence of a universal nucleation mechanism in all crystallization cases. The authors' review represents a substantial leap in our understanding of how protein molecules, scattered homogeneously in the solution, are constrained to form stable crystal nuclei [30] (see also [31]).

3. Bond Selection during Protein Crystallization

Interactions between bio-molecules and protein–substrate interactions play a central role in many complex cellular processes (such as signaling, transcription, inhibition, translation, and regulation), and are responsible for the stability and shelf-life of pharmaceutical formulations. Highly-selective and directional protein-protein interactions govern the protein crystal nucleation as well. The formation of protein crystal lattice contacts (i.e., regions on the surface of a protein that are in contact with regions on the surfaces of neighboring proteins in the crystal lattice) is a typical biochemically-rooted peculiarity of protein crystallization [4].

The study of protein crystal bond formation benefits from the comparison with the physiological protein-protein interactions. Due to their enormous biological importance, the latter have been investigated much more thoroughly than the former. A fundamental postulate for any person working with/on proteins is that only the surface structure of a protein molecule dictates its ability to bind to partners; the protein intramolecular interactions in the bulk do not participate in protein crystal lattice binding, because those interactions are concealed under the amino-acid residues situated at the molecular surface. Although it is impossible to observe the elementary acts of protein crystal bond formation, knowledge about lattice contacts, which are the result of the protein crystallization, are available from Protein Data Bank data, e.g., [32–35]. This information shows the clear differences between physiological protein-protein interactions and biologically non-functional protein crystal bonds. It is well known that the former arise due to hydrophobic areas that occupy relatively large portions on the protein molecule surface. Besides, the physiological protein-protein bonds are extremely specific and strong. In contrast, protein crystal lattice contacts are hydrophilic, polar, and occupy relatively small fractions of the molecule surfaces [32,36]. The intra-crystalline contacts are due to van der Waals, salt bridging, and hydration interactions. These data show that the patches on the protein molecule surface, which can create crystal lattice contacts, are also selected. Observed across many proteins, the most frequently-found residues in crystal contacts are arginine (Arg) and glutamine (Gln) residues, while the most infrequent participation in protein crystal contacts is that of lysine (Lys) and glutamate residues (i.e., the carboxylate anion of glutamic acid, Glu) [32], although both are situated almost exclusively on the surface of the proteins. The hypothesis of Doye et al. [37,38] for an evolutionary negative design (i.e., the evolutionary selection of the surface properties of proteins to prevent any arbitrary association) is that lysine helps to prevent unwanted protein-protein interactions. This protein ability is compulsory because proteins operate within the cellular context, with typical concentrations of up to 300 mg/mL. Therefore, because of millions of years of natural selection, physiological protein-protein bonds are highly specific; any non-specific inter-protein interaction may be fatal.

Information about specific residues participating in the crystal contacts was reported by Gillespie et al. [39]. The contacting residues were identified as participating in either direct residue-residue interactions or in water-mediated interactions, with the interaction free energy of crystal contacts being calculated as well. (For a thorough consideration of the role which play water molecules found at interfaces see [40].) The authors confirmed that the non-specific, long-range electrostatics are not significant in crystal contacts, and that the interactions in crystals are likely dominated by short-range interactions, such as van der Waals, salt bridging, and hydration interactions, which are highly specific.

The big success of the rational site-directed mutagenesis strategy is an (albeit indirect) argument for bond selection during protein crystal nucleation. As early as 1992, rational surface engineering was accomplished by McElroy et al. [41]. It has been established that even single amino acid changes on the surface of the thymidylate synthase can dramatically affect the solubility of a protein and its crystallizability, while not decreasing stability. The latter is of importance because structurally-rigid proteins are more prone to crystallization. Therefore, keeping the protein folded is a prerequisite for protein crystallization ability. It is logical to conclude that the surface mutations mediate novel crystal contacts.

Site-directed mutagenesis has also been applied to crystallize other proteins that are recalcitrant to crystallization [42–44]. Mutation of surface amino acid residues with high conformational entropy to residues with no conformational entropy lead to the enhancement of crystallization; lysine residues were systematically mutated to alanine. Also, modifying the surface properties of a protein by mutating glutamic acid residues by alanine or aspartic acid results in enhanced crystallizability. In fact, any additional CH2 group enhances the conformational entropy [45].

Also, chemical functionalization (a modification which often retains native protein structure), e.g., acetylation of surface lysine groups, coerces bovine carbonic anhydrase to crystallize [46]. The authors have also shown how acetylation altered the organization and composition of crystal contacts—when proteins crystallize, they bury solvent-exposed surface areas in contact regions, which accommodate charged residues through salt bridges and/or hydrogen bonds. Acetylation has little influence on the size and geometry of crystal contacts, but reduces their charge complementarity. Also, contact regions generate and define adjacent non-contact regions. It was concluded that crystal contacts appear to be the principal determinant of the quality of the crystals, as measured by the X-ray diffraction resolution, and that the probability of obtaining crystals with adequate diffraction has been increased substantially due to acetylation of surface lysine groups [46].

Studying the role of charges in protein-protein interactions, Kang et al. [46] show why lysine residues are often excluded from the crystal lattice contacts. The authors conclude that the charge of Lys, but not its conformational flexibility, reduces its propensity to participate in the contact regions of proteins [47]. Kang et al. [46] also concluded that the protein-protein interactions that give rise to crystals are generally weaker than the protein-protein interactions that permit biological function. Indeed, molecules with weak intermolecular interactions can relatively easily reorient to find the preferred orientation needed for the crystal. To compensate for the weak interaction, crystals of proteins form in vitro at concentrations of protein that are much higher than those found inside the cell.

4. Equilibration between Crystal Bonding and Destructive Energies (EBDE)

The intuitive suggestion of Garcia-Ruiz [48] for a balance between the sum of all intra-crystal bond energies (which maintain the integrity of a crystalline cluster) and the sum of surface destructive energies (which tend to tear up it) was our starting platform for further considerations [49], which brings us to the so-called EBDE method [50]. The thermodynamic substantiation of EBDE was presented earlier in this Journal. Let us outline it here.

When the system is undersaturated and no crystallization is possible, the protein 'affinity' to water molecules prevails over the crystallization propensity. To evoke crystallization, it is necessary to impose supersaturation, and the higher the latter, the more thermodynamically-stable the crystal, with respect to the solution. Thus, it is logical to assume that the imposed supersaturation decreases the protein-to-water 'affinity', i.e., supersaturation diminishes the destructive energy (ψ_d) per bond. In other words, the tendency to disintegrate the crystal depends on the degree of supersaturation, in contrast to the cohesive energy per bond in the crystal lattice (ψ_b) which is supersaturation independent. This means that any supersaturation increase will lead to an increase in ψ_b/ψ_d ratio. A rigorous definition of protein-to-water 'affinity', which uses both nucleation process enthalpy and entropy, was provided in [50].

The advantage of the EBDE method is that it can predict the nucleation of crystals of diverse lattice structures (and can be aided by crystallographic computer programs), while the classical mean work of the separation (MWS) method of Stranski–Kaischew [51–53] is applicable only with a Kossel-crystal [54]. However, because EBDE rests on an intuitive suggestion, a question may arise of whether the supersaturation dependent critical nucleus size determined using the EBDE method corresponds to the one from CNT. Proof that the two coincide is presented below.

According to CNT, a free energy change (ΔG) is needed for a homogeneous crystal to occur, e.g., [55]:

$$\Delta G = -n\Delta\mu + S\gamma, \tag{1}$$

where n denotes the number of molecules constituting the cluster and $\Delta\mu$ is supersaturation; S means the total surface of the new phase, and γ is the specific interphase energy.

Classical thermodynamics points out that the cohesive energy ($\Delta G_v = -n\Delta\mu$) maintaining the integrity of a new phase cluster is proportional to its volume, i.e., to n in power three, while the sum of energies ($\Delta G_s = S\gamma$) tending to tear up the cluster is proportional to its surface, i.e., to n in power two. According to EBDE, the balance between cohesive and destructive energies means $-\Delta G_v + \Delta G_s = 0$. Importantly, in the case of under-critically sized clusters when ΔG_v is smaller than ΔG_s, also ΔG is smaller than ΔG_{max}, which is the energy barrier for critical nucleus formation. However, increasing more rapidly with cluster size enlargement (in power three) than ΔG_s, ΔG_v approaches ΔG_s, and thus also ΔG_{max}. Afterwards, when super-critically sized clusters are growing, ΔG_v becomes larger than ΔG_s. Thus, ΔG becomes again smaller than ΔG_{max}. So, it is seen that ΔG is smaller on both sides of ΔG_{max}. This is exactly the mathematical definition for maximum of a function $y = f(x)$, according to which the function maximum appears at the argument value x_o for which:

$$f(x_o + h) < f(x_o) \tag{2}$$

This is an inequality that holds true for every negative and positive small value of h. In other words, $f(x_o)$ is larger than all neighboring function values. In conclusion, ΔG_{max} coincides with the energy balance appearing at $\Delta G_v = \Delta G_s$.

The EBDE method is used here to consider the protein crystal nucleation in pores.

5. Protein Crystal Nucleation in Pores

Experimental studies have shown that porous materials are effective at inducing protein crystal nucleation [56–65]. Remarkable success has been achieved with the use of a mesoporous bioactive gel-glass ($CaO–P_2O_5–SiO_2$) as a nucleating reagent [66]. The bioglass (Naomi's Nucleant') [67] proved to be very successful in producing high quality crystals of both model and target proteins. So, this discovery has set a trend for the use of porous materials as nucleants [68].

Theoretical considerations [69] have shown that a combined diffusion-adsorption effect can increase protein concentration inside pores (and crevices) to a level that is enough for crystal nucleation onset [70]. The reason is that molecular diffusion is the sole mass-transfer mechanism working in pores, and due to translational Brownian motion, which is equally probable in all directions, the probability that large protein molecules will land on pore walls is several times greater than the probability of their escape [71].

The quantitative estimation of the size of a pore which can induce protein crystal nucleation is based on the mean squared diffusion displacement, $<x^2>$:

$$<x^2> = 2Dt, \tag{3}$$

where D is the diffusion coefficient; a typical value for proteins being $D_{prot} = 10^{-6}$ cm^2 s^{-1}.

Regarding the adsorption effect, it was shown that, provided the pore is sufficiently narrow, protein molecules approach their walls and adsorb there more frequently than they can escape. The time span during which protein molecules remain adsorbed at the pore walls is calculated from the desorption rate (R_d), which is an activated first-order rate process:

$$R_d = -dc_a/dt = k_d c_a, \tag{4}$$

where c_a is the surface concentration of the adsorbed molecules and t is time; k_d being the rate constant for desorption. Integration of Equation (4) gives:

$$c_a = c_a{}^0 \exp(-k_d t), \tag{5}$$

where $c_a{}^0$ is the initial concentration of the adsorbed molecules.

The rate constant for desorption (k_d) is $k_d = \theta \exp(-E_d/k_B T)$, where E_d is the desorption activation energy, θ an "attempt frequency" for desorption, k_B is Boltzmann's constant, and T is temperature. On the other hand, the half-life ($\tau_{1/2}$) for adsorption of protein molecules at the pore wall is defined as the time when $c_a = c_a{}^0/2$. Thus:

$$\tau_{1/2} = \ln 2/k_d \sim (E_d/k_B T), \tag{6}$$

It has been found [72] that the apparent activation energy, E_d, for desorption of an isolated protein molecule is extremely low, i.e., within the range of 2–4 kJ/mol, or less than $2k_B T$ per molecule. Therefore, most individual protein molecules exhibited short residence times of about 1 s and even less.

Equation (3) shows that during adsorption time t = 0.5 s, a protein molecule can diffuse from wall to wall of a 10 μm sized pore and adsorb there (the tacit assumption being that the molecule would follow the shortest path). So, the total time during which the protein molecules are in the adsorbed state in pores narrower than several micrometers thus becomes greater than the total time during which they are in the desorbed state. Therefore, despite the increasing amounts of protein in the pores, no concentration gradient favoring back diffusion (from the pores towards bulk solution) can arise. Hence, due to adsorption being more frequent than desorption, sufficiently narrow pores (presumably about 1 μm in size) can become quasi-permanent traps for macromolecules, and thus, accumulate protein [73]. The conclusion is that protein crystal nucleation may be enabled in sufficiently narrow pores, even under conditions where heterogeneous nucleation on flat surfaces (and even more so in bulk solution) is absent [69].

Using the MWS method of Stranski and Kaischew [51–53] and a Kossel-crystal as a model, Nanev et al. [69] have calculated the crystal nucleation energy barriers resulting from pore space confinement and interaction with pore walls; pores having the shape of rectangular prism were considered. Also considered was the case in which the size of the pore opening is large enough to allow a critical nucleus smaller than the pore opening to form inside the pore. It was found, however, that such a crystal nucleus would be larger, and therefore, a smaller pore that is completely filled with the nucleus is more effective [69]. Bearing this in mind, only the most closely-packed monomolecular crystalline layers filling the entire pore orifices are considered here; once formed, the nuclei continue their growth (with a probability 1/2) outside the pore. Evidently, to enable selection of a pore size, which is suitable for crystallization of a given protein molecule, the optimal porous material should possess a broad distribution of pore sizes [69].

Because only idealized pore-shapes are readily liable to be brought to quantitative account, the protein crystallization promoting effect of pores, having hexagonal shapes (Figure 1a), trigonal (Figure 1b) and a rhombic (Figure 1c) prisms, is calculated here by using the EBDE method. Instead of Kossel crystals (existing extremely rarely in nature), closest packings of equal spheres are considered:

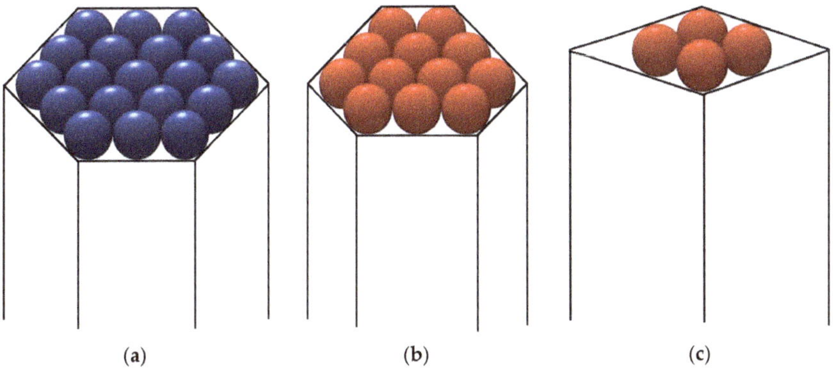

Figure 1. (**a–c**) Prismatic pores having hexagonal, trigonal and rhombic cross-sections. Flat pore walls prevent the nucleating crystal from trying to conform to any curved pore surface and becoming strained [74].

1. Crystallography equations for hexagonal crystal structures (see [50]) are applied for EBDE calculations using pore model of a hexagonal prism (Figure 1a). By denoting the total number of molecules in the hexagonal monolayer by (z), and the number of molecules in its edge by (λ), we have:

$$z = 3\lambda(\lambda - 1) + 1, \tag{7}$$

which gives $z = 7, 19, 37, 61, 91, 127...$ for $\lambda = 2, 3, 4, 5, 6, 7 \ldots$ respectively. Using the crystallographic formula concerning the number of bonds in the hexagonal crystal layer:

$$\Delta G_v{}^h = (3\lambda - 3)(3\lambda - 2), \tag{8}$$

we obtain the balance between cohesive and destructive energies, $-\Delta G_v + \Delta G_s = 0$:

$$(3\lambda - 3)(3\lambda - 2)\psi_b + [12 + 6(\lambda - 2)\psi = [3\lambda(\lambda - 1) + 1]\psi_d, \tag{9}$$

where ψ is the work of separation of one protein molecule from a cavity wall; $\psi \approx E_d$. Note that the molecules at the six crystal apexes are bond to the pore walls by energy of 2ψ. Simultaneously, no crystal apexes and edges (which can be exposed to the enhanced destructive action of water molecules) exist on such crystalline monolayers.

Three different ratios between ψ and ψ_b are used to form an idea of how the energetic interactions between protein and pore materials influence the supersaturation dependence of critical nucleus size:

$$\psi = 0.25\psi_b, \text{ and hence: } [(3\lambda - 3)(3\lambda - 2) + 3 + 1.5(\lambda - 2)]\psi_b = [3\lambda(\lambda - 1) + 1]\psi_d, \tag{10}$$

$$\psi = 0.5\psi_b, \text{ and hence: } [(3\lambda - 3)(3\lambda - 2) + 6 + 3(\lambda - 2)]\psi_b = [3\lambda(\lambda - 1) + 1]\psi_d, \tag{11}$$

$$\psi = 0.75\psi_b, \text{ and hence: } [(3\lambda - 3)(3\lambda - 2) + 9 + 4.5(\lambda - 2)]\psi_b = [3\lambda(\lambda - 1) + 1]\psi_d, \tag{12}$$

The results are presented in Table 1.

Table 1. Supersaturation dependence of the critical nucleus size (thus, also of the suitable pore-opening size). Recall that the higher ψ_b/ψ_d ratio means a higher supersaturation that is needed for formation of the critical crystal nucleus.

	$\psi_b/\psi_d \approx$					
λ =	2	3	4	5	6	7
ψ/ψ_b = 0.25	0.47	0.41	0.39	0.37	0.37	0.36
ψ/ψ_b = 0.5	0.39	0.37	0.36	0.36	0.35	0.35
ψ/ψ_b = 0.75	0.33	0.34	0.34	0.34	0.34	0.34

The results in Table 1 confirm the intuitive expectation that the closer the energetic interaction between protein and pore material (i.e., the larger ψ/ψ_b), the lower supersaturation needed for nucleus formation; and the nucleus size becomes almost supersaturation-independent if $\psi/\psi_d \rightarrow 1$. To all appearances, the 'biocompatibility' of the pore material plays a major role. Perhaps, this is the reason for the effectiveness of the bioactive gel-glass at inducing protein crystal nucleation, as described by Naomi Chayen and coworkers [66].

2. Considering a pore's capacity to increase supersaturation, Nanev et al. [69] have noted that the narrower the pore, the smaller the protein molecule escape probability, and thus, the higher the concentration- (respectively, supersaturation-) increase in the pore. On the other hand, the pore opening is reached, and the protein molecules enter the pore with the same probability with which they reach an equally large flat surface area, meaning that smaller openings are less accessible. The problem is addressed here by considering the results in Table 1 and the model of a smaller crystal (see Figure 1c). In this case, the balance between cohesive and destructive energies, $-\Delta G_v + \Delta G_s = 0$, gives:

$$5\psi_b + 8\psi = 4\psi_d \tag{13}$$

Thus, we obtain:

1. $\psi = 0.25\psi_b$, $\psi_b/\psi_d \approx 0.57$.
2. $\psi = 0.5\psi_b$, $\psi_b/\psi_d \approx 0.44$.
3. $\psi = 0.75\psi_b$, $\psi_b/\psi_d \approx 0.36$.

As seen, the above ψ_b/ψ_d-values are larger than the data for pores having the shape of a hexagonal prism, Figure 1a; see Table 1. In other words, as intuitively expected, the smaller the crystal nucleus, i.e., the narrower the suitable pore-opening size, the higher supersaturation needed for the crystal to nucleate. However, though feasible from an energetic point of view, very large crystal nuclei are less likely to appear for kinetic reasons—bringing together a vast number of molecules via molecule-by-molecule assembly into a crystal nucleus involves very large fluctuations, which, in turn, requires very long waiting times. So, crystals of modest size appear most probable.

3. Besides pore-opening size, pore volume and shape are also of importance. For instance, in intricate pores with many turns and corners, protein molecules can be trapped more readily. To address this point, protein crystal nucleation in a somewhat non-regularly shaped pore (see Figure 1b) is considered. The result for this model is:

$$24\psi_b + 15\psi = 12\psi_d, \tag{14}$$

which gives:

1. $\psi = 0.25\psi_b$, $\psi_b/\psi_d \approx 0.43$.
2. $\psi = 0.5\psi_b$, $\psi_b/\psi_d \approx 0.38$.
3. $\psi = 0.75\psi_b$, $\psi_b/\psi_d \approx 0.34$.

No substantial difference between the two pore kinds (Figure 1a,b) can be established by comparing these results with the data for λ = 2 to 3 in Table 1. So, we might conclude that although pores in actual disordered porous materials (e.g., such as Bioglass) are much more complex, the models considered above provide adequate clues to comprehend why pores and crevices facilitate protein crystal nucleation.

6. Protein Crystallization in Solution Flow

The interest in intensifying drug manufacturing is increasing nowadays [75,76]. Continuous tubular crystallizers which use crystallization in solution flow have received a growing attention. They improve efficiency and enable better product-quality control, thus showing advantages in the process intensification required in the pharmaceutical industry [75,77,78]. Such crystallizers contain periodically-spaced orifice baffles, which allow strong radial solution motions (turbulences) to be created, thus securing a uniform mixing [79,80]. Using a combination of a laboratory-scale, continuous-stirred tank crystallizer and a cooled tubular reactor in bypass, Hekmat et al. [81] have proposed continuous protein crystallization as a viable purification alternative for continuous preparative chromatography. Li and Lakerveld [82] demonstrate that the induction time for protein nucleation can be reduced in continuous flow compared to batch crystallization with electric-field-assisted crystallization and for the control experiment without an electric field.

As already seen [83], the electric field energy affects protein crystal nucleation. A reasonable question here would be whether also the flow kinetic energy may affect protein crystal nucleation. Like any moving object, solution flow has a kinetic energy (E_k). Considering a cylinder of a fluid that is travelling at velocity (u), we have:

$$E_k = \rho A \ell u^2/2, \tag{15}$$

where ρ is the fluid density, A is the tube cross-section area, and ℓ is the tube length.

Unfortunately, there is no evident answer to above question. Shear flow alters the rate at which crystals nucleate from solution, yet the underlying mechanisms remain poorly understood [84]. To the best of the author's knowledge, there are no rigorous experimental indications to the solution flow stimulating or postponing crystal nucleation. Although numerous studies have been devoted to protein crystallization in a flow, they refer predominantly to crystal growth e.g., [85–90], rather than nucleation [84,91–93]. On the other hand, however, the spontaneous (primary) crystal nucleation is technologically-unmanageable and is avoided, while being replaced by easily controllable seeding technology. However, despite the seeding strategy, a significant increase in new crystal number is observed frequently in the presence of an introduced crystalline material. Why new crystals are bred is a fundamental question. Despite its significance, the understanding of this phenomenon remains limited. It is argued that the origin of the secondary nuclei is the crystal seeds themselves. Anwar et al. [94] have considered mechanisms of secondary crystal nucleation at a molecular level. Via a molecular dynamic simulation, the authors stipulate that under-critically sized molecule clusters forming close to the crystal seed can move towards the latter, and if/when contacting the seed's surface, grow like crystal particles. This explanation is quite logical, but the locally-decreased supersaturation in crystal vicinity (due to crystal growth under diffusion matter supply) limits its applicability to some supersaturation interval where sufficiently-large, under-critically sized molecule clusters can arise; the size of the heterogeneous critical nucleus, although smaller than the homogeneously formed nucleus, also depends on the supersaturation.

To assess the impact of flow kinetic energy on crystal nucleation, it is necessary to recall that this energy can be liberated/dissipated by conversion into internal energy, mainly heat, i.e., solution temperature is expected to increase because of flow energy dissipation (the energy needed to raise the temperature of 1 g of water by 1 °C is 1 calorie). Temperature increase disfavors crystallization of substances possessing normal temperature dependent solubility, while in contrast, it favors crystallization of substances possessing retrograde temperature-dependent solubility. Using the mechanical equivalent of heat equal to 4.1868 joules per calorie, the resulting temperature increase is determined simply using E_k.

Occurring at larger flow velocities, a turbulent flow dissipates more energy than a laminar flow. Therefore, the turbulent flow energy can be used as a benchmark for the significance of the E_k-effect. The fluid velocity (u) needed for a turbulent flow to appear is found from the Reynolds number, Re = uL/ν, where L [mm] is a typical length scale in the system, and ν is the kinematic viscosity of the fluid; for water at 20 °C, ν = 1.0034 [mm^2/s] [95]. For a flow in a pipe of diameter D, experimental observations have shown that turbulence occurs when $\mathrm{Re}_D \geq 4000$ [96]. Thus, assuming L = 1 mm, the velocity needed for a turbulent flow to occur is u = 4000 mm/s. Then, if A = 1 mm^2 and ℓ = 1 mm, Equation (15) will give $E_k = 16 \times 10^6$ erg = 1.6 joule, i.e., a negligible temperature increase (less than 0.4 °C) in a solution of 1 mm^3. Like mass mixing however, heat transfer is further enhanced by liquid shear. So, even smaller temperature increases can be expected. Furthermore, the industrial tubular crystallizers are usually tempered.

In conclusion, although the E_k-impact of the fluid flow is negligible, mass mixing may favor crystal nucleation kinetically, e.g., by increasing the microscopic transport rate of a molecule to the cluster in shear flow [97]. For instance, protein crystal nucleation can benefit from enhanced rotation of the huge biomolecules, which assists their proper mutual orientation during colliding; see Section 3.

7. Conclusions

Evidently, it is the biochemically-conditioned peculiarity of the protein crystallization that makes it different from the crystallization of small molecules. Therefore, to assist in the solution of practical problems, e.g., in the pharmaceutical industry, it is obligatory to elucidate all such peculiarities. Pharmaceutical proteins in crystal form are used to treat and prevent a wide range of diseases [98], and crystallization can be an important purification step to remove degraded, aggregated, or misfolded forms of the protein. Furthermore, the native configuration of proteins, which is of vital importance in maintaining their therapeutic function, is best retained in crystalline drug formulations. In conclusion, a profound knowledge of key protein crystallization process details is needed to intensify, and possibly improve, the process in the pharmaceutical industry [75].

Funding: This work was co-financed by the National Science Fund of the Bulgarian Ministry of Science and Education, contract DCOST 01/22.

Acknowledgments: The author would like to acknowledge networking support by the COST Action CM1402 Crystallize.

Conflicts of Interest: The author declares no conflict of interest. The founding sponsor had no role in the design of the study; in the collection, analyses, or interpretation of data; in the writing of the manuscript, and in the decision to publish the results.

References and Notes

1. Giege, R. A historical perspective on protein crystallization from 1840 to the present day. *FEBS J.* **2013**, *280*, 6456–6497. [CrossRef] [PubMed]
2. Bernal, J.D.; Crowfoot, D. X-ray photographs of crystalline pepsin. *Nature* **1934**, *133*, 794–795. [CrossRef]
3. A good example in this respect is the Ritonavir (trade name Norvir, an antiretroviral medication) case; see "Chemburkar, S.R.; Bauer, J.; Deming, K.; Spiwek, H.; Patel, K.; Morris, J.; Henry, R.; Spanton, S.; Dziki, W.; Porter, W.; et al. Dealing with the Impact of Ritonavir Polymorphs on the Late Stages of Bulk Drug Process Development. *Org. Proc. Res. Dev.* **2000**, *4*, 413–417." and "Bauer, J.; Spanton, S.; Henry, R.; Quick, J.; Dziki, W.; Porter, W.; Morris, J. Ritonavir: An Extraordinary Example of Conformational Polymorphism. *Pharm. Res.* **2001**, *18*, 859–866. [CrossRef]"
4. Nanev, C.N. Phenomenological consideration of protein crystal nucleation; the physics and biochemistry behind the phenomenon. *Crystals* **2017**, *7*, 193. [CrossRef]
5. Nanev, C.N. On the slow kinetics of protein crystallization. *Cryst. Growth Des.* **2007**, *7*, 1533–1540. [CrossRef]
6. Nanev, C.N. Protein crystal nucleation: Recent notions. *Cryst. Res. Technol.* **2007**, *42*, 4–12. [CrossRef]
7. Nanev, C.N. On slow protein crystal nucleation: Cluster-cluster aggregation on diffusional encounters. *Cryst. Res. Technol.* **2009**, *44*, 7–12. [CrossRef]

8. Nanev, C.N. Kinetics and intimate mechanism of protein crystal nucleation. *Prog. Cryst. Growth Charact. Mater.* **2013**, *59*, 133–169. [CrossRef]
9. Fusco, D.; Headd, J.J.; De Simone, A.; Wangd, J.; Charbonneau, P. Characterizing protein crystal contacts and their role in crystallization: Rubredoxin as a case study. *Soft Matter* **2014**, *10*, 290–302. [CrossRef] [PubMed]
10. Fusco, D.; Charbonneau, P. Soft matter perspective on protein crystal assembly. *Colloids Surf. B Biointerfaces* **2016**, *137*, 22–31. [CrossRef] [PubMed]
11. Northrup, S.H.; Erickson, H.P. Kinetics of protein-protein association explained by Brownian dynamics computer simulation. *Proc. Natl. Acad. Sci. USA* **1992**, *89*, 3338–3342. [CrossRef] [PubMed]
12. Nanev, C.N. On the elementary processes of protein crystallization: Bond selection mechanism. *J. Cryst. Growth* **2014**, *402*, 195–202. [CrossRef]
13. Malkin, A.J.; Kuznetsov, Y.G.; McPherson, A. In situ atomic force microscopy studies of surface morphology, growth kinetics, defect structure and dissolution in macromolecular crystallization. *J. Cryst. Growth* **1999**, *196*, 471–488. [CrossRef]
14. Chernov, A.A.; Komatsu, H. Topics in crystal growth kinetics. In *Science and Technology of Crystal Growth*; van Eerden, J.P., Bruinsma, O.S.L., Eds.; Kluwer Academic Publishers: Dordrecht, The Netherlands, 1995; pp. 67–80.
15. Akella, S.; Mowitz, A.; Heymann, M.; Fraden, S. Emulsion-based technique to measure protein crystal nucleation rates of lysozyme. *Cryst. Growth Des.* **2014**, *14*, 4487–4509. [CrossRef]
16. Evans, J.S. Polymorphs, Proteins, and Nucleation Theory: A Critical Analysis. *Minerals* **2017**, *7*, 62. [CrossRef]
17. Vekilov, P.G. Dense liquid precursor for the nucleation of ordered solid phases from solution. *Cryst. Growth Des.* **2004**, *4*, 671–685. [CrossRef]
18. Yau, S.-T.; Vekilov, P.G. Quasi-planar Nucleus Structure in Apoferritin Crystallization. *Nature* **2000**, *406*, 494–497. [CrossRef] [PubMed]
19. Sleutel, M.; Lutsko, J.; Van Driessche, A.E.S.; Duran-Olivencia, M.A.; Maes, D. Observing classical nucleation theory at work by monitoring phase transitions with molecular precision. *Nat. Commun.* **2014**, *5*, 5598. [CrossRef] [PubMed]
20. Ten Wolde, P.R.; Frenkel, D. Enhancement of protein crystal nucleation by critical density fluctuations. *Science* **1997**, *277*, 1975–1978. [CrossRef] [PubMed]
21. Vekilov, P.G. Nucleation of protein crystals. *Prog. Cryst. Growth Character. Mater.* **2016**, *62*, 136–154. [CrossRef]
22. Contributing to TSNM appeal is the resemblance it bears to the well-known Ostwald's rule of stages. (This rule stipulates that a thermodynamically less-stable phase appears first, then a polymorphic transition toward a stable phase occurs).
23. Sauter, A.; Roosen-Runge, F.; Zhang, F.; Lotze, G.; Jacobs, R.M.J.; Schreiber, F. Real-Time Observation of Nonclassical Protein Crystallization Kinetics. *J. Am. Chem. Soc.* **2015**, *137*, 1485–1491. [CrossRef] [PubMed]
24. Vivares, D.; Kaler, E.; Lenhoff, A. Quantitative imaging by confocal scanning fluorescence microscopy of protein crystallization via liquid-liquid phase separation. *Acta Crystallogr. D Biol. Crystallogr.* **2005**, *61*, 819–825. [CrossRef] [PubMed]
25. Schubert, R.; Meyer, A.; Baitan, D.; Dierks, K.; Perbandt, M.; Betzel, C. Real-time observation of protein dense liquid cluster evolution during nucleation in protein crystallization. *Cryst. Growth Des.* **2017**, *17*, 954–958. [CrossRef]
26. Yamazaki, T.; Kimura, Y.; Vekilov, P.G.; Furukawa, E.; Shirai, M.; Matsumoto, H.; Van Driessche, A.E.S.; Tsukamoto, K. Two types of amorphous protein particles facilitate crystal nucleation. *Proc. Natl. Acad. Sci. USA* **2017**, *114*, 2154–2159. [CrossRef] [PubMed]
27. Van Driessche, A.E.S.; Van Gerven, N.; Bomans, P.H.H.; Joosten, R.R.M.; Friedrich, H.; Gil-Carton, D.; Sommerdijk, N.A.J.M.; Sleutel, M. Molecular nucleation mechanisms and control strategies for crystal polymorph selection. *Nature* **2018**, *556*, 89–94. [CrossRef] [PubMed]
28. The large-sized protein molecules enable direct observations of how the molecules begin assembly into clusters.
29. Krautwurst, N.; Nicoleau, L.; Dietzsch, M.; Lieberwirth, I.; Labbez, C.; Fernandez-Martinez, A.; Van Driessche, A.E.S.; Barton, B.; Leukel, S.; Tremel, W. Two-step nucleation process of calcium silicate hydrate, the nanobrick of cement. *Chem. Mater.* **2018**, *30*, 2895–2904. [CrossRef]
30. Sleutel, M.; Van Driessche, A.E.S. Nucleation of protein crystals—A Nanoscopic Perspective. *Nanoscale* **2018**, *10*, 12256–12267. [CrossRef] [PubMed]

31. Alberstein, R.G.; Tezcan, F.A. Observations of the birth of crystals. *Nature* **2018**, *556*, 41–42. [CrossRef] [PubMed]
32. Dasgupta, S.; Iyer, G.H.; Bryant, S.H.; Lawrence, C.E.; Bell, J.A. Extent and nature of contacts between protein molecules in crystal lattices and between subunits of protein oligomers. *Proteins Struct. Funct. Genet.* **1997**, *28*, 494–514. [CrossRef]
33. Iyer, G.H.; Dasgupta, S.; Bell, J.A. Ionic strength and intermolecular contacts in protein crystals. *J. Cryst. Growth* **2000**, *217*, 429–440. [CrossRef]
34. Dale, G.E.; Oefner, C.; D'Arcy, A.J. The protein as a variable in protein crystallization. *J. Struct. Biol.* **2003**, *142*, 88–97. [CrossRef]
35. Bahadur, R.P.; Chakrabarti, P.; Rodier, F.; Janin, J. A dissection of specific and non-specific protein–protein interfaces. *J. Mol. Biol.* **2004**, *336*, 943–955. [CrossRef] [PubMed]
36. Janin, J.; Rodier, F.; Chakrabarti, P.; Bahadur, R.P. Macromolecular recognition in the protein data bank. *Acta Crystallogr. D Biol. Crystallogr.* **2007**, *63*, 1–8. [CrossRef] [PubMed]
37. Doye, J.P.K.; Louis, A.A.; Vendruscolo, M. Inhibition of protein crystallization by evolutionary negative design. *Phys. Biol.* **2004**, *1*, 9–13. [CrossRef] [PubMed]
38. Doye, J.P.K.; Louis, A.A.; Lin, I.-C.; Allen, L.R.; Noya, E.G.; Wilber, A.W.; Kok, H.C.; Lyus, R. Controlling crystallization and its absence: Proteins, colloids and patchy models. *Phys. Chem. Chem. Phys.* **2007**, *9*, 2197–2205. [CrossRef] [PubMed]
39. Gillespie, C.M.; Asthagiri, D.; Lenhoff, A.M. Polymorphic Protein Crystal Growth: Influence of Hydration and Ions in Glucose Isomerase. *Cryst. Growth Des.* **2014**, *14*, 46–57. [CrossRef] [PubMed]
40. Derewenda, Z.S.; Vekilov, P.G. Entropy and surface engineering in protein crystallization. *Acta Crystallogr. D Biol. Crystallogr.* **2006**, *62*, 116–124. [CrossRef] [PubMed]
41. McElroy, H.H.; Sisson, G.W.; Schottlin, W.E.; Aust, R.M.; Villafranca, J.E. Studies on engineering crystallizability by mutation of surface residues of human thymidylate synthase. *J. Cryst. Growth* **1992**, *122*, 265–272. [CrossRef]
42. Longenecker, K.L.; Garrard, S.M.; Sheffield, P.J.; Derewenda, Z.S. Protein crystallization by rational mutagenesis of surface residues: Lys to Ala mutations promote crystallization of RhoGDI. *Acta Cryst. D* **2001**, *57*, 679–688. [CrossRef]
43. Mateja, A.; Devedjiev, Y.; Krowarsch, D.; Longenecker, K.; Dauter, Z.; Otlewski, J.; Derewenda, Z.S. The impact of Glu→ Ala and Glu→ Asp mutations on the crystallization properties of RhoGDI: The structure of RhoGDI at 1.3 Å resolution. *Acta Cryst. D* **2002**, *58*, 1983–1991. [CrossRef]
44. Derewenda, Z.S. Application of protein engineering to enhance crystallizability and improve crystal properties. *Acta Crystallogr. D Biol. Crystallogr. D* **2010**, *66*, 604–615. [CrossRef] [PubMed]
45. Nanev, C.N. How do protein lattice contacts reveal the protein crystallization mechanism? *Cryst. Res. Technol.* **2008**, *43*, 914–920.
46. Kang, K.; Choi, J.-M.; Fox, J.M.; Snyder, P.W.; Moustakas, D.T.; Whitesides, G.M. Acetylation of surface lysine groups of a protein alters the organization and composition of its crystal contacts. *J. Phys. Chem. B* **2016**, *120*, 6461–6468. [CrossRef] [PubMed]
47. This result is incompatible with the previous studies, which attributed the infrequent participation of Lys in interfaces to the entropic cost of restricting its highly mobile side chain, see [40] and "Price, W.N., II; Chen, Y.; Handelman, S.K.; Neely, H.; Manor, P.; Karlin, R.; Nair, R.; Liu, J.; Baran, M.; Everett, J.; et al. Understanding the physical properties that control protein crystallization by analysis of large-scale experimental data. *Nat. Biotechnol.* **2009**, *27*, 51–57. [CrossRef]"
48. Garcia-Ruiz, J.M. Nucleation of protein crystals. *J. Struct. Biol.* **2003**, *142*, 22–31. [CrossRef]
49. Nanev, C.N. On some aspects of crystallization process energetics, logistic new phase nucleation kinetics, crystal size distribution and Ostwald ripening. *J. Appl. Cryst.* **2017**, *50*, 1021–1027. [CrossRef]
50. Nanev, C.N. Recent Insights into Protein Crystal Nucleation. *Crystals* **2018**, *8*, 219. [CrossRef]
51. Stranski, I.; Kaischew, R. The theory of the linear rate of crystallisation. *Z. Phys. Chem. A* **1934**, *170*, 295–299.
52. Stranski, I.; Kaischew, R. Über den Mechanismus des Gleichgewichtes kleiner Kriställchen, I. *Z. Phys. Chem. B* **1934**, *26*, 100–113. [CrossRef]
53. Stranski, I.; Kaischew, R. Über den Mechanismus des Gleichgewichtes kleiner Kriställchen II. *Z. Phys. Chem. B* **1934**, *26*, 114–116.

54. The so-called Kossel-crystal is a crystal build by small cubes held together by equal forces in a cubic primitive crystal lattice.
55. Nanev, C.N. Theory of nucleation. In *Handbook of Crystal Growth*, 2nd ed.; Nishinaga, T., Ed.; Elsevier: Amsterdam, The Netherlands, 2016; Volume 1A, pp. 315–358.
56. Chayen, N.E.; Saridakis, E.; El-Bahar, R.; Nemirovsky, Y. Porous silicon: An effective nucleation-inducing material for protein crystallization. *J. Mol. Biol.* **2001**, *312*, 591–595. [CrossRef] [PubMed]
57. Rong, L.; Komatsu, H.; Yoshizaki, I.; Kadowaki, A.; Yoda, S. Protein crystallization by using porous glass substrate. *J. Synchrotron Radiat.* **2004**, *11*, 27–29. [CrossRef] [PubMed]
58. Saridakis, E.; Chayen, N.E. Towards a 'universal' nucleant for protein crystallization. *Trends Biotechnol.* **2009**, *27*, 99–106. [CrossRef] [PubMed]
59. Khurshid, S.; Saridakis, E.; Govada, L.; Chayen, N.E. Porous nucleating agents for protein crystallization. *Nat. Protoc.* **2014**, *9*, 1621–1633. [CrossRef] [PubMed]
60. Asanithi, P.; Saridakis, E.; Govada, L.; Jurewicz, I.; Brunner, E.W.; Ponnusamy, R.; Cleaver, J.A.S.; Dalton, A.B.; Chayen, N.E.; Sear, R.P. Carbon-Nanotube-Based Materials for Protein Crystallization. *Appl. Mater. Interfaces* **2009**, *1*, 1203–1210. [CrossRef] [PubMed]
61. Kertis, F.; Khurshid, S.; Okman, O.; Kysar, J.W.; Govada, L.; Chayen, N.; Erlebacher, J. Heterogeneous nucleation of protein crystals using nanoporous gold nucleants. *J. Mater. Chem.* **2012**, *22*, 21928–21934. [CrossRef]
62. Saridakis, E.; Khurshid, S.; Govada, L.; Phan, Q.; Hawkins, D.; Crichlow, G.V.; Lolis, E.; Reddy, S.M.; Chayen, N.E. Protein crystallization facilitated by molecularly imprinted polymers. *Proc. Natl. Acad. Sci. USA* **2011**, *108*, 11081–11086. [CrossRef] [PubMed]
63. Saridakis, E.; Chayen, N.E. Polymers assisting in protein crystallization. *Trends Biotechnol.* **2013**, *31*, 515–520. [CrossRef] [PubMed]
64. Sugahara, M.; Asada, Y.; Morikawa, Y.; Kageyama, Y.; Kunishima, N. Nucleant-mediated protein crystallization with the application of microporous synthetic zeolites. *Acta Crystallogr. D* **2008**, *64*, 686–695. [CrossRef] [PubMed]
65. Di Profio, G.; Curcio, E.; Ferraro, S.; Stabile, C.; Drioli, E. Effect of supersaturation control and heterogeneous nucleation on porous membrane surfaces in the crystallization of L-glutamic acid polymorphs. *Cryst. Growth Des.* **2009**, *9*, 2179–2186. [CrossRef]
66. Chayen, N.E.; Saridakis, E.; Sear, R.P. Experiment and theory for heterogeneous nucleation of protein crystals in a porous medium. *Proc. Natl. Acad. Sci. USA* **2006**, *103*, 597–601. [CrossRef] [PubMed]
67. Eisenstein, M. The shape of things. *Nat. Methods* **2007**, *4*, 95–102. [CrossRef]
68. A nucleant is a solid substance that has nucleation-inducing properties.
69. Nanev, C.N.; Saridakis, E.; Chayen, N.E. Protein crystal nucleation in pores. *Sci. Rep.* **2017**, *7*, 35821. [CrossRef] [PubMed]
70. A critical supersaturation may arise in pores, provided the system is not too close to equilibrium.
71. Because Brownian motion is equally probable in all directions, the escape probability of a protein molecule from the pore is about 1/6.
72. Langdon, B.B.; Kastantin, M.; Schwartz, D.K. Apparent activation energies associated with protein dynamics on hydrophobic and hydrophilic surfaces. *Biophys. J.* **2012**, *102*, 2625–2633. [CrossRef] [PubMed]
73. Note that such pores are still much larger than the typical nucleus size.
74. Sear, R.P. The non-classical nucleation of crystals: Microscopic mechanisms and applications to molecular crystals, ice and calcium carbonate. *Int. Mater. Rev.* **2012**, *57*, 328–356. [CrossRef]
75. Wang, J.; Li, F.; Lakerveld, R. Process intensification for pharmaceutical crystallization. *Chem. Eng. Process.* **2018**, *127*, 111–126. [CrossRef]
76. Wang, T.; Lu, H.; Wang, J.; Xiao, Y.; Zhou, Y.; Bao, Y.; Hao, H. Recent progress of continuous crystallization. *J. Ind. Eng. Chem.* **2017**, *54*, 14–29. [CrossRef]
77. Neugebauer, P.; Khinast, J.G. Continuous crystallization of proteins in a tubular plug-flow crystallizer. *Cryst. Growth Des.* **2015**, *15*, 1089–1095. [CrossRef] [PubMed]
78. Castro, F.; Ferreira, A.; Teixeira, J.A.; Rocha, F. Protein crystallization as a process step in a novel meso oscillatory flow reactor: Study of lysozyme phase behavior. *Cryst. Growth Des.* **2016**, *16*, 3748–3755. [CrossRef]

79. Turbulence causes the formation of eddies of many different length scales (including macroscopic), and most of the kinetic energy of the turbulent motion is contained in the large-scale structures. Thus, to sustain turbulent flow, a persistent source of energy supply is required.
80. Lawton, S.; Steele, G.; Shering, P.; Zhao, L.; Laird, I.; Ni, X.W. Continuous crystallization of pharmaceuticals using a continuous oscillatory baffled crystallizer. *Org. Process Res. Dev.* **2009**, *13*, 1357–1363. [CrossRef]
81. Hekmat, D.; Huber, M.; Lohse, C.; von den Eichen, N.; Weuster-Botz, D. Continuous crystallization of proteins in a stirred classified product removal tank with a tubular reactor in bypass. *Cryst. Growth Des.* **2017**, *17*, 4162–4169. [CrossRef]
82. Li, F.; Lakerveld, R. Electric-field-assisted protein crystallization in continuous flow. *Cryst. Growth Des.* **2018**, *18*, 2964–2971. [CrossRef]
83. Nanev, C.N. Recent insights into the crystallization process; protein crystal nucleation and growth peculiarities; Processes in the Presence of Electric Fields. *Crystals* **2017**, *7*, 310. [CrossRef]
84. Byington, M.C.; Safari, M.S.; Conrad, J.C.; Vekilov, P.G. Shear flow suppresses the volume of the nucleation precursor clusters in lysozyme solutions. *J. Cryst. Growth* **2017**, *468*, 493–501. [CrossRef]
85. Grant, M.L.; Savile, D.A. The role of transport phenomena in protein crystal growth. *J. Cryst. Growth* **1991**, *108*, 8–18. [CrossRef]
86. Durbin, S.D.; Feher, G. Crystal growth studies of lysozyme as a model for protein crystallization. *J. Cryst. Growth* **1986**, *76*, 583–592. [CrossRef]
87. Pusey, M.; Witherow, W.; Naumann, R. Preliminary investigations into solutal flow about growing tetragonal lysozyme crystals. *J. Cryst. Growth* **1988**, *90*, 105–111. [CrossRef]
88. Nyce, T.A.; Rosenberger, F. Growth of protein crystals suspended in a closed loop thermosyphon. *J. Cryst. Growth* **1991**, *110*, 52–59. [CrossRef]
89. Vekilov, P.G.; Rosenberger, F. Protein crystal growth under forced solution flow: Experimental setup and general response of lysozyme. *J. Cryst. Growth* **1998**, *186*, 251–261. [CrossRef]
90. Roberts, M.M.; Heng, J.Y.Y.; Williams, D.R. Protein crystallization by horced flow through glass capillaries: Enhanced lysozyme crystal growth. *Cryst. Growth Des.* **2010**, *10*, 1074–1083. [CrossRef]
91. Hodzhaoglu, F.V.; Nanev, C.N. Heterogeneous versus bulk nucleation of lysozyme crystals. *Cryst. Res. Technol.* **2010**, *45*, 281–291. [CrossRef]
92. Penkova, A.; Gliko, O.; Dimitrov, I.L.; Hodjaoglu, F.V.; Nanev, C.N.; Vekilov, P.G. Enhancement and suppression of protein crystal nucleation due to electrically driven convection. *J. Cryst. Growth* **2005**, *275*, e1527–e1532. [CrossRef]
93. Parambil, J.V.; Schaepertoens, M.; Williams, D.R.; Heng, J.Y.Y. Effects of oscillatory flow on the nucleation and crystallisation of insulin. *Cryst. Growth Des.* **2011**, *11*, 4353–4359. [CrossRef]
94. Anwar, J.; Khan, S.; Lindfors, L. Secondary crystal nucleation: Nuclei breeding factory uncovered. *Angew. Chem.* **2015**, *127*, 14894–14897. [CrossRef]
95. Viscosity of Water. Available online: https://wiki.anton-paar.com/en/water (accessed on 6 November 2018).
96. Schlichting, H.; Gersten, K. *Boundary-Layer Theory*; Springer: Berlin, Germany, 2017; pp. 416–419.
97. Mura, F.; Zaccone, A. Effects of shear flow on phase nucleation and crystallization. *Phys. Rev. E* **2016**, *93*, 042803. [CrossRef] [PubMed]
98. Angkawinitwong, U.; Sharma, G.; Khaw, P.T.; Brocchini, S.; Williams, G.R. Solid-state protein formulations. *Ther. Deliv.* **2015**, *6*, 59–82. [CrossRef] [PubMed]

crystals

MDPI

Article

pH and Redox Induced Color Changes in Protein Crystals Suffused with Dyes

Alexander McPherson

Molecular Biology and Biochemistry, University of California Irvine, McGaugh Hall, Irvine, CA 92697-3900, USA; amcphers@uci.edu; Tel.: +1-831-717-4247

Received: 21 January 2019; Accepted: 23 February 2019; Published: 1 March 2019

Abstract: Protein crystals, otherwise usually colorless, can be stained a variety of hues by saturating them with dyes, by diffusion from the mother liquor or co-crystallization. The colors assumed by dyes are a function of chemical factors, particularly pH and redox potential. Protein crystals saturated with a pH sensitive dye, initially at one pH, can be exposed to the mother liquor at a second pH and the crystal will change color over time as H_3O^+ ions diffuse through the crystal. This allows diffusion rates of H_3O^+ through the crystal to be measured. Diffusion fronts are often clearly delineated. Similar experiments can be carried out with redox sensitive dyes by adding reductants, such as ascorbic acid or dithionite, or oxidants such as H_2O_2, to the crystal's mother liquor. Presented here are a number of experiments using pH or redox sensitive dye-saturated protein crystals, and some experiments using double dye, sequential redox–pH changes.

Keywords: dyes; diffusion; H_3O^+; reductants; color change; gradients

1. Introduction

In the course of investigations on the diffusion of dyes and stains into protein crystals, and their subsequent, frequent incorporation [1,2], it was noted that a number of the dyes were sensitive to pH and dramatically changed color as a consequence of proton acquisition or release from their peripheral ionizable groups. These dyes are examples of the pH indicator dyes that have been in use for well over two hundred years by chemists, histologists, and biochemists [3–8]. The dyes have an aromatic, conjugated core that serves as a chromophore and attached, solubilizing ionizable groups that are generally amines in the case of basic dyes, or sulfonates in the case of acidic dyes. Some examples are shown in Figure 1.

Other dyes are sensitive to reduction and oxidation [9], and are generally brightly colored when oxidized, but colorless when reduced. In both pH and redox induced color change, the reactions are fully reversible.

Protein crystals can be saturated with a sensitive dye at a specific pH, the pH of its mother liquor, and the mother liquor then replaced with an equivalent at another pH. If the second pH is on the opposite side of a color transition point from the pH of the initial mother liquor, then the crystal changes color. Over time, the entire crystal, from outside to inside, transforms from its initial color to an alternate color. Because the reaction is reversible, replacement of the second mother liquor with that at the original pH causes the color change to proceed in the opposite direction. Multiple cycles of change of a protein crystal between two colors can thereby be realized. The pH sensitive dyes employed in these experiments, their colors, and the pH transition points for their color change are presented in Table 1.

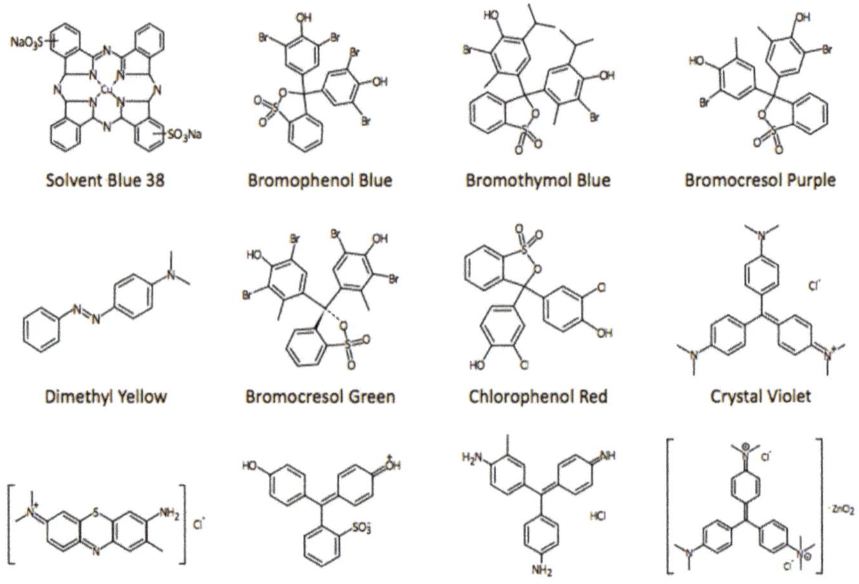

Figure 1. The chemical structures of a selection of the pH and redox sensitive dyes used in the investigation.

Table 1. pH sensitive dyes.

Dye	Low pH Color	Transition pH	High pH Color
Methyl yellow	red	3.5	yellow
Methyl orange	red	3	yellow
Bromophenol blue	yellow	3.8	blue
Chlorophenol red	blue/violet	4.0	red
Bromcresol green	yellow	4.6	blue
Methyl red	red	5.3	yellow
Bromcresol purple	yellow	6.0	purple
m-cresol purple	yellow	6.2	red
Bromothymol blue	yellow	6.8	blue
Phenol red	yellow	7.2	red
Neutral red	red	7.4	yellow

Measuring color transitions of pH sensitive dye-saturated protein crystals as a function of time serves as an indicator of the rates of diffusion of protons (H_3O^+ ions) into and out of the crystals. Furthermore, because pH is defined as the negative log of the H_3O^+ concentration, by knowing the pH of the initial and alternate mother liquors, the H_3O^+ concentrations, inside the crystal and in the surrounding mother liquor are known precisely.

Similar opportunities exist using redox sensitive dyes. A protein crystal can be saturated with such a dye (usually it is green or blue). Addition of an appropriate reductant will cause the dye to lose color and the crystal will become clear. The redox sensitive dyes used in these experiments, along with an appropriate reductant for that dye, are presented in Table 2. Oxygen, present in the environment, including the mother liquor, eventually enters the crystal and re-oxidizes the dye and the original color of the crystal returns. In the experiments described here, the diffusion of a reductant or oxidant into crystals was visualized for various protein crystal/reducing agent combinations.

Table 2. Oxidation–reduction sensitive dyes.

Dye	Color Change on Reduction	Reductant
Basic fuchsin	red > colorless	bisulfite
Methyl green	green > colorless	bisulfite
Crystal violet	violet > colorless	dithionite
Methylene blue	blue > colorless	ascorbic acid
Toluidine blue	blue > colorless	ascorbic acid
Luxol blue or Solvent blue 38	blue > copper	dithionite

2. Materials and Methods

Dyes, their pH transition points, and their colors are listed in Table 1, and redox sensitive dyes and the reductants used in conjunction are shown in Table 2. The crystallization conditions for the proteins on which the experiments were based, along with their unit cell parameters and solvent contents are detailed elsewhere [2]. In general, to provide good observation and recording by photography, it was necessary to use large protein crystals [10–13]. Thus, the crystals studied had dimensions in the range of 0.5 mm to 1.5 mm, or greater edge lengths. Crystals were grown by vapor diffusion [12,13] in sitting drop Cryschem plates with drop sizes of 8 to 12 µL and reservoirs of 0.6 mL. Crystallization was carried out at room temperature, although some crystallization plates were moved to 4 °C to better preserve grown crystals. The mother liquors of canavalin crystals [14–16] were made 15% in 2-methyl 2,4 pentanediol (MPD) after crystals were fully grown, as canavalin crystals, both rhombohedral and orthorhombic modifications [14–16], will otherwise spontaneously degrade over a week's time following their appearance.

Lysozyme in the tetragonal form [17] can be reproducibly grown at a pH between 4.0 and 7.5, canavalin between about pH 5.0 and 7.5 [14–16], and thaumatin [18] between pH 6.0 and 8.0. Although crystals of trypsin, inhibited with benzamidine [19], are usually grown at pH 8.5, they can be transferred to mother liquors at pH 6.0 and even below, without damage. Satellite tobacco mosaic virus (STMV) crystals [20] are stable between pH 4 and 7.5, and concanavalin A and concanavalin B crystals [21,22] between pH 6 and 8. ß-lactoglobulin [23] was grown at pH 3.2 for which bromophenol blue was the only useful dye. Crystals of RNase S were grown as described by Wyckoff et al. [24]. Other dyes in the low pH range that might have been used (e.g. dimethyl yellow) were almost insoluble in water. Some dyes that might appear useful were not, because they were insoluble in concentrated salt (NaCl, $(NH_4)_2SO_4$, etc.) solutions that were used in some crystallizations.

Manipulation of mother liquors in droplets was carried out under a zoom lens dissecting microscope using gel loading tips on 10 µL pipetmen. If necessary, residual mother liquor was drawn off using paper wicks (Hampton Research, Aliso Viejo, CA, USA). Manipulation of crystals was carried out using cryo-mounting tips of 100 µm to 600 µm diameter from MitiGen (Ithaca, NY, USA). Observation of diffusion experiments [2] was with an SZX12 Olympus zoom microscope with a camera adaptor (Olympus Co., Tokyo, Japan). The camera was a Canon EOS digital model. Photos were made with the night–no flash setting. Generally, at timed intervals, pairs of photos were recorded at high and low magnification.

Experiments designed to record the rates of color change in crystals generally proceeded in the following way: Protein crystals were saturated with a pH or redox sensitive dye, or both, *in situ*, by adding a concentrated dye dissolved in the crystal's mother liquor (minus protein). Diffusion of the dye was allowed to continue for at least 48 h or more, as this was necessary in those cases where the dye interacted strongly and positively with the protein comprising the crystals [1,2]. In experiments where multiple dyes (i.e. both pH sensitive dyes and redox sensitive dyes) were investigated sequentially, the crystals were first saturated with the pH sensitive dye and then with the redox sensitive dye. The crystals, upon saturation, exhibited a color appropriate to the dye at the pH of the mother liquor, which would be on one side or the other of the pH transition point. The crystal was first washed with the mother liquor (generally taken from the sitting drop reservoir) to remove extraneous dye from the surroundings. The mother liquor was then replaced with equivalent mother liquor but at a pH on the opposite side of the pH transition point. This meant that the pH of the solvent and the dye inside the crystal was at the growth pH, but that of the new mother liquor was at a pH on the other side of the dye's pH transition point.

Depending on the direction of pH change, H_3O^+ ions either diffused into the crystals (high pH to low pH), or out of the crystals (low pH to high pH). As the pH in a crystal changed, the pH sensitive dye changed color. The progress of H_3O^+ diffusion, therefore, could be followed by recording the color change within the crystal as a function of time. pH is the negative log of the H_3O^+ ion concentration, so that by knowing the initial pH and the replacement pH, the H_3O^+ concentration difference between the inside and outside of a crystal was known.

For experiments with reduction and oxidation, the crystals were saturated with a redox sensitive dye [9] from Table 2 by exposing the crystals to the dye for at least 48 h. Addition of either a reducing compound dissolved in the mother liquor of the crystal (minus protein) to the existing mother liquor, or direct addition of the solid compound to the mother liquor was made. Generally, it was necessary to add a substantial amount of the reducing agent because of competition by oxygen in the system. It was not necessary to remove excess redox sensitive dye from the mother liquor as that was immediately reduced to colorless upon addition of the reducing agent. Fortunately, and perhaps significantly, protein crystals could be exposed to remarkably high concentrations of reducing agents without visible damage.

In experiments where pH and redox dyes were sequentially diffused into crystals to saturation with both, the experiment proceeded by first adding the reducing agent. This caused the color of the redox dye to be lost and as a consequence reveal the color of the pH indicator dye. The pH was then jumped, as described above, and the crystal then observed to transition to the alternate color. In experiments involving reduction of a dye in a crystal, initially the concentration of the reductant in the crystal was zero. Thus, the difference in reductant concentration inside and outside the crystal upon addition of reductant was simply that of the mother liquor after addition.

A caution is advised to others carrying out experiments similar to these. Dyes will occasionally adhere to or accumulate on the surfaces of protein crystals, but not actually penetrate into their interiors. These occurrences are usually apparent on close microscopic inspection, but can nonetheless cause confusion if ignored. Another point to note is that crystals occasionally develop visible cracks and fissures or other significant damage as a consequence of physical manipulation or due to pH or chemical changes to their mother liquors. It might be expected that the faults and defects would provide expedited entry of dyes or ions into crystals. This should be evident by more intense color changes at the defects and cracks. This was not, in general, observed to be the case. Entry through fissures and cracks does not appear to be accelerated beyond the diffusion otherwise prevailing.

3. Results

It is not reasonable to present all of the experiments that were carried out, but a sampling should here be sufficient.

In Figure 2 through 6 are found representative samples of the pH change experiments for a variety of protein crystals. Figure 2 shows thaumatin crystals plus bromophenol blue, Figure 3 is thaumatin crystals plus bromothymol blue, and Figure 4 is lysozyme crystals plus m-cresol purple.

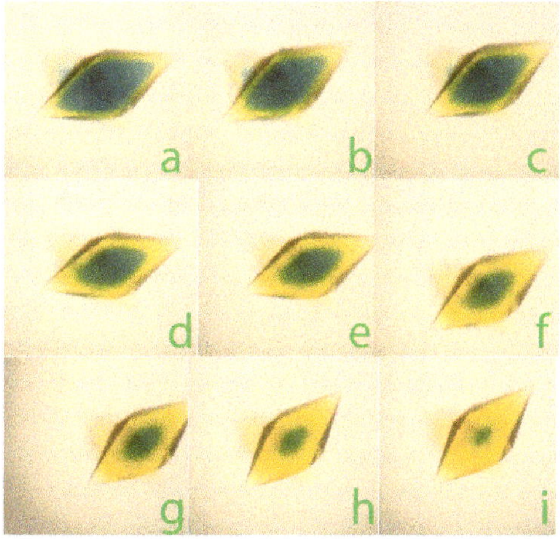

Figure 2. A tetragonal crystal of thaumatin about 1 mm in length was stained to saturation over 48 h with the pH sensitive dye bromophenol blue. The crystal was initially at pH 8.25 so that its color was blue–green. The mother liquor was then removed and replaced with an equivalent mother liquor but at pH 6.5. As H_3O^+ ions diffused into the crystal, the incorporated bromophenol blue dye transformed chemically and yielded a yellow color. The change of color shows a clear leading edge and the rate of penetration of the yellow color provides a measure of the diffusion of H_3O^+ ions into the crystal. The total time required for complete color transformation was about 30 to 40 min. The series of images (**a**–**i**), taken at appropriate time intervals, shows the progress of the dye.

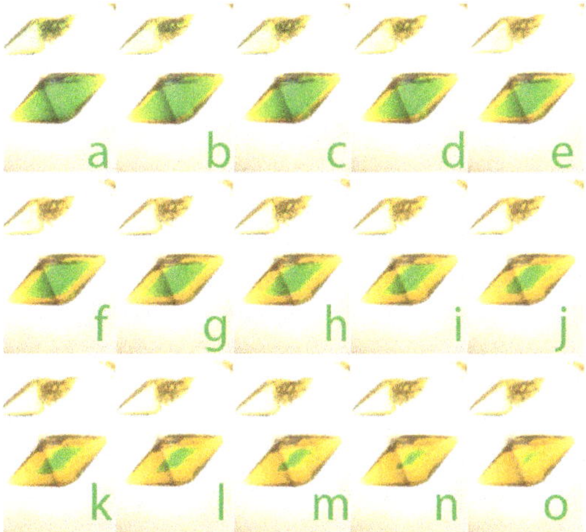

Figure 3. Tetragonal crystals of thaumatin, the larger about 1 mm in length, were stained to saturation over 48 h with the pH sensitive dye bromothymol blue. The crystal was initially at pH 8.25. The mother liquor was then removed and replaced with equivalent mother liquor but at pH 6.5. As H_3O^+ ions diffused into the crystal, the incorporated bromothymol blue dye transformed chemically and yielded a yellow color. The change of color again shows a clear leading edge. The total time required for complete color transformation was about 35 min. The series of images (**a**–**o**), taken at appropriate time intervals, shows the progress of the dye.

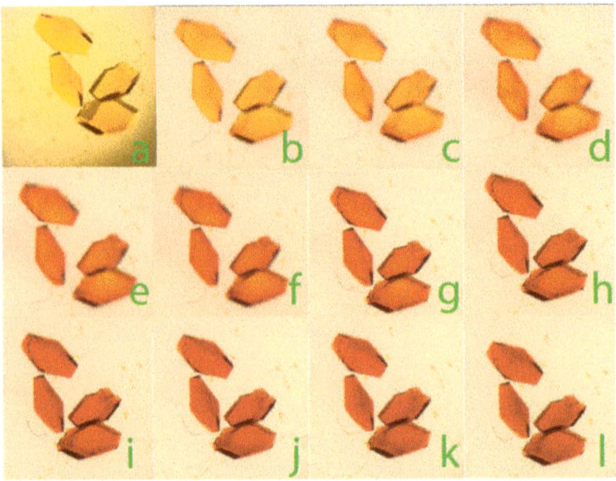

Figure 4. Tetragonal lysozyme crystals were saturated with the dye m-cresol purple at pH 5, which gave the crystals a yellow/orange color. The pH of the mother liquor was changed to pH 7.5 and the conversion of the crystal color to red provided a measure of the diffusion of H_3O^+ ions out of the crystal into the mother liquor. The total time required for the transition was about 30 min. The series of images (**a–l**), taken at appropriate time intervals, shows the progress of the dye.

Figure 5 shows lysozyme crystals plus bromothymol green and Figure 6 shows crystals of the virus STMV, plus m-cresol purple.

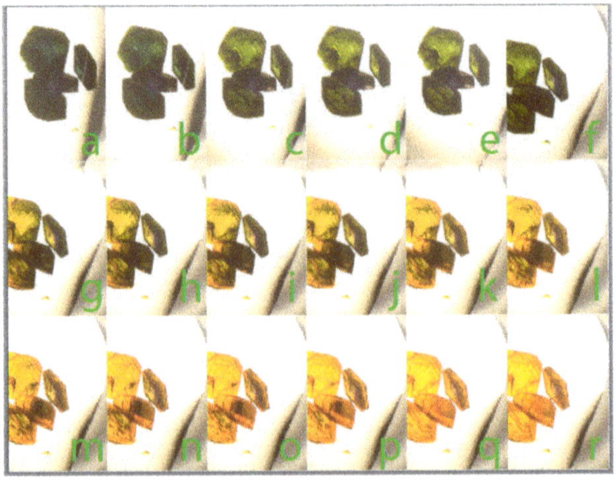

Figure 5. Tetragonal lysozyme crystals saturated with the dye bromocresol green are a dark blue-green color at pH 6.0 in (**a**). When the pH of the mother liquor was altered to pH 3.8, the crystals were seen in (**b**) through (**r**) to change in color to yellow. The series of images (**a–r**), taken at appropriate time intervals, shows the progress of the dye.

Upon pH transformation, the change in color due to dye transition, as H_3O^+ either entered or exited the crystals, appeared immediately at the crystal faces and rapidly proceeded inwards toward the crystal center, generally by isometric approach. The diffusion front was usually clearly evident, particularly when the color contrast was large, as for yellow to blue (or vice versa) transitions.

Those between red and blue, for example, were often less evident. Diffusion was initially swift at the edges and faces of the crystals, but slowed appreciably as the diffusion front approached the crystal center. As diffusion proceeded, a core of color, visible at the crystal center, gradually dissipated and disappeared. This was illustrated particularly well by the examples of the tetragonal thaumatin crystals in Figures 2 and 3 stained with bromophenol blue and bromothymol blue, respectively.

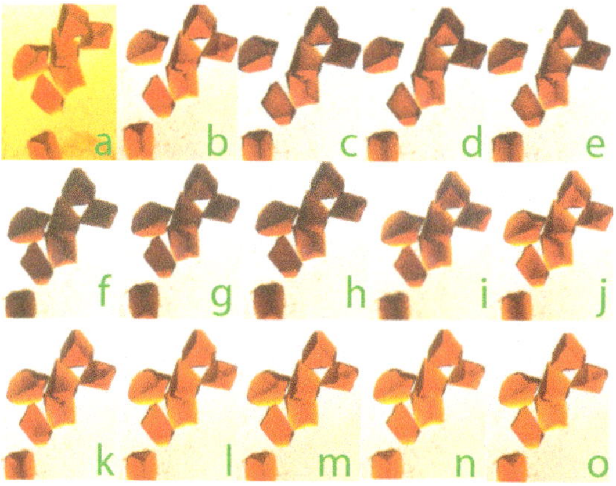

Figure 6. Crystals of satellite tobacco mosaic virus (STMV) were saturated with the dye m-cresol purple at pH 5.5 and appear as in (**a**). The mother liquor was replaced with equivalent mother liquor at pH 8.0 and the crystals changed color to dark red in (**b**) through (**h**). When the original pH was restored in the mother liquor, the crystals (**i**) through (**o**) reverted to the initial color. This experiment, repeated with numerous other protein crystals demonstrates the reversible nature of the color changes and that crystals can experience multiple cycles of pH induced transitions. The series of images (**a**–**o**), taken at appropriate time intervals, shows the progress of the dye.

The total time required for a crystal to completely transition between the initial and replacement pH colors depended to some extent on the protein crystal, its size, and the magnitude of the pH change. For the large crystals used in these experiments, however, the times were noted to be remarkably similar; between 25 and 35 min for pH jumps of, for example pH 8 to pH 6 (10^{-8} mol/L [H_3O^+] to 10^{-6} mol/L [H_3O^+]). If the distances from the edges of the crystals to the center were on the order of 0.25 to 0.50 mm, as they were for most of these experiments, then this represents a diffusion rate in the range of 500 to 1000 μm per hour, or about 8 to 16 μm per minute.

As H_3O^+ is likely the fastest of any ion or molecule diffusing through a protein crystal, then 1000 μm per hour for a two-log difference in concentration inside to outside of the crystal is probably an upper limit for the diffusion of any ion or molecule into a protein crystal. This, it must be emphasized, is for a two-log difference. If, for example, mother liquor is made 10 mM in some compound that is initially absent in a crystal, then the gradient of concentration is appreciably greater. As demonstrated by the redox experiments below, when concentration differences are of that order, diffusion rates substantially exceed those seen for the pH experiments.

Color transitions of dyes are completely reversible when the dye is inside the protein crystal. Using crystals that were particularly stable to dramatic pH change, such as thaumatin and trypsin, it was possible to put a protein crystal-dye complex through multiple cycles of pH and color transition by successive changes of the pH of the mother liquor. This was illustrated in Figure 6 by the satellite tobacco mosaic virus (STMV) crystals. Apparent in many sequences is that the movement of a pH front through a crystal, whether from high to low, or vice versa, is the same in appearance. Absent precise quantitation, it is not, however, certain if the movement of pH fronts, rates of transition from high to low, and low to high pH, are identical.

Table 2 lists a set of dyes that are sensitive to reduction and oxidation, and they include basic fuchsin, methyl green, toluidine blue and methyl blue. Luxol blue, or solvent blue 38, is particularly interesting as it is a soluble porphyrin with a copper atom in the oxidized plus two state that gives the dye its blue color. The dyes are not equally sensitive to the various reducing agents, and in Table 2 are given the appropriate reducing compounds for their transition. Methylene blue, for example, was reduced only by ascorbic acid (vitamin C), while basic fuchsin and methyl green with bisulfite, and crystal violet by dithionite, the latter the most powerful of the reductants.

Experiments with redox sensitive dyes were carried out the same way as for those involving pH change. The protein crystals were first saturated with dye, and reductant was then added to the mother liquor. A noteworthy feature of these experiments was that the reductants, even at quite high concentrations, seemed to inflict no damage on the crystals. This was in contradistinction to jumps in pH that occasionally caused cracking or other damage.

Two, somewhat different kinds of results were recorded in the redox experiments, though they reflect the same chemical process. In the examples shown in Figures 7 and 8 are lysozyme crystals with methyl green and crystal violet, respectively. The color changes as the reductants, bisulfite and dithionite, respectively, diffused into the crystals exhibited a distinct front that progressed from the exterior faces to the center of the crystal, much as the H_3O^+ diffusion fronts in pH experiments did. In some other experiments, however, basic fuchsin or methylene blue infused crystals for example, there was no sharp diffusion front. The crystals simply changed in a more uniform manner from colored to clear simultaneously throughout their volume. An explanation for this latter result may be that the rate of change in color for some of the dyes is slow and comparable to the diffusion rate, or even slower. In that case, the observed result might be expected.

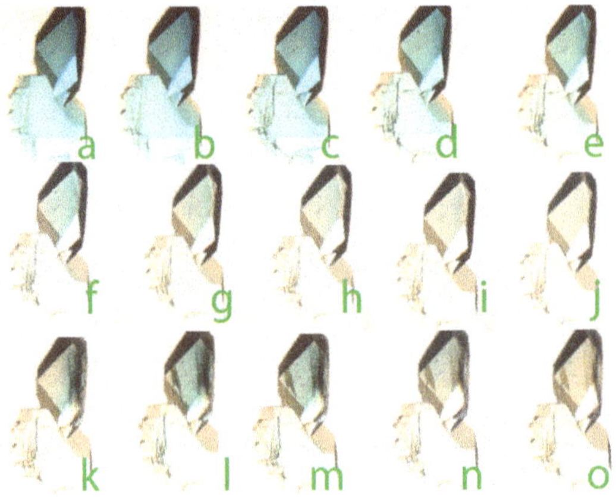

Figure 7. Lysozyme crystals were saturated with the dye methyl green (**a**) and bisulfite added to the mother liquor in (**b**). The green color was progressively lost as the dye was reduced. At (**k**), however, oxygen from the environment competed to re-oxidize the dye, which then experienced a second cycle of reduction in (**n**) and (**o**). The series of images (**a**–**o**), taken at appropriate time intervals, shows the progress of the dye.

As with pH changes in crystals, transitions between oxidized and reduced states are reversible. Addition of a low concentration of an oxidizing agent such as H_2O_2 immediately re-oxidizes the dyes. It was usually not necessary to add any oxidant to reverse the chemical process, however, as this was accomplished by oxygen absorbed into the mother liquor from the air around it. This process of re-oxidation by ambient oxygen was reasonably slow and could be recorded as color returned to crystals.

Figure 8. Lysozyme crystals were saturated with the dye crystal violet (**a**) and dithionite added to the mother liquor in (**b**). The purple color was progressively lost as the dye was reduced. At (**k**), however, oxygen from the environment competed to re-oxidize the dye, which then experienced a second cycle of reduction in (**n**) and (**o**). The series of images (**a**–**o**), taken at appropriate time intervals, shows the progress of the dye.

The rates of change to the redox state of the various dyes were diverse, as the reductants were of different strengths, or chemical activities, and controlling their concentration in mother liquors was technically problematic. Nonetheless, for concentrations of bisulfite in the 10 mM range, reduction of methyl green, for example, was complete throughout a protein crystal in about 10 to 15 min. For crystals saturated with methylene blue, ascorbic acid concentrations of about 40 mM were necessary to achieve the same range of rates. With dyes susceptible to reduction by dithionite, exterior concentrations of the reductant were on the order of 100 mM and complete color conversion was complete in 2 min or less.

In the redox experiments, where a diffusion front was clearly visible and its progress measurable, it was again observed (as in the pH experiments) that the diffusion front initially moved rapidly in from the faces but slowed appreciably as it approached the crystal center. Thus, 80% of a crystal might transition in the first 40% of the total time, while the final 20% of the dye at the center required the last 60%. As already noted, spontaneous reversion to the oxidized state began soon after complete reduction, and in some experiments even before full reduction had completed. Usually, however, re-oxidation by ambient oxygen was graciously slow and required 15 to 20 min. Although no attempt was made to quantitate it, the rates of re-oxidation by ambient oxygen could provide a measure of the rates of diffusion of oxygen through a protein crystal.

Luxol blue, or solvent blue 38, is a sky-blue dye that was of particular interest. It would not enter lysozyme crystals at all, probably due to size or shape, but crystals of thaumatin and STMV were readily infused. The dye is a porphyrin that has two SO_3^- groups attached for solubility. It is redox sensitive because the Cu^{2+} ion at the porphyrin center can be reduced to elemental copper. In our experiments we used dithionite as the reductant, as others failed. This is a strong reductant that is difficult to control in the presence of oxygen. In these experiments grains of powdered dithionite were added directly to the mother liquor at the edge of the droplet as far as possible from the crystals. The grains of dithionite dissolved, and the dithionite then diffused rapidly across the drop before reaching the crystals. This is why in most images the crystals change color, not uniformly from all sides, but directionally as though reduction was passing as a wave through the crystal. The time for reduction of the Luxol blue dye in the crystals was about 1 to 3 min.

As shown in Figures 9 and 10 for thaumatin crystals, when dithionite penetrated the Luxol blue-containing crystals and reduced the dye, the color of the crystals changed from sky blue to the color of a new penny. As with the other redox dyes, this reaction was also reversible and re-oxidation by ambient oxygen could be observed to occur over 5 to 10 min as the crystals were restored to their original color. Because of the ready reversibility of these redox reactions, and the stability of protein crystals in the presence of the reagents, multiple cycles, blue-copper-blue, could be generated with a single crystal.

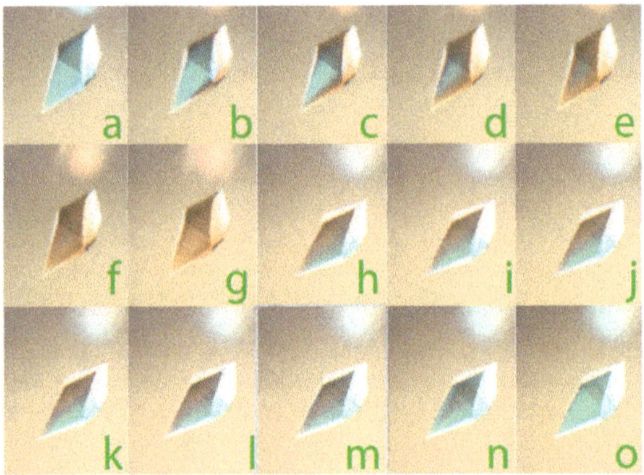

Figure 9. A tetragonal crystal of thaumatin about 1 mm in length was saturated with the redox sensitive dye Luxol blue (Solvent blue 38) giving it an intense blue color. Luxol blue is a porphyrin that contains a copper ion in the plus two state that is responsible for the color. Solid dithionite was added directly to the mother liquor, which reduced the Cu^{2+} ion to elemental copper. Thus, the transformation of color in the crystal from blue to copper shows the progress of the reduction reaction as the dithionite penetrated the crystal. Beginning with image (**h**), oxygen in the environment initiated re-oxidation of the copper and the color transitioned again to blue. This is shown in the final images (**i**) through (**o**). The series of images (**a**–**o**), taken at appropriate time intervals, shows the progress of the dye.

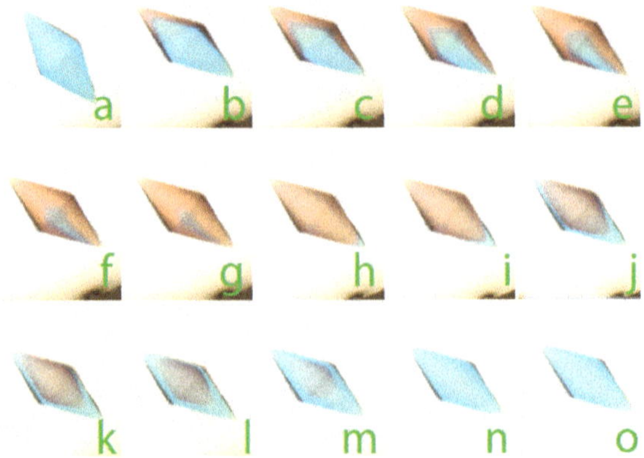

Figure 10. A tetragonal crystal of thaumatin was saturated with the copper porphyrin dye Luxol blue (Solvent blue 38) in (**a**) and dithionite then added to the mother liquor (**b**). The blue color transitioned to elemental copper as the dye was reduced. At about (**k**) oxygen from the environment began to re-oxidize the copper to the Cu^{2+} state and the dye regained its blue color. In both the reduction and oxidation phases the redox front was well defined. The series of images (**a**–**o**), taken at appropriate time intervals, shows the progress of the dye.

Experiments were carried out with crystals of thaumatin, the most stable of the crystals, to investigate possibilities with double staining using two dyes. In these experiments the crystals were first saturated with a pH sensitive dye at a specific pH, and then exposed to a second redox sensitive dye. Crystals then assumed the color of the dye combination. If a pH dye that was yellow was followed by a blue redox dye, then the crystal would be green. If a blue dye was followed by a green dye, the crystal would be blue-green, etc. In the experiment shown in Figure 11, a thaumatin crystal was saturated with bromothymol blue at pH 6 (yellow), and then washed to remove exogenous stain. It was then suffused with methyl green (which is actually blue in color) to give the aquamarine crystals in Figure 11c top panel. Bisulfite was then added. This reduced the methyl green (blue color) to colorless revealing the bright yellow color of the pH dye, bromothymol blue in Figure 11o top panel. In the second series below, the pH was then raised in the mother liquor to pH 8.5 in Figure 11a bottom panel and the crystals can be seen to transition to the high pH blue color in Figure 11o bottom panel.

Figure 11. In the top series, tetragonal crystals of thaumatin were first stained to saturation with the dye bromothymol blue, which at pH 6.5 created a yellow color. This was followed by saturation with the dye toluidine blue that gave the crystals a green (yellow + blue) color. Dithionite was then added to the mother liquor and in (**a**) of the bottom series, as the dithionite penetrated the crystal, the toluidine blue was reduced thereby yielding the underlying yellow color due to the pH sensitive bromothymol blue at low pH. Reduction proceeded until (**o**) at which time virtually all of the redox sensitive dye had been reduced. Starting in (**e**) of the bottom panel the pH of the mother liquor was raised to pH 8.2 and the bromothymol blue dye in the crystal transitioned to the blue chemical state. The series of images (**a**–**o**) in both the upper and lower panels, taken at appropriate time intervals, shows the progress of the dye changes with time, for both pH and redox sensitive dyes.

Aside from demonstrating the feasibility of multiple, successive chemical reactions within the confines of protein crystals, the experiments involving multiple dyes have a second implication. One dye saturating a protein crystal does not exclude another from entering and sharing the space. The dyes, therefore, are not necessarily competitive for a few select sites, but seem to find for themselves new available space.

4. Conclusions

The experiments described here involving protein crystals, dyes, and changes to chemical states can serve as systems for the precise measurement of the penetration of H_3O^+, redox effectors, and oxygen into and out of macromolecular crystals. With proper controls and the use of more sophisticated instruments, such as micro-spectrophotometers, rates of penetration could be determined quantitatively as a function of numerous parameters. These might include concentrations of dyes and diffusing species in the mother liquor, crystal solvent content, dye size or charge, and undoubtedly others. The work presented here yields only rough estimates, but they may be sufficient to at least put succeeding investigations in the "experimental ballpark" with regard to initial trials.

The chemical reactions that were here induced inside the channels and cavities of protein crystals represent only a small number of the many that might be carried out. They were chosen because they were readily visualized and recorded using a simple microscope and camera. Presumably, more interesting and challenging experiments could be designed or contrived that yield more information about the physical and chemical properties that characterize the interiors of macromolecular crystals.

Conflicts of Interest: The authors declare no conflict of interest.

References

1. Larson, S.B.; McPherson, A. Investigation into the binding of dyes in protein crystals. *Acta Crystallogr. F* **2018**, *74*, 593–602.
2. McPherson, A. Penetration of dyes into protein crystals. *Acta Crystallogr. F* **2019**, *75*, 132–140. [CrossRef] [PubMed]
3. Horobin, R.W.; Kiernan, J.A. (Eds.) *Conn's Biological Stains: A Handbook of Dyes, Stains and Fluorchromes for Use in Biology and Medicine*, 10th ed.; BIOS Scientific Publishers: Oxford, UK, 2002.
4. Gurr, E. *Synthetic Dyes*; Academic Press: New York, NY, USA, 1971.
5. Harris, D.C. *Quantitative Chemical Analysis*, 7th ed.; Freeman and Co.: New York, NY, USA, 2007.
6. Skoog, D.A.; West, D.M.; Holler, F.J. *Analytical Chemistry: An Introduction*, 7th ed.; Emily Barrosse: Orlando, FL, USA, 2000; pp. 265–305.
7. Yusuf, M.; Shabbir, M.; Mohammed, F. Natural Colorants; Historical, Processing and Sustainable Prospects. *Nat. Prod. Bioprospects* **2017**, *7*, 123–145. [CrossRef] [PubMed]
8. Zumdahl, S.S. *Chemical Principles*, 6th ed.; Houghten-Mifflin Co.: Boston, MA, USA, 2009; pp. 319–324.
9. Michaelis, L.; Eagle, H. Some Redox Indicators. *J. Biol. Chem.* **1930**, *87*, 713–727.
10. McPherson, A.; Gavira, J.A. Introduction to Protein Crystallization. *Acta Crystllogr. F* **2014**, *70*, 2–20. [CrossRef] [PubMed]
11. McPherson, A.; Cudney, B. Optimization of crystallization conditions for biological macromolecules. *Acta Crystallogr. F* **2014**, *70*, 1445–1467. [CrossRef] [PubMed]
12. McPherson, A. *Preparation and Analysis of Protein Crystals*; John Wiley Co.: New York, NY, USA, 1982; p. 371.
13. McPherson, A. *Crystallization of Biological Macromolecules*; Cold Spring Harbor Laboratory Press: Cold Spring Harbor, NY, USA, 1999; p. 586.
14. McPherson, A.; Spencer, R. Preliminary structure analysis of canavalin from Jack bean. *Arch. Biochem. Biophys.* **1975**, *169*, 650–661. [CrossRef]
15. Ko, T.-P.; Ng, J.D.; McPherson, A. The three dimensional structure of canavalin from jack bean. *Plant Physiol.* **1993**, *101*, 729–744. [CrossRef] [PubMed]

16. Ko, T.-P.; Ng, J.D.; Day, J.; Greenwood, A.; McPherson, A. X-ray structure determination of three crystal forms of canavalin by molecular replacment. *Acta Crystallogr. D* **1993**, *49*, 478–489. [CrossRef] [PubMed]
17. Alderton, G.; Fevold, H.L. Direct crystallization of lysozyme from egg white and some crystalline salts of lysozyme. *J. Biol. Chem.* **1946**, *164*, 1–8. [PubMed]
18. Ko, T.-P.; Day, J.; Greenwood, A.; McPherson, A. The Structures of Three Crystal Forms of the Sweet Protein Thaumatin. *Acta Crystallogr. D* **1994**, *50*, 813–825. [CrossRef] [PubMed]
19. McPherson, A.; Cudney, B. Searching for Silver Bullets: An Alternative Strategy for Crystallizing Macromolecules. *J. Struct. Biol.* **2006**, *156*, 387–406. [CrossRef] [PubMed]
20. Koszelak, S.; Dodds, J.A.; McPherson, A. Preliminary analysis of crystals of satellite tobacco mosaic virus (STMV). *J. Mol. Biol.* **1989**, *209*, 323–326. [CrossRef]
21. Sumner, J.B. The globulins of the jack bean, Canavalia Ensiformis. *J. Biol. Chem.* **1919**, *37*, 137–145.
22. McPherson, A.; Geller, J.; Rich, A. Crystallographic studies on concanavalin B. *Biochem. Biophys. Res. Commun.* **1974**, *57*, 494–499. [CrossRef]
23. Sawyer, L.; Kontopidis, G. The core lipocalin, bovine β-lactoglobulin. *Biochem. Biophys. Acta* **2000**, *1482*, 136–148. [CrossRef]
24. Wyckoff, H.W.; Tsernoglou, D.; Hanson, A.W.; Knox, J.R.; Lee, B.; Richards, F.M. The three dimensional structure of RNase S: Interpretation of an electron density map at a nominal resolution of 2 A. *J. Biol. Chem.* **1970**, *245*, 305–325. [PubMed]

crystals

MDPI

Article

From Initial Hit to Crystal Optimization with Microseeding of Human Carbonic Anhydrase IX—A Case Study for Neutron Protein Crystallography

Katarina Koruza [1,*], Bénédicte Lafumat [1], Maria Nyblom [1], Wolfgang Knecht [1] and Zoë Fisher [1,2,*]

1 Department of Biology and Lund Protein Production Platform, Lund University, Sölvegatan 35, 22362 Lund, Sweden; benedicte.lafumat@gmail.com (B.L.); maria.gourdon@biol.lu.se (M.N.); wolfgang.knecht@biol.lu.se (W.K.)

2 Scientific Activities Division, European Spallation Source ERIC, Box 176, 22100 Lund, Sweden

* Correspondence: katarina.koruza@biol.lu.se (K.K.); zoe.fisher@esss.se (Z.F.)

Received: 22 October 2018; Accepted: 17 November 2018; Published: 20 November 2018

Abstract: Human carbonic anhydrase IX (CA IX) is a multi-domain membrane protein that is therefore difficult to express or crystalize. To prepare crystals that are suitable for neutron studies, we are using only the catalytic domain of CA IX with six surface mutations, named surface variant (SV). The crystallization of CA IX SV, and also partly deuterated CA IX SV, was enabled by the use of microseed matrix screening (MMS). Only three drops with crystals were obtained after initial sparse matrix screening, and these were used as seeds in subsequent crystallization trials. Application of MMS, commercial screens, and refinement resulted in consistent crystallization and diffraction-quality crystals. The crystallization protocols and strategies that resulted in consistent crystallization are presented. These results demonstrate not only the use of MMS in the growth of large single crystals for neutron studies with defined conditions, but also that MMS enabled re-screening to find new conditions and consistent crystallization success.

Keywords: crystallization; microseed matrix screening; seeding; optimization; human carbonic anhydrase IX; neutron protein crystallography

1. Introduction

Protein crystallization is based on creating a supersaturated solution of a macromolecule with the addition of precipitants (neutral salts, high molecular weight polymers, organic solvents, polyalcohols) and by manipulation of crystallization conditions, including pH, incubation temperature, increased ionic strength, alteration of the dielectric constant of the medium, volume exclusion by polymers, and chemical/biochemical modification of proteins. Different approaches have been developed to promote crystallization, and among the most widely used are sitting or hanging drop vapor diffusion, batch, dialysis, and counter-diffusion techniques [1,2].

Growing large crystals for neutron protein crystallography (NPX) presents a major challenge for structural biologists interested in using neutron scattering. Current neutron instruments for NPX are limited in neutron flux and instrument geometry. This imposes limitations on the size of the crystallographic unit cell parameters, asymmetric unit volume, and it requires a very large overall crystal volume compared to what is sufficient for X-ray crystallographic studies. The general recommendation is to aim for crystals ~1 mm^3 in volume with ~30 out of 160 examples coming from crystals smaller than this [3,4]. A survey of deposited NPX structures in the Protein Data Bank (PDB) reveals that most large crystals were grown using vapor diffusion. A drop of protein is mixed with precipitant and equilibrated against the reservoir in a sealed environment. In the sitting drop format it can be scaled up to almost any volume and has been reported to be successful in 100–1000 µL drops [5–7]. Batch crystallization has also been used, and here the protein is mixed with precipitant and optionally covered with a layer of oil. In batch, the supersaturation point is reached at the moment of drop preparation, meaning the process does not rely on evaporation [8]. In both techniques, it is challenging to control nucleation and subsequent crystal growth in a consistent and repeatable way.

Seeding is a very powerful strategy, as it allows optimization of crystal growth conditions independently from conditions needed for initial nucleation. Examples in the literature show that seeding in its variations can increase the number of crystallization hits and can significantly shorten the crystallization time [9–11]. It is also possible to incorporate seeding into any crystallization set-up. Seeding is a technique where crystals are crushed in mother liquor or reservoir solution and diluted to create a seed stock that is added to crystallization solutions to aid with nucleation [12,13]. As illustrated in Figure 1, spontaneous nucleation cannot occur in an undersaturated solution, below the solubility line; therefore, seeds introduced in this phase dissolve. Spontaneous homogeneous nucleation occurs in the supersaturation zone, but it can be a long process that takes several months. The supersaturation is low and suitable for crystal growth in the metastable zone, but here nucleation does not occur spontaneously. This zone is ideal for seeding and crystal growth, and seeding can remarkably speed up the crystallization trajectory and provide more consistent control and results [10,11].

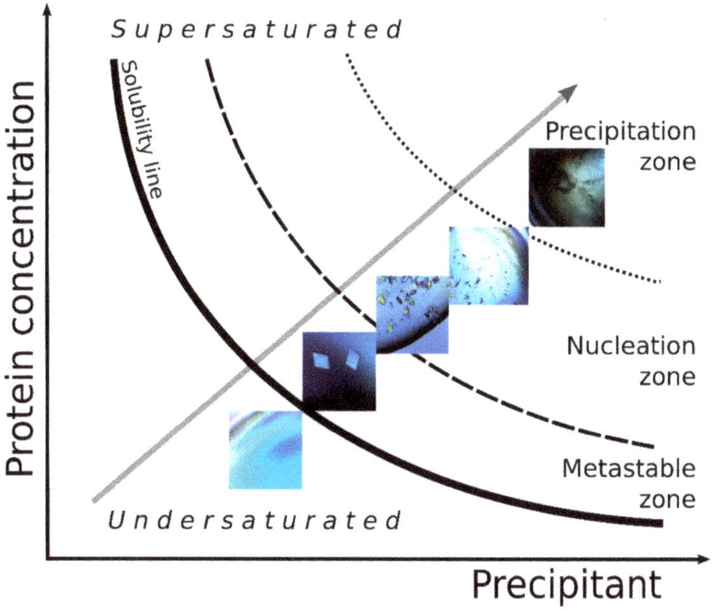

Figure 1. A schematic representation of a phase diagram showing the solubility of a protein in solution as a function of the concentration of the precipitant. The grey arrow illustrates the estimated path of crystallization. The path leads from the undersaturated zone below the solubility curve, where no crystals can grow, to the precipitation zone, where supersaturation is too high and protein precipitates without forming crystals. Below the precipitation zone is the nucleation zone, where the supersaturation is high and nucleation is observed, but the crystal growth is slow. In the metastable zone, the nucleation does not occur spontaneously, but the supersaturation is low and suitable for crystal growth [14].

In the nucleation and precipitation zone where spontaneous nucleation occurs, seeding often results in excessive (over)nucleation and can result in the formation of showers of microcrystals or amorphous precipitate [14]. Additionally, the size and number of crystals can be manipulated by systematically using a dilution series of seed stock until an ideal amount of seedstock for a given batch can be identified [15]. This technique can not only help to grow crystals where none grew before, but it can also generate better-diffracting crystals, since crystals are more likely to grow in the metastable zone [14].

Our work is focused on NPX, and for that we need optimized methods to reliably grow large crystals from large volume drops (typically >100 μL). This is due to the inherent low flux of neutron sources, and crystals have to be 2–3 orders of magnitude larger in volume than what is typical for X-ray crystallography (e.g., ~1 vs. ~0.01 mm^3). Neutrons are sensitive to H atoms and to its isotope, deuterium (D), whose scattering lengths differ in both magnitude and phase [16]. Partial (including H/D exchange) or full (per)deuteration of proteins presents its own challenges, but it reduces the incoherent scattering background from H, significantly increases the signal-to-noise ratio of the diffraction data, and also makes it possible to use up to an order of magnitude smaller crystals [3,17]. In addition to the requirement of having large, deuterated protein crystals for NPX, there are also limitations to the size of the unit cell that current neutron macromolecular beamlines can resolve. Instruments have different design features that enable some to resolve unit cell parameters of 100–150 Å on edge (IMAGINE, LADI-III, BioDIFF, iBIX), while MaNDi is designed to resolve unit cells up to 300 Å on edge [4,18]. This limitation remains a challenge, and while a large unit cell can be overcome by having a large crystal, there are very few examples of neutron structures determined from crystals with unit cells over 150 Å on any edge. Unit cell parameters are then an important consideration when pursuing neutron studies of a given system. Just as a relatively large unit cell can be overcome by having a large crystal, the converse is also true, in that it becomes more feasible to collect neutron data from a much smaller crystal if the unit cell is also small [3,16,19].

In the work presented here, we worked with both hydrogenous and deuterated versions of the same protein. As large crystals can take a long time to grow, it is necessary to use set-ups that are stable for long incubation and equilibration periods. It is desirable to control the thermodynamic and kinetic contribution to the supersaturated state where the protein crystal nucleates, and this can be done by controlling temperature, evaporation rates, and pH, protein, and precipitant concentrations. Automation and systematic screening gives better control of the supersaturation state by having the possibility to screen a large area around crystallization hits. Seeding, in conjunction with large drop volume, provides a way to speed up crystal growth, as well as a way to control the number of crystals in a given drop.

The protein we are working with is human carbonic anhydrase IX (CA IX). CA IX is implicated in cancer metastasis and has emerged as a target for cancer detection, imaging, and treatment [20]. We are using neutrons to obtain details of the enzyme active site with regards to hydrogen bonding, water organization, and ligand binding interactions that will enable rational drug design efforts [16]. However, the native protein is 459 amino acids (UniProtKB–Q16790) with multiple domains, including a membrane spanning domain [20]. There are two reports on the structure of the catalytic domain of native CA IX, but the unit cell and parameters are unsuitable for neutron studies [21,22]. The most recent is from CA IX prepared by expression in yeast, and the space group was H3 with unit cell parameters: a = b = 152.7 Å, c = 170.7 Å; $\alpha = \beta = 90°$, $\gamma = 120°$ (PDB ID 6fe2) [22]. The first structure that was reported was also for the catalytic domain of CA IX, but produced in insect cells and crystallized in $P6_1$ with unit cell parameters: a = b = 144.2 Å, c = 208.9 Å; $\alpha = \beta = 90°$, $\gamma = 120°$ (PDB ID 3iai) [21]. Due to the previously mentioned challenges regarding size and unit cell volumes, we are, therefore, working with a construct composed of residues 140–395 of human CA IX, that also contains six mutations engineered at the surface of the protein (hence called surface variant: CA IX SV). The mutated residues C174S, L180S, A210K, A258K, F259Y, and M360S were designed to make the catalytic domain more soluble, suitable for expression in *E. coli*, and also to be more amenable to crystallization [23].

While surface mutations are introduced away from the highly conserved active site of the enzyme, the catalytic efficiency of the SV enzyme is decreased in comparison with native protein, as described by Mahon et al. [23]. The CA IX SV was previously reported to crystallize in $P2_12_12_1$ with unit cell parameters ~100 Å on edge using 8% PEG 8000, 0.1 M Tris pH 8.5, and as such is more suitable for neutron studies than working with the native protein [21–23]. Despite extensive efforts to reproduce these conditions, we were unable to obtain any crystals. Here we present the details of starting over with nanoliter drops in high-throughput set-ups with commercial screens and going from a single initial hit to optimization for consistent growth of diffraction-quality single crystals. A crystallization strategy of microseed matrix screening (MMS), as described by Ireton and Stoddard, was used [24]. In this method, crystal seeds or nuclei from one crystallization condition are added systematically to a matrix of various conditions to screen for new conditions that promote crystal growth [24].

2. Materials and Methods

2.1. Protein Preparation

The catalytic domain of human CA IX was expressed in *E. coli* BL21(DE3) under hydrogenous and deuterated conditions according to a protocol described in detail elsewhere [23,25]. In this previous work the focus was to optimize deuteration strategies to produce different levels of D incorporation and to measure biophysical effects on the resulting proteins. To measure how deuteration affects crystallization behavior, we chose a single condition from this work to set-up side-by-side crystallization of H and D versions of the same proteins [25]. The protein was previously engineered to have 6 surface mutations that provide a more soluble and stable enzyme, which we refer to as the CA IX surface variant (CA IX SV) [23]. CA IX SV was purified using para-aminobenzenesulfonamide resin (pAMBS, Sigma-Aldrich, Stockholm, Sweden), followed by size exclusion chromatography (Superdex200 16/600, GE Healthcare, Uppsala, Sweden). The protein elutes in 50 mM Tris pH 8.5, 100 mM NaCl in 2 peaks. Both peaks 1 and 2 were tested for crystallization, and only peak 2 yielded any crystals [25]. Fractions containing peak 2 were pooled and concentrated to 12–17 mg/mL, depending on the preparation batch.

2.2. Crystallization

Published crystallization conditions around 8% PEG 8000, 0.1 M Tris pH 8.5 for concentrations of CA IX SV ranging from 10.8 mg/mL to 17.5 mg/mL, were used to set up crystallization screens at different temperatures (4, 18 °C, ambient) in vapor diffusion experiments [23]. To find new conditions, high-throughput screening was performed with a Mosquito (TTP Labtech, Melbourn, UK) crystallization robot for droplets with a final volume of 300 nL (protein:precipitant ratio of 1:1, reservoir volume 40 μL), or with an Oryx8 (Douglas instruments, Hungerford, UK) in vapor diffusion, or under-oil microbatch drops (protein:precipitant:seeds ratio of 3:2:1, final drop volume 0.6 μL, reservoir volume 40 μL). Four commercial crystallization screens were used in the initial rounds: JCSG+ (Molecular Dimensions, Newmarket, UK), Morpheus (Molecular Dimensions, Newmarket, UK), PurePEGs (Anatrace, Maumee, OH, USA), and TOP96 (Anatrace, Maumee, OH, USA). MRC crystallization plates from Molecular Dimensions were used for sitting drop format, while for the microbatch under-oil set-up we used hydrophilic plasma treated plates (Douglas Instruments, Hungerford, UK). Oils used were paraffin oil, silicon oil, and Al's oil (1:1 mix of paraffin and silicon oil). In manual set-ups for larger hanging and sitting drops (protein:precipitant:seeds ratio of 3:2:1, final drop volume 3–24 μL, reservoir volume 1 mL), 24-well Linbro plates (Hampton Research, Aliso Viejo, CA, USA) were used. Crystallization drops were always mixed in order, precipitant–protein–seeds, using freshly prepared buffers and PEG solutions (*w*/*v*).

Seed stocks for microseeding experiments were prepared according to Seed Bead™ Kit instructions (https://www.hamptonresearch.com; collected 2018-10-06) using reservoir solution from the crystallization condition. For seeding optimization the stock was tested in a dilution series. In between crystallization experiments the seed stocks were stored in a −80 °C freezer. Cross-seeding was attempted with human CA isoform II crystal seeds [26]. Seeding was used for both manual and automated experiments with Oryx8 (Douglas instruments). All crystallization plates were incubated at 20 °C and inspected under polarizing light microscope, or with the Minstrel HT UV imaging system (Rigaku). Hydrogenous and deuterated proteins were crystallized using the same hydrogenous precipitant solutions with no adaptation. Table 1 and Figure 2 show summaries of all conditions eventually identified. Crystals were visually inspected and evaluated based on relative size and quality (single, not aggregated, clean edges, no or little precipitation in the drops).

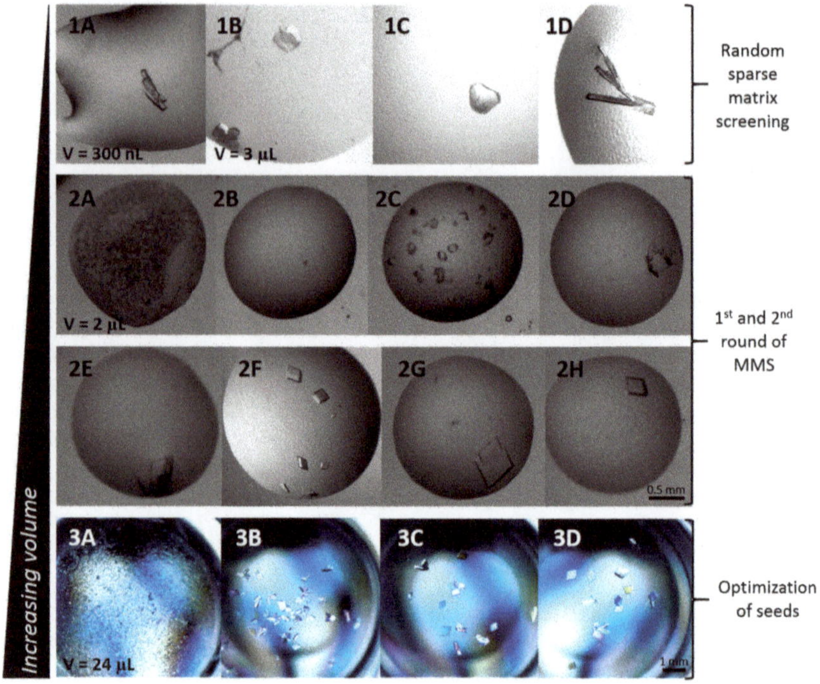

Figure 2. Crystallization of CA IX SV from initial hit to optimized crystals used for diffraction. Drop volumes are indicated per row. The initial screening hit appeared in (**1A**) 20% PEG 3350, 0.2 M ammonium formate (JCSG+). Systematic alternation of the initial conditions was performed in hanging drop set-up. Observed crystals were used for two seed stocks. The first seed stock was prepared from (**1B**) 0.1 M Tris pH 7.5, 22% PEG 3350, 0.2 M NaCl, and the second seed stock was from a mixture of (**1C**) 0.1 M Tris pH 7.0, 20% PEG 4000, 0.2 NaCl and (**1D**) 0.1 M Tris pH 7.5, 20% PEG 6000, 0.2 M NaCl. The two seed stocks were diluted 10× and seeded into JCSG+, Morpheus, PurePEGs, and TOP96 screens for a second round of MMS. The two rounds of MMS resulted in numerous new conditions (**2A–H**): (**2A**) 0.1 M sodium cacodylate pH 6.5, 1 M sodium citrate, (**2B**) 0.1 M ammonium citrate pH 5, 22.5% PurePEGs Cocktail, 0.3 M cesium chloride, (**2C**) 0.1 M Tris pH 5.5, 25% PEG 3350, 0.2 M magnesium chloride, (**2D**) 0.1 M HEPES pH 7.5, 25% PEG 3350, 0.2 M ammonium acetate, (**2E**) 0.1 M HEPES pH 7.5, 20% PEG 4000, 10% 2-propanol, (**2F**) 0.1 M Tris pH 8.5, 30% PEG 4000, 0.2 M sodium acetate, (**2G**) 0.1 M Tris pH 8.5, 25% PEG 3350 and (**2H**) 0.1 M HEPES pH 7.5, 25% PEG 3350. The crystallization condition that was used to prepare crystals of the H/D exchanged and deuterated (D/D) CA IX SV protein corresponds to (**2F**). When scaling up the drop size, seed concentration had to be optimized. Seeds stocks used were diluted 10× (**3A**), diluted 100× (**3B**), 1000× (**3C**), and 10,000× (**3D**) in 24 μL drops.

Table 1. Summary of crystallization conditions that yielded crystals.

Precipitant	Buffer	pH	Additive/Extra	Crystallization Set-Up	Correlation to Figure 2
20% PEG 3350	None	n/a	0.2 M Ammonium Formate	Hanging drop	1A [1]
20–24% PEG 3350	0.1 M Tris	7.0	None	Sitting drop	n/a
25% PEG 3350	0.1 M HEPES	7.5	None	Sitting drop	2H
20–24% PEG 3350	0.1 M Tris	8.0	None	Sitting drop	n/a
25% PEG 3350	0.1 M Tris	8.5	None	Sitting drop	2G
25% PEG 3350	0.1 M Tris	5.5	0.2 M Magnesium Chloride	Sitting drop	2C
20–26% PEG 3350	0.1 M Tris	7.0	4–10% 2-propanol	Sitting drop	n/a
25% PEG 3350	0.1 M HEPES	7.5	0.2 M Ammonium Acetate	Sitting drop	2D
22% PEG 3350	0.1 M Tris	7.5	0.2 M Sodium Chloride	Hanging drop, microbatch	1B [1]
20–26% PEG 3350	0.1 M Tris	8.0	4–10% 2-propanol	Sitting drop	n/a
18–22% PEG 4000	0.1 M Tris	7.0	None	Sitting drop	n/a
18–22% PEG 4000	0.1 M Tris	7.0	0.2 M Sodium Chloride	Hanging drop, microbatch	1C [1]
20% PEG 4000	0.1 M HEPES	7.5	10% 2-propanol	Sitting drop	2E
22–26% PEG 4000	0.1 M Tris	8.0	10% 2-propanol	Sitting drop	n/a
30% PEG 4000	0.1 M Tris	8.5	0.2 M Sodium Acetate	Sitting drop, microbatch	2F, 3A–D
18–22% PEG 6000	0.1 M Tris	7.5	None	Sitting drop	n/a
18–22% PEG 6000	0.1 M Tris	7.5	0.2 M Sodium Chloride	Hanging drop	1D [1]
22.5% PurePEGs Cocktail	0.1 M Citric Acid	3.5	0.3 M Sodium Formate	Sitting drop	n/a
22.5% PurePEGs Cocktail	0.1 M Ammonium Citrate	5.0	0.3 M Cesium Chloride	Sitting drop	2B
22.5% PurePEGs Cocktail	0.1 M Sodium Cacodylate	6.5	0.3 M Magnesium Chloride	Sitting drop	2A
1.0 M Sodium Citrate	0.1 M Sodium Cacodylate	6.5	None	Sitting drop	n/a

[1] These are the only crystallization conditions where spontaneous nucleation occurred.

2.3. Deuterium Labeling and X-ray Analysis

Crystals of hydrogenous CA IX SV were grown under hydrogenous conditions and were kept in hydrogenous buffers during the experiment (H/H CA IX SV). In parallel, crystals of H/H CA IX SV were subjected to H/D exchange by exchanging the well solution for deuterated solutions and resealed for several weeks to allow labile hydrogen atoms to exchange for deuterium (H/D CA IX SV). Deuterated CA IX SV was crystallized with hydrogenous buffers and then later exchanged to regain lost D atoms in labile positions (D/D CA IX SV). Crystals were cryoprotected by quick dipping in reservoir solution supplemented with glycerol (20% *v*/*v* final concentration) and flash frozen in liquid nitrogen. Crystals were tested for diffraction at the BioMAX beamline at the MAX IV laboratory. Diffraction data statistics are shown in Table 2. Structure refinement is ongoing.

Table 2. X-ray diffraction data set statistics for CA IX SV.

	H/H	H/D	D/D
Wavelength (Å)	0.979	0.979	0.979
Beamline, source	BM-30 (FIP), ESRF	BioMAX, MAX IV laboratory	BioMAX, MAX IV laboratory
Space group, unit cell	$P2_1$; a = 44.3, b = 65.1, c = 46.7; β = 115.1	$P2_1$; a = 44.5, b = 65.4, c = 46.7; β = 115.1	$P2_1$; a = 44.4, b = 65.1, c = 46.6; β = 114.7
Resolution range (Å)	50.0–1.77 (1.88–1.77)	40.0–1.39 (1.42–1.39)	40.0–1.49 (1.51–1.49)
Total No. of reflections	63,296 (9199)	327,912 (15,926)	265,165 (13,184)
No. of unique reflections	22,187 (3463)	48,352 (2406)	39,461 (1948)
Redundancy	2.8 (2.6)	6.8 (6.6)	6.7 (6.8)
Completeness (%)	95.1 (92.9)	99.8 (99.1)	99.9 (99.3)
$\langle I/\sigma(I)\rangle$	10.5 (2.2)	19.9 (2.2)	14.3 (2.2)
R_{merge} † (%)	6.3 (48.6)	3.6 (78.1)	6.0 (78.1)

† $R_{merge} = (\Sigma \mid I - \langle I \rangle \mid / \Sigma \langle I \rangle) \times 100$.

3. Results & Discussion

3.1. Initial Screening

Narrow screens for CA IX SV, ranging from 6 to 10% PEG 8000 (0.1 M Tris pH 8.5) around published crystallization conditions (8% PEG 8000, 0.1 M Tris pH 8.5), were used to set up hanging drops at different temperatures (4, 12, 18 °C, ambient) [23]. Despite extensive efforts to recreate the reported crystallizations, we had no success. Protein expression and purification was attempted exactly as previously published but did not yield any crystals [23]. The CA IX SV construct shares 36% of its sequence identity with CA II; however, an attempt to do cross-seeding with human CA isoform II crystals did not result in any hits. In a recent study, Abuhammad et al. reported successful crystallization, increased number of hits, and shorter crystallization times when seeds from protein crystals with sequence identities as low as 24% were used [9]. The search for new conditions started with random sparse matrix screening using two commercial screens, JCSG+ and Morpheus (both from Molecular Dimensions). Small drops were prepared with a Mosquito (TTP Labtech) crystallization robot in a sitting drop format using 96-well MRC crystallization plates. The first and only crystallization hit was from JCSG+ (20% PEG 3350, 0.2 M ammonium formate pH 6.5) and was observed after 30 days (Figure 2(1A)). This hit condition formed the basis of systematic alternation of the initial conditions where 16–24% PEG 3350 (or 4000 or 6000 or 8000) with either 0.2 M NaCl or 0.2 M ammonium formate were included. The pH was also varied in 0.5 unit increments from pH 7.0–9.0. The drops were hanging drop and had a final volume of 3 µL and a protein:precipitant ratio of 1:1. Crystals appeared in three of these drops after ~65 days from PEG 3350, 4000, and 6000 (20–22%) supplemented with 0.2 M NaCl and 0.1 M Tris pH 7.0/7.5 (Figure 2(1B–1D)). There were no spontaneous crystals appearing at the reported higher pH of 8.5 or with PEG 8000 in these initial screens.

3.2. Microseed Matrix Screening (MMS)

The drop shown in Figure 2(1B) was sacrificed for seed stock preparation with its reservoir solution (0.1 M Tris pH 7.5, 22% PEG 3350, 0.2 M NaCl), and this stock was used in the first round of MMS using four commercial screens: JCSG+, Morpheus, PurePEGs, and TOP96. The volume of seeds added to the screens was 0.1 µL. The crystals from Figure 2(1C,1D) were later used to prepare a second seed stock using 0.1 M Tris pH 7.0, 20% PEG 4000, 0.2 NaCl, and were also used as an additive for automated microseeding experiments with the Oryx8 (Douglas instruments). Seeding experiments resulted in the identification of numerous new conditions out of the 384 tested (summarized in Table 1; some shown in Figure 2(2A–2H)).

The expanded hits covered a broad range of PEG sizes (Table 1). PEG 3350 appeared on ten occasions, PEG 4000 on five, PEG 6000 on two, and PurePEGs cocktail on three occasions, and ranged between 18% and 30% (w/v). A number of additional additives, including magnesium chloride, sodium acetate, ammonium acetate, and 2-propanol also appeared, and seven hits did not contain any (Table 1; Figure 2(2G)). We also found a non-PEG condition with sodium citrate, but due to the crystal appearance, we did not explore it further (Figure 2(2A)). A broad range of pHs also became accessible from the expanded screen results, with CA IX SV crystallizing between pH 3.5–8.5 (Table 1).

The distribution of the number of crystallizations of all the PEG conditions is shown in Figure 3 as a function of pH, PEG size, PEG concentration, and additives. The highest number of hits appears to group around conditions with pH above 7, in range of 18–26% PEG 3350 or 4000, without additive or with NaCl or isopropanol. We did not observe differences in conditions yielding crystals between batch or vapor diffusion set-ups in the smaller volumes (0.3–15 µL) (Figure 4). The outcome of the automated seeding experiments was firstly a control over consistent nucleation at lower levels of supersaturation. Secondly, the conditions where crystals grew varied from the conditions where crystal nucleation occurred and actually resulted in new hits yielding crystals that had different shapes. In terms of time, optimization without seeds would, for some batches of fresh protein, never yield crystals, while seeding in small volumes reduced the time to get crystals from weeks or even months to only days.

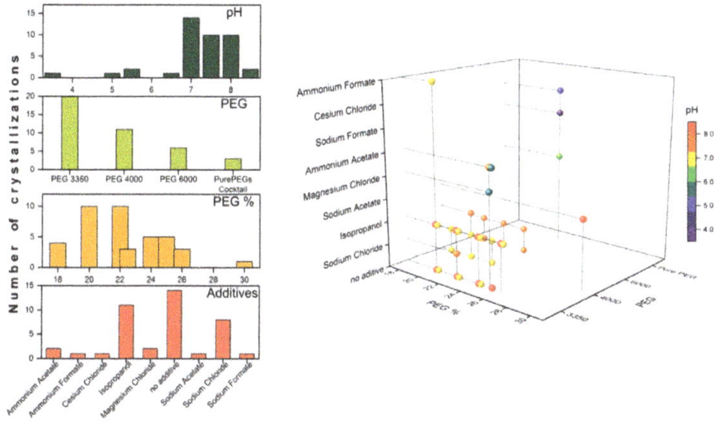

Figure 3. Distribution of the number of successful protein crystallizations of all PEG conditions found in systematic screening as a function of pH, size of the PEG, PEG %, and which additives were included. A 3D graph that displays all of the components of the PEG conditions is shown on the right.

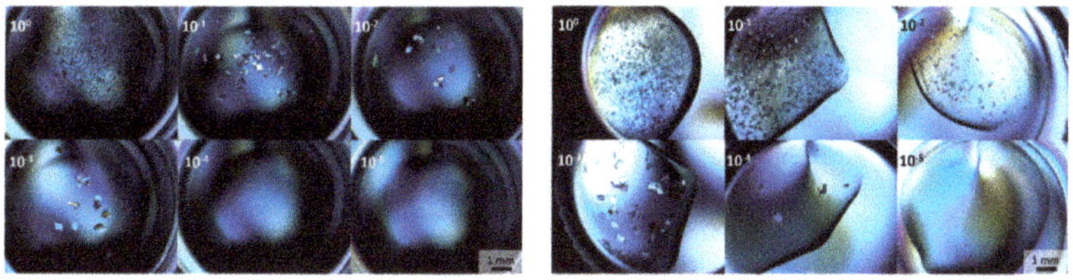

(**a**) (**b**)

Figure 4. The effect of diluting and testing the seed stock solution. Crystallization drops were set up in 0.1 M Tris pH 8.5, 30% PEG 4000, 0.2 M sodium acetate in a ratio of 3:2:1 (protein:precipitant:seeds) in (**a**) vapor diffusion sitting drop set-up (total drop volume is 24 μL) and (**b**) microbatch with paraffin oil set-up (total drop volume is 15 μL).

3.3. Optimization with Seed Concentration

The conditions summarized in Table 1 and Figure 2 that resulted in visually the best crystals were scaled up in volumes from 3 to 24 μL, initially. Crystal quality was ranked based on being single, having clear edges, and containing no inclusions. The Hampton Research recommended ratio of seed stock to use is 3:2:1 (protein:precipitant:seeds), and while this worked well in smaller drops, when we scaled up to 24 μL we observed overnucleation when using 10× diluted seed stocks (Figure 2(3A)). The high number of seeds introduced into the crystallization drop resulted in numerous small crystals, but yielded improved results, i.e., fewer and larger crystals per drop (Figure 2(3C,3D)) when prepared in a 1000× and 10,000× dilution series. The condition that produced the first good crystals selected for X-ray diffraction testing was 22% PEG 3350, 0.2 M NaCl, 0.1 M Tris pH 7.5 (Figure 2(1B)). However, it was later observed that this condition, when using different protein batches, was not consistently successful in giving any crystals, let alone for large crystal growth. After some additional trials, the condition that consistently produced the largest crystals, regardless of the protein batch and length of protein storage, was 30% PEG 4000, 0.2 M sodium acetate, 0.1 M Tris pH 8.5 (Figure 2(2F)).

This condition was successfully used to prepare crystals of the H/D exchanged and deuterated (D/D) CA IX SV protein. Crystals from numerous conditions were selected for X-ray diffraction, but this condition gave the best diffraction data sets for H/H, H/D, and D/D CA IX SV. X-ray diffraction data collection statistics are shown in Table 2, and the refinement and structures are ongoing and will be reported elsewhere.

Through our MMS approach, we were not only able to gain consistent control over nucleation and crystal growth, we also obtained a new space group ($P2_1$) with smaller unit cell parameters than the previously reported $P2_12_12_1$; a = 44.5, b = 65.4, c = 46.7 Å, β = 115.1° vs. a = 57.9, b = 102.7, c = 109.0 Å (Table 2) [23]. The monoclinic space group reported here has a monomer in the asymmetric unit (ASU) compared to a non-crystallographic dimer in the orthorhombic space group. This fortuitous discovery is very beneficial for planned NPX experiments where large crystals with smaller unit cells are required.

3.4. Deuterated CA IX SV and Large Volume Crystallization

It has been demonstrated for different proteins, including different human CA proteins, that side-by-side crystallization with the same conditions produced variable outcomes depending on the deuteration level of the protein [25–27]. Initial structural analysis shows that the resulting X-ray crystal structures are unchanged, but the optimal crystallization conditions can be affected and may have to be adjusted for deuterated protein crystal growth.

For neutron protein crystallography we need substantially larger crystals, one to two orders of magnitude larger in volume than what is routinely used for X-ray diffraction. We attempted larger drop set-ups with the best conditions used for X-rays (e.g., Figure 2(2E–2H)); however, scaling up to 50–150 µL with diluted seeds did not reproduce the results from the smaller drops, and conditions had to be re-explored and refined in larger drops. We were finally able to grow a large crystal of H/H CA IX SV from a 150 µL drop using 24% PEG 4000, 10% (*v*/*v*) 2-propanol, 0.1 M Tris pH 8.0 (Figure 5). We can speculate that, in addition to the chemical parameters (precipitant, pH, and salts), this condition promotes crystallization in large volume by manipulating the dielectric properties of the solution by presence of 2-propanol. All of these factors together successfully created a supersaturated solution and resulted in growth of large single crystals of CA IX SV in nine to 12 months.

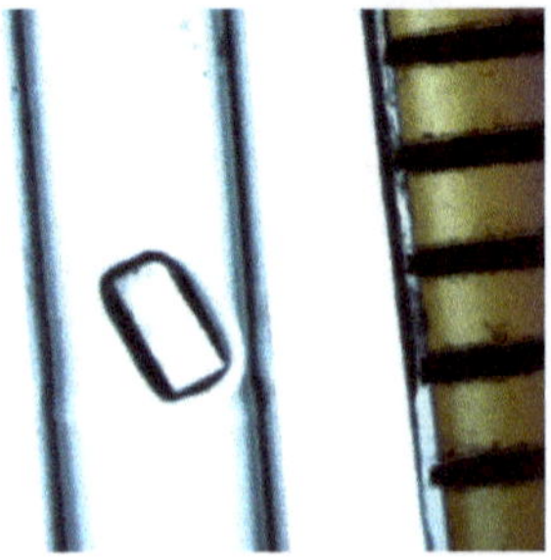

Figure 5. CA IX SV crystal mounted in a capillary. Approximate crystal dimensions are 1.3 × 0.8 × 0.8 mm (volume ~0.8 mm^3), lined ruler visible is graduated in mm increments.

4. Conclusions

For this challenging protein, the combination of microseeding, where crystal growth is induced at low levels of supersaturation, and random sparse matrix screening yielded numerous conditions over a broad range of PEGs, additives, and pH. Additionally, the microseeding allowed us to get diffraction quality crystals for different batches of H/H, H/D, and D/D CA IX SV with a limited amount of material and in a reasonable time frame. Conditions that appeared as the result of automated high-throughput screening in small volumes (300 nL) did not scale as expected. Some screening and refining was necessary when volumes were increased >10-fold from the initial hit volume. The amount and size of the PEG varied greatly, from 18–30% PEG and sizes 3350, 4000, and 6000 all gave good crystals. Throughout our screening efforts, we did not find a condition with PEG larger than 6000. The type of additive did not seem to affect crystal morphology or how fast they appeared in small volumes (0.3–6 μL). However, for scaling up the crystallization drop volume beyond 50 μL, isopropanol was a better additive than the others. An unexpected bonus coming out of these studies was that we were able to find new conditions that gave us consistent control over crystallization. We prepared well-diffracting crystals in a space group with smaller unit cell parameters than are suitable for neutron diffraction experiments. In short, our work here demonstrates the necessity of broad screening and subsequent continuous optimization while scaling up crystallization volumes as a pre-requisite for preparing CA IX SV crystals suitable for NPX. This approach may be beneficial and generally applicable to other challenging targets, as well, especially where larger crystals are required to enable X-ray or neutron diffraction experiments.

Author Contributions: Conceptualization: K.K., W.K. and Z.F.; formal analysis: K.K., B.L., M.N. and Z.F.; investigation: K.K., B.L. and M.N.; methodology: K.K., B.L. and M.N.; supervision: Z.F.; writing—original draft: K.K.; writing review and editing: K.K., M.N., W.K. and Z.F.

Funding: This research was partly funded by: Integrated Infrastructure Initiative No. 262348 European Soft Matter Infrastructure, SINE2020. We thank Lund University, the Royal Physiographic Society of Lund, Interreg/MAX4ESSFUN, The Crafoord Foundation No. 20160528, and BioCARE (a strategic research area at Lund University) for financial support.

Acknowledgments: The authors would like to thank the MAX IV laboratory (BioMAX) beamline scientists for expert assistance, and the Lund Protein Production Platform (LP3) staff for providing technical support for experiments and for data collection. We would also like to extend sincere thanks to Motoyasu Adachi from the National Institutes for Quantum and Radiological Science and Technology (Tokai, Japan) for useful discussions and advice.

Conflicts of Interest: The authors declare no conflict of interest. The funders had no role in the design of the study; in the collection, analyses, or interpretation of data; in the writing of the manuscript, or in the decision to publish the results.

References

1. Ng, J.D.; Gavira, J.A.; García-Ruíz, J.M. Protein crystallization by capillary counterdiffusion for applied crystallographic structure determination. *J. Struct. Biol.* **2003**, *142*, 218–231. [CrossRef]
2. McPherson, A.; Gavira, J.A. Introduction to protein crystallization. *Acta Crystallogr. F Struct. Biol. Cryst. Commun.* **2014**, *70*, 2–20. [CrossRef] [PubMed]
3. Blakeley, M.P.; Hasnain, S.S.; Antonyuk, S.V. Sub-atomic resolution X-ray crystallography and neutron crystallography: Promise, challenges and potential. *IUCrJ* **2015**, *2*, 464–474. [CrossRef] [PubMed]
4. Meilleur, F.; Coates, L.; Cuneo, M.; Kovalevsky, A.; Myles, D. The neutron macromolecular crystallography instruments at Oak Ridge National Laboratory: Advances, challenges, and opportunities. *Crystals* **2018**, *8*, 388. [CrossRef]
5. O'Dell, W.B.; Swartz, P.D.; Weiss, K.L.; Meilleur, F. Crystallization of a fungal lytic polysaccharide monooxygenase expressed from glycoengineered Pichia pastoris for X-ray and neutron diffraction. *Acta Crystallogr. F Struct. Biol. Cryst. Commun.* **2017**, *73*, 70–78. [CrossRef] [PubMed]
6. Ohlin, M.; von Schantz, L.; Schrader, T.E.; Ostermann, A.; Logan, D.T.; Fisher, S.Z. Crystallization, neutron data collection, initial structure refinement and analysis of a xyloglucan heptamer bound to an engineered carbohydrate-binding module from xylanase. *Acta Crystallogr. F Struct. Biol. Cryst.* **2015**, *71*, 1072–1077. [CrossRef] [PubMed]
7. Manzoni, F.; Saraboji, K.; Sprenger, J.; Kumar, R.; Noresson, A.-L.; Nilsson, U.J.; Leffler, H.; Fisher, S.Z.; Schrader, T.E.; Ostermann, A.; et al. Perdeuteration, crystallization, data collection and comparison of five neutron diffraction data sets of complexes of human galectin-3C. *Acta Crystallogr. D Struct. Biol.* **2016**, *72*, 1194–1202. [CrossRef] [PubMed]
8. Gerlits, O.O.; Coates, L.; Woods, R.J.; Kovalevsky, A. Mannobiose binding induces changes in hydrogen bonding and protonation states of acidic residues in Concanavalin A as revealed by neutron crystallography. *Biochemistry* **2017**, *56*, 4747–4750. [CrossRef] [PubMed]

9. Abuhammad, A.; McDonough, M.A.; Brem, J.; Makena, A.; Johnson, S.; Schofield, C.J.; Garman, E.F. "To Cross-Seed or Not To Cross-Seed": A Pilot Study Using Metallo-β-lactamases. *Cryst. Growth Des.* **2017**, *17*, 913–924. [CrossRef]
10. Obmolova, G.; Malia, T.J.; Teplyakov, A.; Sweet, R.W.; Gilliland, G.L. Protein crystallization with microseed matrix screening: Application to human germline antibody Fabs. *Acta Crystallogr. F Struct. Biol. Cryst.* **2014**, *70*, 1107–1115. [CrossRef] [PubMed]
11. D'Arcy, A.; Mac Sweeney, A.; Haber, A. Using natural seeding material to generate nucleation in protein crystallization experiments. *Acta Crystallogr. D Biol. Crystallogr.* **2003**, *59*, 1343–1346. [CrossRef] [PubMed]
12. Blakeley, M.P.; Kalb, A.J.; Helliwell, J.R.; Myles, D.A. The 15-K neutron structure of saccharide-free Concanavalin A. *Proc. Natl. Acad. Sci. USA* **2004**, *101*, 16405–16410. [CrossRef] [PubMed]
13. Gavira, J.A.; Hernandez-Hernandez, M.A.; Gonzalez-Ramirez, L.A.; Briggs, R.A.; Kolek, S.A.; Shaw Stewart, P.D. combining counter-diffusion and microseeding to increase the success rate in protein crystallization. *Cryst. Growth Des.* **2011**, *11*, 2122–2126. [CrossRef]
14. Asherie, N. Protein crystallization and phase diagrams. *Methods* **2004**, *34*, 266–272. [CrossRef] [PubMed]
15. Luft, J.R.; DeTitta, G.T. A method to produce microseed stock for use in the crystallization of biological macromolecules. *Acta Crystallogr. D Biol. Crystallogr.* **1999**, *55*, 988–993. [CrossRef] [PubMed]
16. Oksanen, E.; Chen, J.C.; Fisher, S.Z. Neutron crystallography for the study of hydrogen bonds in macromolecules. *Molecules* **2017**, *22*, 596. [CrossRef] [PubMed]
17. Fisher, Z.; Jackson, A.; Kovalevsky, A.; Oksanen, E.; Wacklin, H. Biological structures. In *Experimental Methods in the Physical Sciences*; Felix, F.-A., David, L.P., Eds.; Elsevier: San Diego, CA, USA, 2017; Volume 49, pp. 1–75.
18. Blum, M.-M.; Tomanicek, S.J.; John, H.; Hanson, B.L.; Rüterjans, H.; Schoenborn, B.P.; Langan, P.; Chen, J.C.-H. X-ray structure of perdeuterated diisopropyl fluorophosphatase (DFPase): Perdeuteration of proteins for neutron diffraction. *Acta Crystallogr. F Struct. Biol. Cryst.* **2010**, *66*, 379–385. [CrossRef] [PubMed]
19. Tanaka, I.; Kusaka, K.; Hosoya, T.; Niimura, N.; Ohhara, T.; Kurihara, K.; Yamada, T.; Ohnishi, Y.; Tomoyori, K.; Yokoyama, T. Neutron structure analysis using the IBARAKI biological crystal diffractometer (iBIX) at J-PARC. *Acta Crystallogr. D Struct. Biol.* **2010**, *66*, 1194–1197. [CrossRef] [PubMed]
20. Pastorek, J.; Pastorekova, S. Hypoxia-induced carbonic anhydrase IX as a target for cancer therapy: From biology to clinical use. *Semin. Cancer Biol.* **2015**, *31*, 52–64. [CrossRef] [PubMed]
21. Alterio, V.; Hilvo, M.; Di Fiore, A.; Supuran, C.T.; Pan, P.; Parkkila, S.; Scaloni, A.; Pastorek, J.; Pastorekova, S.; Pedone, C.; et al. Crystal structure of the catalytic domain of the tumor-associated human carbonic anhydrase IX. *Proc. Natl. Acad. Sci. USA* **2009**, *106*, 16233–16238. [CrossRef] [PubMed]
22. Kazokaite, J.; Niemans, R.; Dudutiene, V.; Becker, H.M.; Leitans, J.; Zubriene, A.; Baranauskiene, L.; Gondi, G.; Zeidler, R.; Matuliene, J.; et al. Novel fluorinated carbonic anhydrase IX inhibitors reduce hypoxia-induced acidification and clonogenic survival of cancer cells. *Oncotarget* **2018**, *9*, 26800–26816. [CrossRef] [PubMed]
23. Mahon, B.P.; Bhatt, A.; Socorro, L.; Driscoll, J.M.; Okoh, C.; Lomelino, C.L.; Mboge, M.Y.; Kurian, J.J.; Tu, C.; Agbandje-McKenna, M.; et al. The Structure of Carbonic Anhydrase IX Is Adapted for Low-pH Catalysis. *Biochemistry* **2016**, *55*, 4642–4653. [CrossRef] [PubMed]
24. Ireton, G.C.; Stoddard, B.L. Microseed matrix screeningto improve crystals of yeast cytosine deaminase. *Acta Crystallogr. D Struct. Biol.* **2004**, *60*, 801. [CrossRef]
25. Koruza, K.; Lafumat, B.; Végvári, Á.; Knecht, W.; Fisher, S.Z. Deuteration of human carbonic anhydrase for neutron crystallography: Cell culture media, protein thermostability, and crystallization behavior. *Arch. Biochem. Biophys.* **2018**, *645*, 26–33. [CrossRef] [PubMed]
26. Budayova-Spano, M.; Fisher, S.Z.; Dauvergne, M.-T.; Agbandje-McKenna, M.; Silverman, D.N.; Myles, D.A.A.; McKenna, R. Production and X-ray crystallographic analysis of fully deuterated human carbonic anhydrase II. *Acta Crystallogr. F Struct. Biol. Cryst. Commun.* **2006**, *62*, 6–9. [CrossRef] [PubMed]
27. Di Costanzo, L.; Moulin, M.; Haertlein, M.; Meilleur, F.; Christianson, D.W. Expression, purification, assay, and crystal structure of perdeuterated human arginase I. *Arch. Biochem. Biophys.* **2007**, *465*, 82–89. [CrossRef] [PubMed]

crystals

MDPI

Article

Cry Protein Crystal-Immobilized Metallothioneins for Bioremediation of Heavy Metals from Water

Qian Sun, Sze Wan Cheng, Kelton Cheung, Marianne M. Lee * and Michael K. Chan *

School of Life Sciences and Center of Novel Biomaterials, The Chinese University of Hong Kong, Shatin, New Territories, Hong Kong SAR, China; sunqian@link.cuhk.edu.hk (Q.S.); 1155034893@link.cuhk.edu.hk (S.W.C.); skykelton@gmail.com (K.C.)

* Correspondence: mariannemmlee@cuhk.edu.hk (M.M.L.); michaelkchan88@cuhk.edu.hk (M.K.C.)

Received: 8 May 2019; Accepted: 30 May 2019; Published: 1 June 2019

Abstract: Cry proteins have been the subject of intense research due to their ability to form crystals naturally in *Bacillus thuringiensis (Bt)*. In this research we developed a new strategy that allows for the removal of cadmium and chromium from wastewater by using one Cry protein, Cry3Aa, as a framework to immobilize tandem repeats of the cyanobacterial metallothionein SmtA from *Synechococcus elongatus* (strain PCC 7942). SmtA is a low molecular weight cysteine-rich protein known to bind heavy metals. A series of Cry3Aa-SmtA constructs were produced by the fusion of one, three, or six tandem repeats of SmtA to Cry3Aa. Overexpression of these constructs in *Bt* resulted in the production of pure Cry3Aa-SmtA fusion crystals that exhibited similar size, crystallinity, and morphology to that of native Cry3Aa protein crystals. All three Cry3Aa-SmtA constructs exhibited efficient binding to cadmium and chromium, with the binding capacity correlated with increasing SmtA copy number. These results suggest the potential use of Cry3Aa-SmtA crystals as a novel biodegradable and cost-effective approach to the removal of toxic heavy metals from the environment.

Keywords: Cry protein crystals; metallothioneins; bioremediation; heavy metal contamination

1. Introduction

Cry proteins are a family of proteins that are produced and directly crystallized within the bacterium *Bacillus thuringiensis* (*Bt*) [1–4]. The resulting crystals are very stable and can be isolated with minimal purification. In many cases, simple washing is all that is needed to remove soluble impurities. Functionally, Cry proteins are most famous for their role as biological insecticides. They have been incorporated into many genetically modified crops as a way to make the crops resistant to insect and nematode pests [5]. Due to their use in genetically modified crops that are consumed by humans, Cry proteins are generally considered to be safe for organisms [6]. Our group was the first to demonstrate that various reporter proteins could be fused to one such Cry protein, Cry3Aa, to produce isolatable Cry3Aa fusion crystals without apparent loss of reporter function. Isolated Cry3Aa-GFP or Cry3Aa-mCherry crystals were fluorescent, indicating the proper folding of GFP and mCherry, while Cry3A-luciferase crystals remained functionally active and could be used for in vivo imaging studies in mice [7]. It should be noted that while the crystalline nature of some Cry protein crystals produced in *Bt* has been confirmed by powder diffraction [8], and in the case of Cry3Aa, its structure determination [4], we have not been able to use diffraction to confirm this for our Cry3Aa fusions. Nevertheless, their morphology, uniformity, and insolubility are similar to those of native Cry3Aa crystals, and as such, we loosely use the term crystals to reflect this similarity. Consistent with these properties, we recently demonstrated that the lipase lipA from *Bacillus subtilis* could be immobilized by genetically fusing it to Cry3Aa and the resultant Cry3Aa-lipA fusion crystals were found to be more stable against temperature and solvent than the free lipA protein, thus enabling multiple cycles of biodiesel conversion [9]. Herein, we extend the application of this Cry3Aa fusion technology for use in heavy metal sequestration.

The release of heavy metals such as cadmium (Cd) or chromium (Cr) into natural water ecosystems has increased with expanding industrialization [10], with industrial wastewater from mining, metal processing, tanneries, pharmaceuticals, pesticides, organic chemicals, rubber and plastics, lumber and wood products being common sources of heavy metal pollutants [11,12]. The exposure of these heavy metals to humans can be harmful, leading to chronic or acute health conditions [13], including reduced growth and development, cancer, organ damage, nervous system damage, and in extreme cases, death [14–16]. Due to the mobility and toxicity of heavy metal ions in water, effective methods for their removal are needed. Several different physicochemical methods such as chemical precipitation and membrane filtration have been utilized for this purpose, but there are disadvantages to each of these methods, including high cost, energy requirement, or production of harmful by-products [17–20].

To produce our biologically inspired immobilized chelator, we chose to genetically fuse Cry3Aa to the metallothionein SmtA from the cyanobacterium *Synechococcus elongates* (strain PCC 7942). Metallothioneins are low molecular weight cysteine-rich proteins with a high binding capacity for heavy metals due to the presence of multiple cysteine residues that facilitate the formation of metal-thiolate clusters at sub-micromolar concentrations [21–25]. SmtA is arguably the best characterized bacterial metallothionein. The NMR structure of a Zn^{2+}-bound SmtA suggests that its cysteine residues form a pocket for binding four metal ions [26]. Like other metallothioneins, this use of cysteines to ligate the metal ions provides SmtA with selectivity towards soft metals such as Hg^{2+}, Cd^{2+}, Zn^{2+}, Cu^{2+}, and Co^{2+}, allowing it to be used as a metal chelator of toxic heavy metals such as Cd^{2+}—even in the presence of high concentrations of sodium, calcium, and magnesium [27,28]. This property makes them potentially useful for multiple purposes [15,29–33], including as binding agents for the selective removal of heavy atoms from contaminated water [27]. Here we describe the in vivo production and physical characterization of three different Cry3Aa-SmtA fusion crystals, and the evaluation of their ability to remove Cd^{2+} and Cr^{3+} ions from heavy metal-containing solutions.

2. Materials and Methods

2.1. Construction of the Cry3Aa-SmtA Fusion Plasmids

The gene encoding three tandem repeats of SmtA ($[SmtA]_3$) was synthesized and cloned into a standard plasmid by GeneArt (ThermoFisher). A single copy of SmtA ([SmtA]) or $[SmtA]_3$ was amplified by PCR using Kapa HiFi Hot Start Ready Mix (Kapa Biosystems) from the aforementioned plasmid while the gene fragment encoding six repeats of SmtA ($[SmtA]_6$) was generated by overlap PCR of two $[SmtA]_3$ fragments. Each gene was subcloned into an existing pHT315 vector harboring the *cry3Aa* gene (pHT315-Cry3Aa) [7]. Briefly, the pHT315-Cry3Aa vector was linearized using the restriction endonucleases BamHI and KpnI, and the gene fragment $[smtA]_x$ was then cloned between the BamHI and KpnI sites downstream of the *cry3Aa* gene using Gibson Assembly® Master Mix (New England Biolabs) following the manufacturer's instructions, yielding the plasmids for *Bt* expression of Cry3Aa-[SmtA], Cry3Aa-$[SmtA]_3$, or Cry3Aa-$[SmtA]_6$ protein crystals.

2.2. Expression and Purification of Cry3Aa and Cry3Aa-SmtA Fusion Crystals in Bt

All *Bt* Cry3Aa and Cry3Aa-SmtA fusion protein crystals were produced by transforming the corresponding plasmids into *Bt407* cells. The cells were overexpressed at 25 °C for 72 h, after which the pellet was harvested using centrifugation at 8000 rpm for 10 min. The pellet was washed with autoclaved double distilled water (ddH_2O) and purified by centrifugation using a discontinuous sucrose gradient (40% 4 mL, 55% 7 mL, 65% 7 mL, 72% 4 mL) at 5000 rpm for 30 min. The crystals were extracted from the 55% and 65% layer and washed with autoclaved ddH_2O to remove the sucrose. SDS-PAGE samples were prepared by resuspending crystals in autoclaved ddH_2O, adding 5× SDS dye, boiling for 5 min, and then loading onto a 10% TGX Stain-Free gel (BioRad) to verify the presence and purity of the corresponding crystals. Protein concentrations were determined using the Bradford standard assay (BioRad).

2.3. *Dynamic Light Scattering of Cry3Aa and Cry3Aa-SmtA Fusion Crystals*

The size and dispersity of Cry3Aa and the three Cry3Aa-SmtA fusion crystals were measured by dynamic light scattering (DLS) at 25 °C by resuspending 100 μg of crystals in 1 mL of autoclaved ddH_2O and then applying them to a Malvern Zetasizer Nano ZS90 (Malvern Instruments Ltd., Malvern, UK).

2.4. *Scanning Electron Microscopy of Cry3Aa and Cry3Aa-SmtA Fusion Crystals*

Scanning electron microscopy (SEM) imaging samples were prepared by resuspending 0.1 mg of Cry3Aa or Cry3Aa-SmtA fusion crystals in 1 mL autoclaved ddH_2O. The samples were sonicated for 5 min and 2 μL of this solution was added to a copper stub and allowed to dry overnight. Samples were coated with Au using a Sputter Coater S150B (Edwards) and imaged in a SU8000 (Hitachi) operated at 5 kV and a working distance of 8.0 mm to 8.2 mm. The length and width of crystals were measured using software ImageJ Version 1.52a (NIH, Bethesda, MD, USA).

2.5. *Metal Binding Capacity Studies by Atomic Absorption Spectrophotometer (AAS)*

To ascertain the effect of fusing different numbers of repeat units of SmtA to Cry3Aa on the binding capacity of the corresponding fusion crystals to Cd^{2+} and Cr^{3+} ions, 1 nmol of Cry3Aa (control), Cry3Aa-[SmtA], Cry3Aa-$[SmtA]_3$, or Cry3Aa-$[SmtA]_6$ was incubated with 2 mL of 0.1 ppm of Cd standard for AAS (Sigma) or Cr standard for AAS (Sigma) solution at room temperature overnight. The samples were filtered with a 0.22 μm filter and the remaining Cd^{2+} or Cr^{3+} ions in the supernatant were measured by atomic absorption spectrometry (AAS) (Hitachi Z2300 flame). The metal concentrations were determined by comparing the measured absorbance against a series of calibration standards from 0 to 1 ppm prepared from a Cd standard solution (Sigma) or Cr standard solution (Sigma), and then used to derive the % bound by subtracting from the initial amount (0.1 ppm).

2.6. *Statistical Analysis*

GraphPad Prism software (version 8.0.2, GraphPad, San Diego, CA, USA) was used for statistical analysis. An unpaired two-tailed student's *t*-test was used for size comparison of the Cry3Aa-SmtA fusions with native Cry3Aa. One-way analysis of variance (ANOVA) was used for the multi-group comparison of metal binding by the different constructs. Data are presented as mean ± standard error of the mean.

3. Results

3.1. *Production of Cry3Aa and Cry3Aa-SmtA Fusion Crystals*

Previous work by our group demonstrated the utility of Cry3Aa as an immobilization platform to generate crystalline particles with functional fusion partners such as green fluorescent protein, mCherry, luciferase, and lipase A [7,9]. To further expand the utility of the Cry3Aa protein crystal platform, we decided to explore whether these crystals could be used to accommodate multiple copies of a small protein, such as a metallothionein, and in so doing, improve their utility—in this case, for increasing metal binding capacity. As such, a series of plasmids containing the gene encoding Cry3Aa-SmtA with different numbers of SmtA repeats were constructed and the corresponding Cry3Aa-SmtA fusion protein crystals were produced in *Bt*407. The identity and purity of each of the Cry3Aa-SmtA fusion crystals were confirmed by SDS-PAGE (Figure 1) based on their molecular weights.

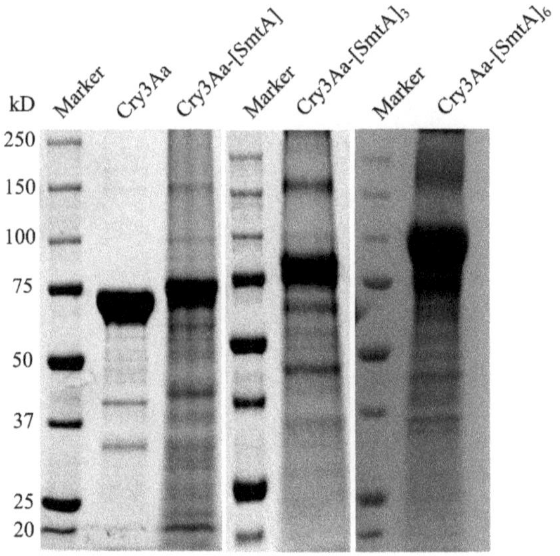

Figure 1. SDS-PAGE of Cry3Aa and Cry3Aa-SmtA fusion crystals produced in *Bacillus thuringiensis* (*Bt*). The theoretical molecular weights are: Cry3Aa, 73.1 kDa; Cry3Aa-[SmtA], 78.8 kDa; Cry3Aa-[SmtA]$_3$, 90.3 kDa; and Cry3Aa-[SmtA]$_6$, 107.5 kDa.

3.2. Characterization of Cry3Aa and Cry3Aa-SmtA Fusion Protein Crystals

The size and morphology of each Cry3Aa-SmtA fusion crystal were investigated by DLS and SEM and compared to those of Cry3Aa. Based on the DLS measurements, the size distributions of the different Cry3Aa-SmtA fusion crystals were similar to that of native Cry3Aa crystals (Figure 2a–d). The mean hydrodynamic diameters and polydispersity index (PDI) of Cry3Aa-[SmtA], Cry3Aa-[SmtA]$_3$, and Cry3Aa-[SmtA]$_6$ were found to be 860.8 nm (PDI = 0.147), 862.7 nm (PDI = 0.226), and 895.1 nm (PDI = 0.015), respectively, whereas that of the native Cry3Aa was 754.1 nm (PDI = 0.035). In agreement with the DLS results, SEM images of Cry3Aa and the three Cry3Aa-SmtA fusion crystals revealed that the morphologies of the Cry3Aa-SmtA fusion crystals were also similar to that of Cry3Aa crystals (Figure 3a–h)—a rod-like shape of similar length and width (Table 1). These findings suggest that the fusion of SmtA and its repeats to Cry3Aa did not appear to alter the crystal-forming properties of Cry3Aa with respect to both its size and morphology.

Table 1. The mean length and width of Cry3Aa and Cry3Aa-SmtA fusion crystals based on scanning electron microscopy (SEM) images.

Construct	Length (nm) *	Width (nm) *
Cry3Aa	1215 ± 197	912 ± 97
Cry3Aa-[SmtA]	1378 ± 86	871 ± 138
Cry3Aa-[SmtA]$_3$	1471 ± 195	1014 ± 151
Cry3Aa-[SmtA]$_6$	1405 ± 112	913 ± 48

* Any difference between Cry3Aa-SmtA fusion and Cry3Aa crystals was not statistically significant ($p > 0.05$), as confirmed by unpaired two-tailed student's t-test.

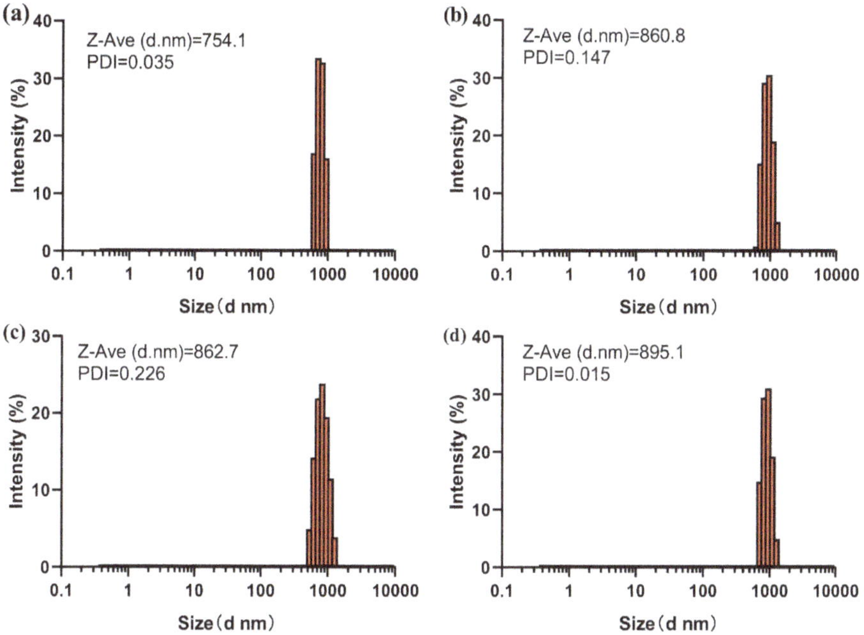

Figure 2. Size comparison of Cry3Aa and Cry3Aa-SmtA fusion crystals by dynamic light scattering (DLS). (**a**) Cry3Aa, (**b**) Cry3Aa-[SmtA], (**c**) Cry3Aa-[SmtA]$_3$, and (**d**) Cry3Aa-[SmtA]$_6$. All samples were measured at a concentration of 100 μg/mL of crystal.

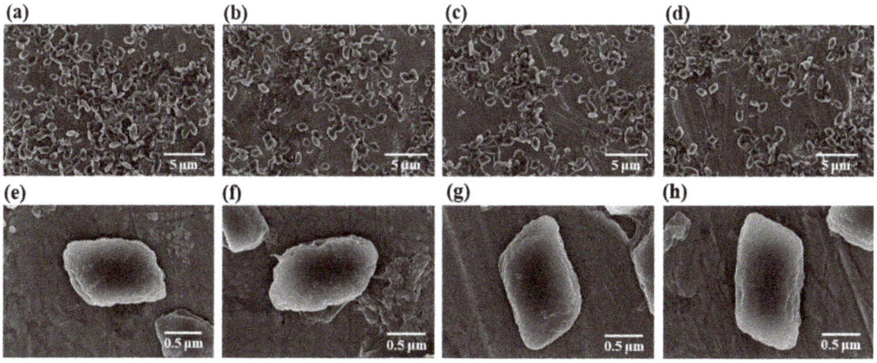

Figure 3. Size and morphological comparison of Cry3Aa and Cry3Aa-SmtA fusion crystals by scanning electron microscopy (SEM). Global SEM images of purified crystals of (**a**) Cry3Aa, (**b**) Cry3Aa-[SmtA], (**c**) Cry3Aa-[SmtA]$_3$, and (**d**) Cry3Aa-[SmtA]$_6$ at 5000× magnification. SEM images of single crystals of (**e**) Cry3Aa, (**f**) Cry3Aa-[SmtA], (**g**) Cry3Aa-[SmtA]$_3$, and (**h**) Cry3Aa-[SmtA]$_6$ at 45,000× magnification.

3.3. Cadmium and Chromium Binding by Cry3Aa and Cry3Aa-SmtA Fusion Crystals

To evaluate the heavy metal binding capacity of the Cry3Aa-SmtA fusion constructs, 1 nmol each of Cry3Aa-SmtA fusion and Cry3Aa (as a control) crystals were incubated with 2 mL of 0.1 ppm of Cd standard solution or Cr standard solution (Figure 4). As revealed by the AAS assay, Cd binding to the Cry3Aa-SmtA constructs increased with increasing SmtA repeats. Of note, Cry3Aa-[SmtA]$_6$ crystals bound nearly 100% of the Cd^{2+} ions in 0.1 ppm solution, suggesting that these fusion crystals were able to remove Cd^{2+} ions even at trace levels (Figure 4a). Similar trends were also observed for binding of Cr^{3+} ions, with the Cry3Aa-[SmtA]$_6$ crystals showing the highest level of Cr binding (Figure 4b).

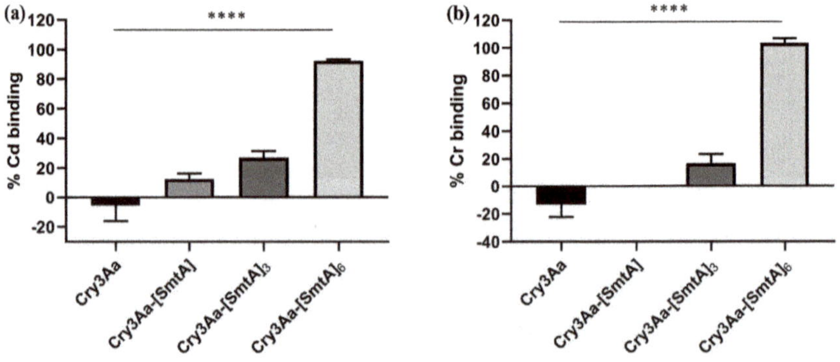

Figure 4. Atomic absorption spectrometry (AAS) analysis of the metal binding capacity of Cry3Aa (control) and Cry3Aa-SmtA fusion crystals. Percentage of (**a**) Cd^{2+} and (**b**) Cr^{3+} ions bound to crystals. All measurements were performed in triplicate. The error bars show the standard deviation of the mean. Repeated measurements of Cry3Aa-[SmtA] showed no binding of Cr^{3+} ions. **** $p < 0.0001$ by one-way ANOVA.

4. Discussion

There has been long-term interest in developing metallothioneins for use in the selective removal of heavy metals from wastewater [27]. Their selectivity allows for the binding of toxic metal ions such as Cd^{2+} or Cr^{3+} without removing other metal ions that might be important for the environment [28]. A major challenge, however, is the generation of an immobilized metallothionein for this purpose in a cheap and easy fashion that allows for its potential recycling.

As demonstrated here, the Cry3Aa crystal as an immobilization platform is particularly attractive for this purpose. A plasmid for the expression of Cry3Aa fused to tandem repeats of the SmtA can be readily produced and used for the production of Cry3Aa-SmtA fusion crystals directly in *Bt*. These crystalline particles can easily be purified without the use of columns by density gradient centrifugation, greatly simplifying the isolation step and lowering the cost of production.

While the fusion of one SmtA to Cry3Aa gives only limited binding of heavy metals, we show that up to six tandem repeats of SmtA can be incorporated, resulting in enhanced metal binding without significantly changing the size and morphology of the biologically produced crystal. As an aside, the Cry3Aa-[SmtA]$_9$ construct with nine copies of SmtA was also cloned, but the yields were low and the resultant crystalline particles seemed less stable. This may suggest that the solvent channels within Cry3Aa crystals can accommodate six copies of SmtA, but not nine. Increasing the number of copies of SmtA is important for metal binding; however, the metal binding capacity for either Cd^{2+} or Cr^{3+} is strongly correlated with the number of SmtA repeats incorporated. The results affirm the potential use of biologically produced Cry3Aa-SmtA fusion crystals for the bioremediation of heavy metals from wastewater.

While there are a number of conventional treatment processes such as ion exchange, activated charcoal, chemical precipitation, and chemical reduction and adsorption, these approaches have certain limitations including their high cost, production of secondary waste, and limited effectiveness at low concentrations of metal [34,35]. In contrast, Cry3Aa-SmtA fusion crystals can be produced directly in bacteria, are easy to isolate, and for the Cry3Aa fusion construct with six copies of SmtA, Cry3Aa-[SmtA]$_6$ can efficiently remove Cd^{2+} or Cr^{3+} ions from solutions even at trace levels of 0.1 ppm. Thus, Cry3Aa-SmtA fusion protein crystals offer an environmentally friendly, effective, and economical alternative method for the removal of heavy metals from water—highlighting yet another use of biologically produced Cry3Aa fusion crystals for practical application.

Author Contributions: Conceptualization, M.K.C. and M.M.L.; investigation, Q.S., S.W.C., K.C.; writing—original draft preparation, Q.S., M.K.C. and M.M.L.; supervision, M.K.C. and M.M.L.; funding acquisition M.K.C. and M.M.L.

Funding: This research was funded by the Hong Kong Research Grants Council GRF grant 14300414 and the Center of Novel Biomaterials, the Chinese University of Hong Kong.

Acknowledgments: We would like to thank Josie Lai (CUHK SBS) for her assistance in collecting the SEM images, and Didier Lereclus (Institut Pasteur in Paris) and Daniel Ziegler (*Bacillus* Genetic Stock Center, Ohio State University, USA) for kindly providing us with the *Bt* cells.

Conflicts of Interest: The authors declare no conflict of interest.

References

1. Whiteley, H.R.; Schnepf, H.E. The Molecular Biology of Parasporal Crystal Body Formation in Bacillus thuringiensis. *Annu. Rev. Microbiol.* **1986**, *40*, 549–576. [CrossRef] [PubMed]
2. Schnepf, E.; Crickmore, N.; Van Rie, J.; Lereclus, D.; Baum, J.; Feitelson, J.; Zeigler, D.R.; Dean, D.H. Bacillus thuringiensis and Its Pesticidal Crystal Proteins. *Microbiol. Mol. Boil. Rev.* **1998**, *62*, 775–806.
3. Park, H.-W.; Ge, B.; Bauer, L.S.; Federici, B.A. Optimization of Cry3A Yields in Bacillus thuringiensis by Use of Sporulation-Dependent Promoters in Combination with the STAB-SD mRNA Sequence. *Appl. Environ. Microbiol.* **1998**, *64*, 3932–3938. [PubMed]
4. Sawaya, M.R.; Cascio, D.; Gingery, M.; Rodriguez, J.; Goldschmidt, L.; Colletier, J.-P.; Messerschmidt, M.M.; Boutet, S.; Koglin, J.E.; Williams, G.J. Protein crystal structure obtained at 2.9 Å resolution from injecting bacterial cells into an X-ray free-electron laser beam. *Proc. Natl. Acad. Sci. USA* **2014**, *111*, 12769–12774. [CrossRef]
5. Shimada, N.; Miyamoto, K.; Kanda, K.; Murata, H. Bacillus thuringiensis insecticidal Cry1Ab toxin does not affect the membrane integrity of the mammalian intestinal epithelial cells: An *in vitro* study. *In Vitro Cell. Dev. Biol.-Anim.* **2006**, *42*, 45–49. [CrossRef] [PubMed]
6. Chowdhury, E.H.; Shimada, N.; Murata, H.; Mikami, O.; Sultana, P.; Miyazaki, S.; Yoshioka, M.; Yamanaka, N.; Hirai, N.; Nakajima, Y. Detection of Cry1Ab protein in gastrointestinal contents but not visceral organs of genetically modified Bt11-fed calves. *Vet. Hum. Toxicol.* **2003**, *45*, 72–75. [PubMed]
7. Nair, M.S.; Lee, M.M.; Bonnegarde-Bernard, A.; Wallace, J.A.; Dean, D.H.; Ostrowski, M.C.; Burry, R.W.; Boyaka, P.N.; Chan, M.K. Cry Protein Crystals: A Novel Platform for Protein Delivery. *PLoS ONE* **2015**, *10*, e0127669. [CrossRef]
8. Holmes, K.; Monro, R.; Holmes, K. Studies on the structure of parasporal inclusions from Bacillus thuringiensis. *J. Mol. Boil.* **1965**, *14*, 572-IN25. [CrossRef]
9. Heater, B.S.; Lee, M.M.; Chan, M.K. Direct production of a genetically-encoded immobilized biodiesel catalyst. *Sci. Rep.* **2018**, *8*, 12783. [CrossRef] [PubMed]
10. George, S.G.; Carpene, E.; Coombs, T.L.; Overnell, J.; Youngson, A. Characterisation of cadmium-binding proteins from mussels, Mytilus edulis (L), exposed to cadmium. *Biochim. Biophys. Acta (BBA)-Protein Struct.* **1979**, *580*, 225–233. [CrossRef]
11. Lakherwal, D. Adsorption of heavy metals: A review. *Int. J. Environ. Res. Dev.* **2014**, *4*, 41–48.
12. Srivastava, N.; Majumder, C. Novel biofiltration methods for the treatment of heavy metals from industrial wastewater. *J. Hazard. Mater.* **2008**, *151*, 1–8. [CrossRef] [PubMed]
13. Chowdhury, S.; Mazumder, M.J.; Al-Attas, O.; Husain, T. Heavy metals in drinking water: Occurrences, implications, and future needs in developing countries. *Sci. Total Environ.* **2016**, *569*, 476–488. [CrossRef] [PubMed]
14. Bailey, S.E.; Olin, T.J.; Bricka, R.; Adrian, D. A review of potentially low-cost sorbents for heavy metals. *Water Res.* **1999**, *33*, 2469–2479. [CrossRef]
15. Demirbaş, A. Heavy metal adsorption onto agro-based waste materials: A review. *J. Hazard. Mater.* **2008**, *157*, 220–229. [CrossRef]
16. Barakat, M. New trends in removing heavy metals from industrial wastewater. *Arab. J. Chem.* **2011**, *4*, 361–377. [CrossRef]

17. Roy, D.; Greenlaw, P.N.; Shane, B.S. Adsorption of heavy metals by green algae and ground rice hulls. *J. Environ. Sci. Health Part A Environ. Sci. Eng. Toxicol.* **2008**, *28*, 37–50. [CrossRef]
18. Pansini, M.; Colella, C.; De Gennaro, M. Chromium removal from water by ion exchange using zeolite. *Desalination* **1991**, *83*, 145–157. [CrossRef]
19. Pérez-Candela, M.; Martín-Martínez, J.; Torregrosa-Maciá, R. Chromium(VI) removal with activated carbons. *Water Res.* **1995**, *29*, 2174–2180. [CrossRef]
20. Rengaraj, S.; Yeon, K.-H.; Moon, S.-H. Removal of chromium from water and wastewater by ion exchange resins. *J. Hazard. Mater.* **2001**, *87*, 273–287. [CrossRef]
21. Vašák, M. Advances in metallothionein structure and functions. *J. Trace Elem. Med. Boil.* **2005**, *19*, 13–17. [CrossRef]
22. Romero-Isart, N.; Vašák, M. Advances in the structure and chemistry of metallothioneins. *J. Inorg. Biochem.* **2002**, *88*, 388–396. [CrossRef]
23. Blindauer, C.A. Bacterial metallothioneins: Past, present, and questions for the future. *JBIC J. Boil. Inorg. Chem.* **2011**, *16*, 1011–1024. [CrossRef]
24. Capdevila, M.; Atrian, S. Metallothionein protein evolution: A miniassay. *JBIC J. Boil. Inorg. Chem.* **2011**, *16*, 977–989. [CrossRef] [PubMed]
25. Capdevila, M.; Bofill, R.; Palacios, Ò.; Atrian, S. State-of-the-art of metallothioneins at the beginning of the 21st century. *Coord. Chem. Rev.* **2012**, *256*, 46–62. [CrossRef]
26. Blindauer, C.A.; Harrison, M.D.; Parkinson, J.A.; Robinson, A.K.; Cavet, J.S.; Robinson, N.J.; Sadler, P.J. A metallothionein containing a zinc finger within a four-metal cluster protects a bacterium from zinc toxicity. *Proc. Natl. Acad. Sci. USA* **2001**, *98*, 9593–9598. [CrossRef] [PubMed]
27. Esser-Kahn, A.P.; Iavarone, A.T.; Francis, M.B. Metallothionein-Cross-Linked Hydrogels for the Selective Removal of Heavy Metals from Water. *J. Am. Chem. Soc.* **2008**, *130*, 15820–15822. [CrossRef] [PubMed]
28. Freisinger, E.; Vašák, M. Cadmium in metallothioneins. In *Cadmium: From Toxicity to Essentiality*; Sigel, A., Sigel, H., Sigel, R.K., Eds.; Springer: Dordrecht, The Netherlands, 2013; Volume 11, pp. 339–371.
29. Verma, N.; Singh, M. Biosensors for heavy metals. *BioMetals* **2005**, *18*, 121–129. [CrossRef] [PubMed]
30. Martín-Betancor, K.; Rodea-Palomares, I.; Muñoz-Martín, M.A.; Leganés, F.; Fernández-Piñas, F.; Palomares, I.M.R. Construction of a self-luminescent cyanobacterial bioreporter that detects a broad range of bioavailable heavy metals in aquatic environments. *Front. Microbiol.* **2015**, *6*, 186. [CrossRef] [PubMed]
31. Bontidean, I.; Berggren, C.; Johansson, G.; Csöregi, E.; Mattiasson, B.; Lloyd, J.R.; Jakeman, K.J.; Brown, N.L. Detection of Heavy Metal Ions at Femtomolar Levels Using Protein-Based Biosensors. *Anal. Chem.* **1998**, *70*, 4162–4169. [CrossRef]
32. Wu, C.-M.; Lin, L.-Y. Immobilization of metallothionein as a sensitive biosensor chip for the detection of metal ions by surface plasmon resonance. *Biosens. Bioelectron.* **2004**, *20*, 864–871. [CrossRef] [PubMed]
33. Gonzalezbellavista, A.; Atrian, S.; Muñoz, M.; Capdevila, M.; Fàbregas, E. Novel potentiometric sensors based on polysulfone immobilized metallothioneins as metal-ionophores. *Talanta* **2009**, *77*, 1528–1533. [CrossRef]
34. Lim, A.P.; Aris, A.Z. A review on economically adsorbents on heavy metals removal in water and wastewater. *Rev. Environ. Sci. Bio/Technol.* **2013**, *13*, 163–181. [CrossRef]
35. Babel, S. Low-cost adsorbents for heavy metals uptake from contaminated water: A review. *J. Hazard. Mater.* **2003**, *97*, 219–243. [CrossRef]

crystals

MDPI

Communication

Over-Expression, Secondary Structure Characterization, and Preliminary X-ray Crystallographic Analysis of *Xenopus tropicalis* Ependymin

Jeong Kuk Park, Yeo Won Sim and SangYoun Park *

School of Systems Biomedical Science, Soongsil University, Seoul 06978, Korea; water1028@naver.com (J.K.P.); rato0001@naver.com (Y.W.S.)

* Correspondence: psy@ssu.ac.kr; Tel.: +82-2-820-0456

Received: 19 June 2018; Accepted: 9 July 2018; Published: 11 July 2018

Abstract: The gene encoding frog (*Xenopus tropicalis*) ependymin without the signaling sequence was gene-synthesized, and the protein successfully over-expressed in ~mg quantities adequate for crystallization using insect cell expression. Circular dichroism (CD) analysis of the protein purified with >95% homogeneity indicated that ependymin contains both α-helix and β-strand among the secondary structure elements. The protein was further crystallized using polyethylene glycol 8000 as the precipitating reagent, and X-ray diffraction data were collected to 2.7 Å resolution under cryo-condition at a synchrotron facility. The crystal belongs to a hexagonal space group $P6_122$ (or $P6_522$) having unit cell parameters of $a = b = 61.05$ Å, $c = 234.33$ Å. Matthews coefficient analysis indicated a crystal volume per protein mass (V_M) of 2.76 $Å^3$ Da^{-1} and 55.4% solvent content in the crystal when the calculated molecular mass of the protein only was used. However, the apparent SDS-PAGE molecular mass of ~33 kDa (likely resulting from *N*-glycosylation) suggested V_M of 1.90 $Å^3$ Da^{-1} and 35.4% solvent content instead. In both cases, the asymmetric unit of the crystal likely contains only one subunit of the protein.

Keywords: ependymin (EPN); ependymin-related protein (EPDR); mammalian ependymin-related protein (MERP)

1. Introduction

Ependymin (EPN) was first discovered in the ependymal zone of goldfish brain upon emergence of a learning behavior [1–3]. Additionally, EPN being one of the abundant glycoproteins in the brain extracellular fluid and cerebrospinal fluid in the teleost fish, has been suggested to have various roles in memory consolidation, neuronal regeneration, brain calcium homeostasis [4,5] as well as in cold adaptation, and even in determining aggressiveness [6,7]. Interestingly, orthologues of fish EPN, also exist in other animals such as sea urchin, frog, and even in mammals, and have been also named as EPDR (ependymin-related) protein or MERP (mammalian ependymin-related protein). Humans, in particular, have one isoform of EPN and its expression has been reported in various human normal tissues [8,9]. More interestingly, the transcription level of human EPN was elevated in colorectal tumor cells, and hence is also called UCC1 (upregulated in colorectal cancer gene 1) [10]. Hereafter, all EPN and EPN-like proteins will be mentioned as EPN for simplicity.

EPNs in general contain endoplasmic reticulum-targeting signal sequences at the N-terminus, which are likely cleaved by signal peptidases and further processed for secretion [9,11]. The sequences also hold four cysteine residues that may form either intramolecular disulfide crosslinking for stabilizing protein conformation or intermolecular crosslinking for mediating dimeric interaction

(Figure 1). Furthermore, studies on fish EPN suggest *N*-glycosylations at two asparagine residues [1,12], and other EPNs also have predicted *N*-glycosylation sites (Figure 1). An isolated digestive fragment of fish EPN has been shown to activate c-Jun N-terminal kinase, and components of AP-1 (c-Jun and c-Fos) via the PKC and MAPK pathways [13,14]. Because oxidative stresses induce the expression of AP-1 by the MAP kinase pathway, upregulation of superoxide dismutase (SOD), catalase, and glutathione peroxide have been implied as the downstream targets of at least fish EPN to prevent damages resulting from reactive oxygen species [15].

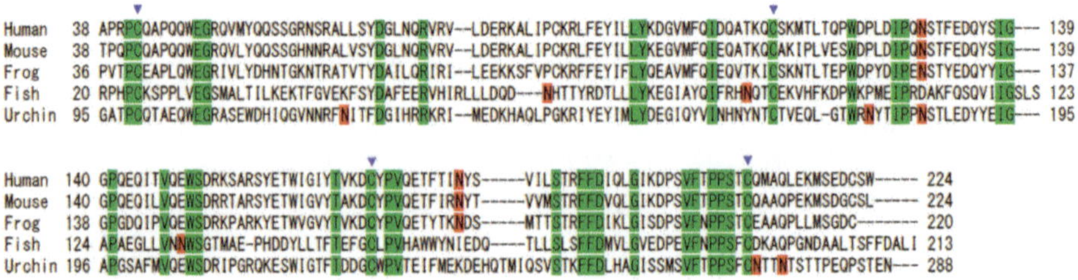

Figure 1. Protein sequence alignment of EPNs (without their N-terminal signal sequence) from various organisms. Homologous residues in all species are in green background. Predicted *N*-glycosylation sites are in red background. The four conserved cysteines are indicated with triangles on top. The mouse sequence is from *Mus musculus* EPDR2, the fish sequence is from *Oncorhynchus mykiss* (Rainbow trout), the frog sequence is from *Xenopus tropicalis*, and the sea urchin sequence is from *Strongylocentrotus purpuratus* (purple sea urchin). In comparison to the frog EPN sequence expressed in this study, the EPNs of fish (29%), sea urchin (43%), human (66%), and mouse (65%) all show sequence conservations (identity percent in parenthesis).

Despite the fact that EPN is similar from sea urchins to humans [16], studies leading to its multi-functional role are limited to EPN of the fish. Furthermore, the detailed mechanism of EPN's action remains yet to be unveiled. Since no known structure of any EPN exists so far, we believe that the future structure of EPN would give directions in predicting the function of EPNs in general. In this attempt, frog (*Xenopus tropicalis*) EPN without the signal sequence was recombinantly expressed in insect cells and purified. The protein was further crystallized, and the diffraction data obtained in a synchrotron facility.

2. Materials and Methods

2.1. Macromolecule Production

DNA encoding frog (*X. tropicalis*) EPN (residues 38~224), which excludes the N-terminal signal sequence was gene-synthesized with an addition of N-terminal His_8-tag sequence and *Bam*HI/*Not*I restriction enzyme sites (Bioneer, Daejeon, Korea) for cloning into pAcGP67A vector (BD Biosciences, Franklin Lakes, NJ, USA) (Table 1). The final plasmid generated was sequence verified before using it for transfection into insect cells. The media used for the insect cell culture was Corning® Insectagro® (Thermo Fisher Scientific, Waltham, MA, USA) supplemented with ×1 of Gibco® (Thermo Fisher Scientific, Waltham, MA, USA) Antibiotic-Antimycotic [penicillin 100 units/mL, streptomycin 100 μg/mL, Fungizone® (Amphotericin B) 250 ng/mL]. The media was pre-warmed in 27 °C water bath for 1 h, and the insect cell culture was performed in 27 °C benchtop or shaking incubators. Transfection of frog EPN containing plasmid and baculovirus DNA into SF9 (*Spodoptera frugiperda*) cells, and the subsequent virus amplification through multistep infections were performed according to the manufacturer's method. Briefly, a mixture of 0.1 μg plasmid and 5 μL baculovirus DNA (ProEasy™, AB vector, San Diego, CA, USA) was pre-mixed and diluted with water to 50 μL. Into the pre-mix, 50 μL of 10% Profectin™ (AB vector, San Diego, CA, USA) solution was added dropwise, and incubated at room temperature for 20 min. For transfection into SF9 cells, the resulting 100 μL mixture was added dropwise to the 50% confluent cells in a 6-well plate. The cells were harvested after four days of infection to obtain the initial P0 EPN virus stock. All subsequent infections for virus amplification were made by using the virus stock as ×20 stock. Hence, the P1 EPN virus stock was obtained by infecting the P0 stock into 50% confluent SF9 cells in T75 flask and harvesting after 4 days. Further infection and harvest after 4 days in 50% confluent SF9 cells in T150 flask generated the P2 EPN virus stock. Subsequently, another infection was made into a suspension culture in 50 mL SF9 cells (2 × 10^6 cells/mL) for the P3 EPN virus stock which was harvested in 4 days. The final P4 EPN virus stock was obtained by infecting the P3 EPN virus stock into 200 mL SF9 cells (2 × 10^6 cells/mL). For large-scale protein expression, 50 mL of P4 EPN virus stock was infected into 1 L of 2 × 10^6 cells/mL SF9 cells with 140 RPM shaking at 27 °C. The cells were harvested after 2 days to obtain only the supernatant in which the secreted EPN was found.

Table 1. Macromolecule production information for frog EPN.

Source Organism	***Xenopus tropicalis***
DNA source	Synthesized DNA
Cloning sites	*Bam*HI and *Not*I
Cloning vector	pAcGP67A
Expression vector	pAcGP67A
Expression host	SF9 (*S. frugiperda*)
Complete amino acid sequence of the construct produced [1]	ADPHHHHHHHHPVTPCEAPLQWEGRIVLYDH NTGKNTRATVTYDAILQRIRILEEKKSFVPCK RFFEYIFLYQEAVMFQIEQVTKICSKNTLTEPWDPY DIPENSTYEDQYYIGGPGDQIPVQEWSDRKPARK YETWVGVYTVKDCYPVQETYTKNDSMTTSTRFFDI KLGISDPSVFNPPSTCEAAQPLLMSGDC

[1] Non-native His_6-tag is underlined.

Stock solutions of Tris pH 7.5 and NaCl were added into the harvested supernatant to a final concentration of 50 mM Tris pH 7.5 and 200 mM NaCl, and further adjusted to pH 7 optimal for Ni-NTA binding. For ~1 L of supernatant, 20 mL Ni-NTA resin (Qiagen, Hilden, Germany) was added and rocked at 100 RPM (20 °C, 1 h) for bead binding. The mixture was incubated in 4 °C for 1 h to allow the separation of resin from the supernatant. The supernatant was carefully removed until ~100 mL and the re-suspended mixture was packed in a glass column. The resin was further washed with ~100 mL wash buffer (20 mM imidazole, 25 mM Tris pH 7.5, and 500 mM NaCl), and the frog EPN eluted with ~100 mL elution buffer (200 mM imidazole, 25 mM Tris pH 7.5, and 500 mM NaCl) by collecting in fractions of ~10 mL. The frog EPN in the collection fraction was checked on SDS-PAGE, and concentrated to ~10 mL using Amicon® Ultra-15 centrifugal filter (Merck, Kenilworth, NJ, USA). The concentrated crude protein was further purified using size-exclusion chromatography (SEC) on a ÄKTA FPLC (GE Healthcare, Little Chalfont, UK) connected to Superdex® 200 HR26/60 (GE Healthcare, Little Chalfont, UK) pre-equilibrated with gel filtration buffer (50 mM Tris pH 7.5 and 150 mM NaCl). FPLC eluate peak of EPN showed one major peak, and was concentrated to yield ~13 mg/mL (~200 μL) of frog EPN protein. Absorptivity coefficient of $\varepsilon = 1.8\ (\text{mg/mL})^{-1}\ \text{cm}^{-1}$ at λ = 280 nm was calculated by using the numbers of tyrosine and tryptophan in the protein [17]. The homogeneity of the final frog EPN was checked using SDS-PAGE under both reducing (with 5 mM DTT) and non-reducing (without DTT) conditions (Figure 2) and the aliquoted protein flash-frozen in liquid nitrogen for storage in −80 °C.

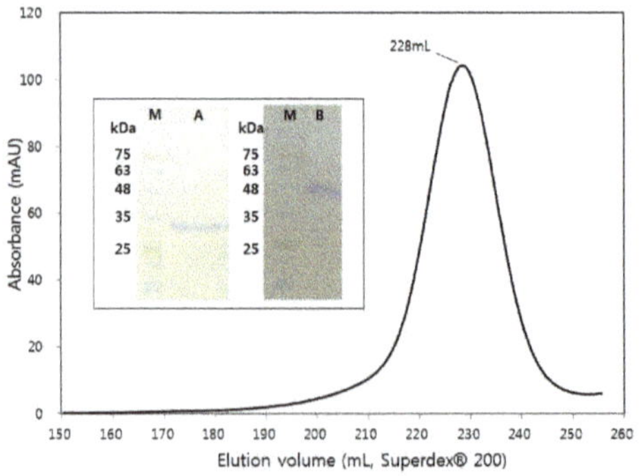

Figure 2. Size-exclusion chromatogram of frog EPN near 228 mL elution volume with the SDS-PAGE results on the concentrated fractions under the peak (*inset*). Labels for inset are as follows: Protein standard marker (M); SDS-PAGE under reducing condition (A); SDS-PAGE under non-reducing condition (B).

2.2. Circular Dichroism (CD)

Proper folding and further secondary structure element content of the purified frog EPN protein were analyzed using circular dichroism (CD). Ellipticity was scanned over 205~240 nm wavelength on a JASCO spectropolarimeter (Model J-810, Tokyo, Japan) where a concentration of 0.2 mg/mL frog EPN in 0.1 cm path-length cuvette was used. The ellipticity data was fitted using a secondary structure estimation program K2D3 [18] to approximate the secondary structure content (Figure 3).

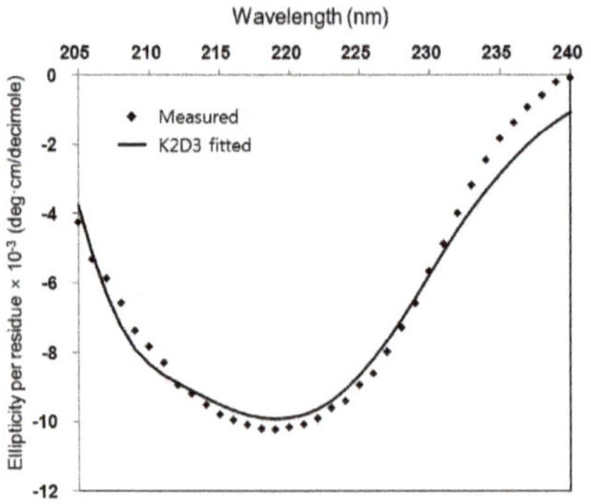

Figure 3. Circular dichroism study on the frog EPN.

2.3. Crystallization

Commercial screening solutions (Hampton Research, Aliso Viejo, CA, USA) were used to screen for frog EPN crystals. Single crystals appeared in a well reservoir of 0.2 M Calcium acetate hydrate, 0.1 M Sodium cacodylate trihydrate pH 6.5, and 18% (*w*/*v*) Polyethylene glycol 8000 in ~2 days over a hanging drop at 22 °C (Figure 3). A single crystal was transferred to a cryo-protectant solution which was made by adding glycerol to the reservoir solution to final 20% concentration, and flash-cooled in liquid nitrogen for storage. The frozen crystal was transported to a synchrotron facility where diffraction experiment was performed (Figure 4).

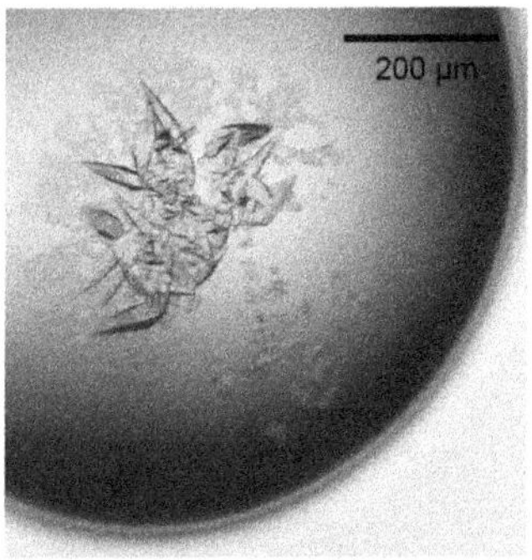

Figure 4. Crystals of frog EPN.

2.4. Data Collection and Processing

X-ray diffraction data were collected under a liquid nitrogen stream (100 K) on a CCD detector (ADSC Quantum Q270) at beamline 7A of PLS (Pohang, Korea). The crystal was rotated with 1.0° oscillation per frame to a total of 180° for complete data collection. Data were processed using HKL2000 [19] to a space group of $P6_122$ (or $P6_522$) (Table 2).

Table 2. Data collection and processing.

Diffraction Source	Pohang Light Source (PLS 7A) (Pohang, Korea)
Wavelength (Å)	0.97934
Temperature (K)	100
Detector	ADSC Quantum Q270
Crystal-detector distance (mm)	350
Rotation range per image (°)	1
Total rotation range (°)	180
Exposure time per image (s)	1
Space group	$P6_122$ (or $P6_522$)
a, b, c (Å)	61.05, 61.05, 234.33
α, β, γ (°)	90.0, 90.0, 120.0
Mosaicity (°)	0.496
Resolution range (Å)	50.0–2.70 (2.75–2.70) [1]
Total No. of reflections	129,713
No. of unique reflections	7871
Completeness (%)	99.6 (100.0)
Redundancy	16.6 (17.0)
$\langle I/\sigma(I)\rangle$	43.8 (3.7)
R_{merge}	0.118 (1.136)
$R_{p.i.m.}$	0.029 (0.271)
CC1/2	(0.886)
Overall B factor from Wilson plot (Å^2)	58.9

[1] Values for the outer shell are given in parentheses.

3. Results and Discussion

Although EPN is a protein that is similar among various species (Figure 1), neither the functional role (other than in fish) nor the detailed mechanism of action are known. In this study, the DNA encoding the frog (*X. tropicalis*) EPN without the signal sequence was generated by gene-synthesis and cloned into a pAcGP67A vector for secreted recombinant expression in SF9 insect cells. Of note, all attempts in using *Escherichia coli* bacterial expression of the same protein in various vectors and competent cells failed, which led us to depend on the eukaryotic expression system of using insect cells. The EPN protein was successfully over-expressed in a soluble form and was secreted into the media. The SEC elution profile of initial Ni-NTA purified frog EPN on HiLoad® 26/60 Superdex® 200 showed a single peak (Figure 2). The final SDS-PAGE analysis of the concentrated EPN protein under the peak indicated successful over-expression and purification of the EPN protein (Figure 2 *inset*). When looking at the size of the expressed frog EPN protein under reducing (5 mM DTT) condition, the molecular mass estimated based on standard protein markers seems to be ~33 kDa (Figure 2 *inset*) while the expected size calculated from the protein sequence is 22.8 kDa. This increased size in SDS-PAGE may be due to the *N*-glycosylations that are possible during expression within the insect cells. Also, unlike SDS-PAGE ran under reducing condition, non-reducing SDS-PAGE of the protein showed a single band corresponding to the size of a dimer. Hence, the frog EPN is likely associated into a dimer by at least one intermolecular disulfide bridge. The overall yield of the purified frog EPN was ~3 mg per 1 L of culture.

Due to the lack of functional studies that can be performed on the frog EPN, circular dichroism (CD) was used to determine that the recombinant protein had proper folding. Furthermore, secondary structure contents of frog EPN were estimated using the CD data (Figure 3). When the experimental ellipticity values of 205~240 nm range were fitted for secondary structure estimation, the result indicated that ~19% of the protein are α-helical and ~33% of the protein are of β-strands (Figure 3).

Despite the fact that the purified frog EPN may contain *N*-glycosylation which often result in difficulties in crystallizing proteins [20], crystals of frog EPN were obtained by screening against commercial solutions (Figure 4). Single crystals were grown and optimized in a reservoir solution of 0.2 M Calcium acetate hydrate, 0.1 M Sodium cacodylate trihydrate pH 6.5, and 18% (*w*/*v*) Polyethylene glycol 8000 at 22 °C. In ~2 days, the EPN crystal grew to approximately 10 μm × 10 μm × 100 μm (Figure 4) in size, which was sufficiently sized for X-ray diffraction experiments. The crystal diffracted well in a synchrotron X-ray radiation maintaining isotropic diffraction throughout all 180° rotation (Figure 5). A total of 7871 unique reflections were measured throughout 180° rotation and merged in the space group $P6_122$ (or $P6_522$) with unit cell parameters of $a = b = 61.05$ Å, and $c = 234.33$ Å. The merged dataset was overall 99.6% complete with R_{merge} of 11.8% and $R_{p.i.m.}$ of 2.9% to 2.70 Å resolution. Further diffraction statistics for the data collected are shown in Table 2. Matthews coefficient analysis [21] using the unit cell parameters, space group and the calculated molecular mass of protein only (22.8 kDa) indicated that the crystal volume per protein mass (V_M) of the EPN crystal was 2.76 Å^3 Da^{-1} with 55.4% solvent content. Instead of the calculated molecular mass, using the apparent molecular mass of ~33 kDa from SDS-PAGE analysis, which likely results from *N*-glycosylation, led to V_M of 1.90 Å^3 Da^{-1} and 35.4% solvent content. In both cases, these values suggested that the asymmetric unit of EPN crystal likely contains only one subunit of the monomeric frog EPN. Because no model for EPN exists in the Protein Databank (PDB), attempts for phasing via molecular replacement couldn't be made.

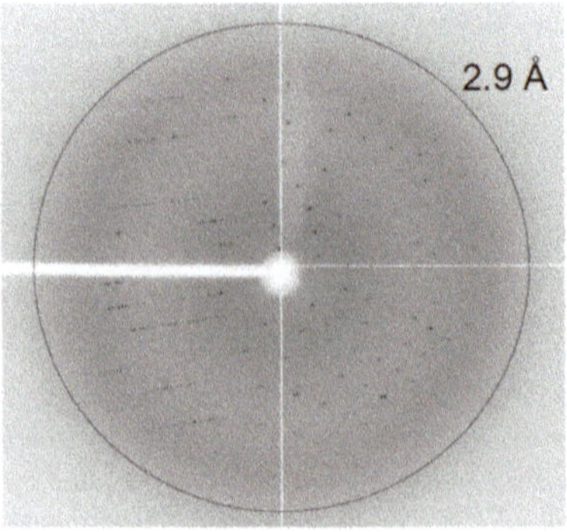

Figure 5. Isotropic X-ray diffraction image of frog EPN crystal.

In the future, we plan to improve the current 2.7 Å resolution crystal for better diffraction, and determine the high resolution structure of frog EPN by direct phasing. During the process, we plan to exploit conventional metal soaking [22] and other well-known techniques to work such as the use of $(Ta_6Br_{12})^{2+}$ [23] or 5-amino-2,4,6-triiodoisophthalic acid [24], and even sulfur-SAD [25,26] or ultraviolet radiation damage-induced phasing (UV-RIP) [27] methods. Since no structural information on any EPN exists in databases, the structure would give us insight into understanding its function.

Author Contributions: S.Y.P. and J.K.P. conceived and designed the experiments; J.K.P. and Y.W.S. performed the experiments; J.K.P. analyzed the data; S.Y.P. wrote the paper.

Funding: This research was supported by the Basic Science Program through the National Research Foundation of Korea (NRF) funded by the Ministry of Science, ICT & Future Planning (2016R1D1A1A09918187).

Acknowledgments: The authors would like to thank the staff at PAL 7A beamline for their support and beam time.

Conflicts of Interest: The authors declare no conflict of interest.

References

1. Shashoua, V.E. Brain metabolism and the acquisition of new behaviors. I. Evidence for specific changes in the pattern of protein synthesis. *Brain Res.* **1976**, *111*, 347–364. [CrossRef]
2. Benowitz, L.I.; Shashoua, V.E. Localization of a brain protein metabolically linked with behavioral plasticity in the goldfish. *Brain Res.* **1977**, *136*, 227–242. [CrossRef]
3. Shashoua, V.E. Brain protein metabolism and the acquisition of new patterns of behavior. *Proc. Natl. Acad. Sci. USA* **1977**, *74*, 1743–1747. [CrossRef] [PubMed]
4. Shashoua, V.E. Ependymin, a brain extracellular glycoprotein, and CNS plasticity. *Ann. N. Y. Acad. Sci.* **1991**, *627*, 94–114. [CrossRef] [PubMed]
5. Schmidt, R. Cell-adhesion molecules in memory formation. *Behav. Brain Res.* **1995**, *66*, 65–72. [CrossRef]
6. Tang, S.J.; Sun, K.H.; Sun, G.H.; Lin, G.; Lin, W.W.; Chuang, M.J. Cold-induced ependymin expression in zebrafish and carp brain: Implications for cold acclimation. *FEBS Lett.* **1999**, *459*, 95–99. [CrossRef]
7. Sneddon, L.U.; Schmidt, R.; Fang, Y.; Cossins, A.R. Molecular correlates of social dominance: A novel role for ependymin in aggression. *PLoS ONE* **2011**, *6*, e18181. [CrossRef] [PubMed]
8. Apostolopoulos, J.; Sparrow, R.L.; McLeod, J.L.; Collier, F.M.; Darcy, P.K.; Slater, H.R.; Ngu, C.; Gregorio-King, C.C.; Kirkland, M.A. Identification and characterization of a novel family of mammalian ependymin-related proteins (MERPs) in hematopoietic, nonhematopoietic, and malignant tissues. *DNA Cell Biol.* **2001**, *20*, 625–635. [CrossRef] [PubMed]
9. Gregorio-King, C.C.; McLeod, J.L.; Collier, F.M.; Collier, G.R.; Bolton, K.A.; Van Der Meer, G.J.; Apostolopoulos, J.; Kirkland, M.A. MERP1: A mammalian ependymin-related protein gene differentially expressed in hematopoietic cells. *Gene* **2002**, *286*, 249–257. [CrossRef]
10. Nimmrich, I.; Erdmann, S.; Melchers, U.; Chtarbova, S.; Finke, U.; Hentsch, S.; Hoffmann, I.; Oertel, M.; Hoffmann, W.; Müller, O. The novel ependymin related gene UCC1 is highly expressed in colorectal tumor cells. *Cancer Lett.* **2001**, *165*, 71–79. [CrossRef]
11. Müller-Schmid, A.; Rinder, H.; Lottspeich, F.; Gertzen, E.M.; Hoffmann, W. Ependymins from the cerebrospinal fluid of salmonid fish: Gene structure and molecular characterization. *Gene* **1992**, *118*, 189–196. [CrossRef]
12. Königstorfer, A.; Sterrer, S.; Hoffmann, W. Biosynthesis of ependymins from goldfish brain. *J. Biol. Chem.* **1989**, *264*, 13689–13692. [PubMed]
13. Shashoua, V.E.; Adams, D.; Boyer-Boiteau, A. CMX-8933, a peptide fragment of the glycoprotein ependymin, promotes activation of AP-1 transcription factor in mouse neuroblastoma and rat cortical cell cultures. *Neurosci. Lett.* **2001**, *312*, 103–107. [CrossRef]
14. Adams, D.S.; Hasson, B.; Boyer-Boiteau, A.; El-Khishin, A.; Shashoua, V.E. A peptide fragment of ependymin neurotrophic factor uses protein kinase C and the mitogen-activated protein kinase pathway to activate c-Jun N-terminal kinase and a functional AP-1 containing c-Jun and c-Fos proteins in mouse NB2a cells. *J. Neurosci. Res.* **2003**, *72*, 405–416. [CrossRef] [PubMed]
15. Kaska, J. Ependymin Mechanism of Action: Full Length EPN vs Peptide CMX-8933. Master's Thesis, Worcester Polytechnic Institute, Worcester, MA, USA, 28 May 2003.

16. Suárez-Castillo, E.C.; García-Arrarás, J.E. Molecular evolution of the ependymin protein family: A necessary update. *BMC Evol. Biol.* **2007**, *7*, 23. [CrossRef] [PubMed]
17. Gill, S.C.; von Hippel, P.H. Calculation of protein extinction coefficients from amino acid sequence data. *Anal. Biochem.* **1989**, *182*, 319–326. [CrossRef]
18. Louis-Jeune, C.; Andrade-Navarro, M.A.; Perez-Iratxeta, C. Prediction of protein secondary structure from circular dichroism using theoretically derived spectra. *Proteins* **2012**, *80*, 374–381. [CrossRef] [PubMed]
19. Otwinowski, Z.; Minor, W. Processing of X-ray Diffraction Data Collected in Oscillation Mode. *Methods Enzymol.* **1997**, *276*, 307–326. [CrossRef] [PubMed]
20. Chang, V.T.; Crispin, M.; Aricescu, A.R.; Harvey, D.J.; Nettleship, J.E.; Fennelly, J.A.; Yu, C.; Boles, K.S.; Evans, E.J.; Stuart, D.I.; et al. Glycoprotein Structural Genomics: Solving the Glycosylation Problem. *Structure* **2007**, *15*, 267–273. [CrossRef] [PubMed]
21. Matthews, B.W. Solvent content of protein crystals. *J. Mol. Biol.* **1968**, *33*, 491–497. [CrossRef]
22. Pike, A.C.; Garman, E.F.; Krojer, T.; von Delft, F.; Carpenter, E.P. An overview of heavy-atom derivatization of protein crystals. *Acta Crystallogr. D Struct. Biol.* **2016**, *72*, 303–318. [CrossRef] [PubMed]
23. Pasternak, O.; Bujacz, A.; Biesiadka, J.; Bujacz, G.; Sikorski, M.; Jaskolski, M. MAD phasing using the (Ta6Br12)2+ cluster: A retrospective study. *Acta Crystallogr. D Struct. Biol.* **2008**, *64*, 595–606. [CrossRef] [PubMed]
24. Sippel, K.H.; Robbins, A.H.; Reutzel, R.; Domsic, J.; Boehlein, S.K.; Govindasamy, L.; Agbandje-McKenna, M.; Rosser, C.J.; McKenna, R. Structure determination of the cancer-associated Mycoplasma hyorhinis protein Mh-p37. *Acta Crystallogr. D Struct. Biol.* **2008**, *64*, 1172–1178. [CrossRef] [PubMed]
25. El Omari, K.; Iourin, O.; Kadlec, J.; Fearn, R.; Hall, D.R.; Harlos, K.; Grimes, J.M.; Stuart, D.I. Pushing the limits of sulfur SAD phasing: De novo structure solution of the N-terminal domain of the ectodomain of HCV E1. *Acta Crystallogr. D Struct. Biol.* **2014**, *70*, 2197–2203. [CrossRef] [PubMed]
26. Gorgel, M.; Bøggild, A.; Ulstrup, J.J.; Weiss, M.S.; Müller, U.; Nissen, P.; Boesen, T. Against the odds? De novo structure determination of a pilin with two cysteine residues by sulfur SAD. *Acta Crystallogr. D Struct. Biol.* **2015**, *71*, 1095–1101. [CrossRef] [PubMed]
27. Foos, N.; Seuring, C.; Schubert, R.; Burkhardt, A.; Svensson, O.; Meents, A.; Chapman, H.N.; Nanao, M.H. X-ray and UV radiation-damage-induced phasing using synchrotron serial crystallography. *Acta Crystallogr. D Struct. Biol.* **2018**, *74*, 366–378. [CrossRef] [PubMed]

crystals

MDPI

Communication

Crystal Structure of the Catalytic Domain of MCR-1 (cMCR-1) in Complex with D-Xylose

Zhao-Xin Liu [1,2,†], Zhenggang Han [1,†], Xiao-Li Yu [1], Guoyuan Wen [3] and Chi Zeng [1,3,*]

1 Hubei Province Engineering Research Center of Healthy Food, School of Biology and Pharmaceutical Engineering, Wuhan Polytechnic University, Wuhan 430023, China; jiangxinklzx@163.com (Z.-X.L.); zhengganghan@whpu.edu.cn (Z.H.); yxll268@126.com (X.-L.Y.)
2 College of Life Sciences, Wuhan University, Wuhan 430072, China
3 Key Laboratory of Prevention and Control Agents for Animal Bacteriosis (Ministry of Agriculture), Institute of Animal Husbandry and Veterinary, Hubei Academy of Agricultural Sciences, Wuhan 430064, China; wgy_524@163.com
* Correspondence: czeng@whpu.edu.cn
† These authors contributed equally to this work.

Received: 4 March 2018; Accepted: 14 April 2018; Published: 17 April 2018

Abstract: The polymyxin colistin is known as a "last resort" antibacterial drug toward pandrug-resistant enterobacteria. The recently discovered plasmid-encoded *mcr-1* gene spreads rapidly across pathogenic strains and confers resistance to colistin, which has emerged as a global threat. The *mcr-1* gene encodes a phosphoethanolamine transferase (MCR-1) that catalyzes the transference of phosphoethanolamine to lipid A moiety of lipopolysaccharide, resulting in resistance to colistin. Development of effective MCR-1 inhibitors is crucial for combating MCR-1-mediated colistin resistance. In this study, MCR-1 catalytic domain (namely cMCR-1) was expressed and co-crystallized together with D-xylose. X-ray crystallographic study at a resolution of 1.8 Å found that cMCR-1-D-xylose co-crystals fell under space group $P2_12_12_1$, with unit-cell parameters a = 51.6 Å, b = 73.1 Å, c = 82.2 Å, α = 90°, β = 90°, γ = 90°. The asymmetric unit contained a single cMCR-1 molecule complexed with D-xylose and had a solvent content of 29.13%. The structural model of cMCR-1-D-xylose complex showed that a D-xylose molecule bound in the putative lipid A-binding pocket of cMCR-1, which might provide a clue for MCR-1 inhibitor development.

Keywords: polymyxin resistance; colistin resistance; MCR-1

1. Introduction

Antimicrobial resistance among Gram-negative bacteria, especially the multidrug-resistant enterobacteria which are the leading cause of human clinical infections, is a global healthcare concern [1]. The carbapenemase-producing carbapenem-resistant *Enterobacteriaceae* (CRE), such as *Klebsiella pneumoniae* strains expressing the KPC-2 enzyme and *Enterobacteriaceae* strains expressing the NDM-1 enzyme, are of special clinical importance [1].

Polymyxin is often employed as the final therapeutic option to treat CRE-caused clinical infections because of its low resistance and high efficiency among CRE [2]. Polymyxins (colistin, polymyxin B) are cationic polypeptides which could bind the lipid A moiety of bacterial lipopolysaccharide and disrupt the bacterial cytomembrane subsequently [2]. Bacterial polymyxin resistance was considered to be very low and primarily caused by genomic mutations associated with specific two-component regulatory systems, which either modify lipid A or lead to complete loss of the lipopolysaccharide [2].

Recently, a novel mobile colistin resistance mechanism, led by a protein named MCR-1 (a phosphoethanolamine (PEA) transferase that confers colistin resistance by catalyzing the transference of phosphoethanolamine to lipid A moiety of lipopolysaccharide), has been discovered [3]. The gene encoding MCR-1 (*mcr-1*) has been shown plasmid-located and self-transmittable between various bacterial strains [2]. Until now, *mcr-1* has already been detected within a broad range of pathogenic isolates from humans and animals worldwide, which poses a huge threat to the sustaining effectiveness of colistin against CRE-caused clinical infections [2]. Development of effective MCR-1 inhibitors might be the only way to extend the usage of colistin as a reserved antibacterial drug to treat CRE infections [4].

Although several structures of MCR-1 catalytic domain (namely cMCR-1) have been determined [5–8], few effective inhibitors for MCR-1 are known. A recent co-crystallization study [9] showed that two substrate analogues of MCR-1, ethanolamine and D-glucose, could specifically bind to cMCR-1. Here, the crystallization and primary structure analysis of cMCR-1 complexed with D-xylose is reported. The structure determined showed that a D-xylose molecule bound in the putative lipid A-binding pocket of cMCR-1, which might provide a clue for MCR-1 inhibitor development.

2. Materials and Methods

2.1. Recombinant cMCR-1 Production

The sequence of *mcr-1* gene is available in GenBank (GenBank accession no. KY685070). Based on the secondary structure predictions, the MCR-1 catalytic domain (namely cMCR-1) includes 326 amino acids, from Pro216 to Arg541. The partial *mcr-1* gene sequence encoding cMCR-1 with *Nco*I/*Xho*I restriction sites incorporated at the 5′/3′ ends was commercially synthesized and cloned into *Nco*I/*Xho*I restriction sites of the expression vector pET-28a(+) (Novagen), creating pET-28a(+)-mcr-1. In construct pET-28a(+)-mcr-1, a histidine tag (HHHHHH) was fused to the C-terminus of cMCR-1 (Table 1).

Table 1. Production specifics for cMCR-1.

Source	*Escherichia coli*
DNA	Synthesized DNA
Forward primer [1]	5′-CATGCCATGGCCAAAAGATACCATTTATCAC-3′
Reverse primer [2]	5′-CCCTCGAGGCGGATGAATGCGGTGCGGTC-3′
Expression vector	pET-28a(+)
Host	*E. coli* BL21(DE3)pLysS
Recombinant protein sequence [3]	MGPKDTIYHAKDAVQATKPDMRKPRLVVF VVGETARADHVSFNGYERDTFPQLAKIDGVTNF SNVTSCGTSTAYSVPCMFSYLGADEYDVDTAK YQENVLDTLDRLGVSILWRDNNSDSKGVMDKLPKA QFADYKSATNNAICNTNPYNECRDVGMLVGLDDFV AANNGKDMLIMLHQMGNHGPAYFKRYDEKFAKFT PVCEGNELAKCEHQSLINAYDNALLATDDFIAQSIQ WLQTHSNAYDVSMLYVSDHGESLGENGVYLHGMP NAFAPKEQRSVPAFFWTDKQTGITPMATDTVLTHD AITPTLLKLFDVTADKVKDRTAFIRLEHHHHHH

[1] The *Nco*I site noted. [2] The *Xho*I site noted. [3] The cloning artifacts are underlined.

Escherichia coli BL21(DE3)pLysS was transformed with pET-28a(+)-mcr-1 and grown at 310 K, 200 rpm rotation in LB liquid medium containing 50 μg mL^{-1} kanamycin for cMCR-1 expression. Confluent cultures (OD_{600}~0.6) were then treated with 0.3 mM (final concentration) IPTG at 298 K with shaking (180 rpm) for 20 h. Cells were collected by 20 min of centrifugation (4500 g, 277 K) and pellets were kept at 193 K for subsequent use.

The cell pellets were lysed with 10 mM Tris-HCl pH 8.0, 200 mM NaCl, 5% (*v/v*) glycerol, 0.3% (*v/v*) Triton X-100, 1 mM DTT, and 0.1 mM PMSF in a French press. Cell wastes were excluded by centrifuging the lysates at 12,000 g for 30 min at 277 K, and the supernatant was clarified using a

0.45 μm filter and then passed through a pre-equilibrated Ni-NTA affinity column (GE Healthcare). The affinity column was washed thoroughly using 10 mM Tris-HCl pH 8.0 containing 50 mM imidazole to remove the miscellaneous proteins. The target proteins were eluted using 10 mM Tris-HCl pH 8.0 containing 200 mM imidazole. Concentrated protein was then loaded onto a MonoQ 5/50 GL anion exchange column (GE Healthcare) and chromatographed at 1 mL min^{-1} using a linear NaCl gradient generated with 10 mM Tris-HCl pH 8.0 (buffer A) and 10 mM Tris-HCl pH 8.0, 1 M NaCl (buffer B). Peak fractions were pooled and run through a Superdex 200 10/300 GL column (GE Healthcare) equilibrated with buffer A at a flow rate of 1 mL min^{-1}. Peak fractions were recovered and concentrated to 10 mg mL^{-1} for crystallization. The purity of the final protein (cMCR-1) was checked by SDS-PAGE. All specifics for recombinant cMCR-1 production are present in Table 1.

2.2. Crystallization

The cMCR-1 was crystallized at 277 K using the sitting-drop vapor-diffusion method as described by Wei et al. [9]. The 0.5 μL sitting drops consisting of 0.25 μL cMCR-1 solution and 0.25 μL reservoir solution were equilibrated against 30 μL reservoir solution in 96-well MRC plates (Molecular Dimensions). The best crystals were achieved in 10% (*w/v*) PEG 1000, 5% (*w/v*) PEG 8000.

2.3. Data Collection, Structure Solution, and Refinement

The cMCR-1 crystals were incubated in mother liquor containing 100 mM D-xylose for 10 s to form the cMCR-1-D-xylose complex. The cMCR-1-D-xylose co-crystals used for diffraction data collection were incubated in cryoprotectant (mother liquor containing 20% (*v/v*) glycerol) for 10 s before flash cooling in streams of liquid nitrogen. Data for cMCR-1-D-xylose complex were acquired at 100 K using an ADSC Q315r detector at beamline BL17U1 of Shanghai Synchrotron Radiation Facility (SSRF), China; 360 frames were taken with 1.0° oscillations. The data were indexed, integrated, and scaled using *HKL-2000* (HKL Research, Inc., Charlottesville, VA, USA) [10] and *iMosflm* programs [11]. The structure was solved by molecular replacement with *Phaser* [12] using a single monomer of cMCR-1 (PDB entry 5GRR [7]) as the search model. The structure model was constructed using alternating manual building in *Coot* [13] and restrained refinement in *PHENIX* [14]. The final model was optimized on PDB_REDO web server [15] and validated by *MolProbity* [16]. All figures were prepared by *PyMOL* (Schrödinger). Table 2 summarizes data-collection and crystallographic statistics of cMCR-1-D-xylose complex. Coordinates and structure factors of the cMCR-1-D-xylose complex have been deposited in the Protein Data Bank (PDB) under accession code 5ZJV.

Table 2. Data-collection and crystallographic statistics of cMCR-1-D-xylose complex.

Diffraction Source	BL17U1, SSRF
Wavelength (Å)	0.9792
Temperature (K)	100
Detector	ADSC Q315r
Crystal-to-detector distance (mm)	350
Total rotation range (°)	360
Rotation range per image (°)	1.0
Exposure time per image (s)	0.5
Space group	$P2_12_12_1$
a, b, c (Å)	51.6, 73.1, 82.2
α, β, γ (°)	90, 90, 90
Resolution range (Å)	43.71–1.82 (1.88–1.82) [1]
Total number of reflections	56014 (5514)
Number of unique reflections	28258 (2789)
Mosaicity (°)	0.5
Multiplicity	2.0 (2.0)
Completeness (%)	99.0 (99.0)
Mean $I/\sigma(I)$	10.61 (4.94)
R_{merge} (%)	3.5 (11.1)
$CC_{1/2}$	0.997 (0.949)
Wilson plot overall B factor (Å^2)	13.74
Reflection number, working set	28223 (2785)
Reflection number, test set	1997 (196)
R_{work}	0.139
R_{free}	0.178
Ramachandran favored region (%)	98
Ramachandran allowed region (%)	1.75
Ramachandran outliers (%)	0.25
Rotamer outliers (%)	0.65
R.m.s.d. bond lengths (Å)	0.006
R.m.s.d. bond angles (°)	0.86
Average B factor (Å^2)	17.55

[1] Outer shell values.

3. Results and Discussion

As stated in the Introduction, development of effective MCR-1 inhibitors is crucial for combating the threat of colistin resistance mediated by MCR-1. A recent co-crystallization study [9] showed that two substrate analogues of MCR-1 (ethanolamine and D-glucose) could specifically bind to MCR-1 catalytic domain (cMCR-1). Both D-glucose and lipid A are hexacyclic compounds. Thus, we tried many other hexacyclic compounds for co-crystallization with cMCR-1 (unpublished). The only other co-crystal structure was that obtained for the complex formed between cMCR-1 and D-xylose at a resolution of 1.8 Å.

MCR-1 belongs to the phosphoethanolamine (PEA) transferase family. It contains 541 amino acids with an N-terminal five-helix transmembrane domain (amino acid residues 1–215) and a C-terminal periplasmic catalytic domain (amino acid residues 216–541) [9]. In order to investigate the potential interactions between D-xylose (and other hexacyclic compounds) and MCR-1, cMCR-1 (MCR-1 catalytic domain) was expressed and purified using a combine of affinity, anion exchange and gel filtration chromatography (Figure 1a), as stated in Section 2.1. The purity of the purified cMCR-1 was confirmed with SDS-PAGE (Figure 1b) and subsequent Western blot analysis (Figure 1c). The cMCR-1 was effectively crystallized using the sitting-drop vapor-diffusion method as described by Wei et al. [9], which generated diffraction-quality crystals with a longest dimension of 0.2 mm (Figure 2). SDS-PAGE showed that the obtained crystals were protein crystals and had the same molecular weight as the purified cMCR-1 protein (Figure 1b). The cMCR-1 crystals were incubated in the mother liquor supplemented with 100 mM D-xylose for 10 s to form the cMCR-1-D-xylose complex.

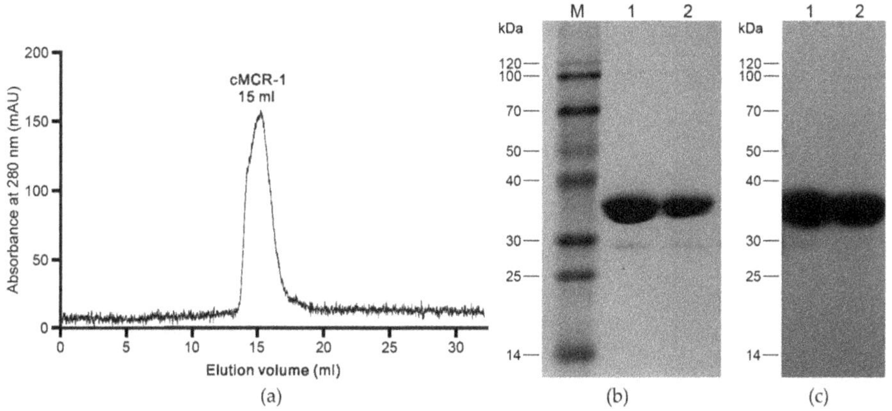

Figure 1. Purification and purity analysis of cMCR-1. (**a**) Gel filtration chromatography of cMCR-1. (**b**) SDS-PAGE of the final purified cMCR-1 and cMCR-1 crystals. Molecular-weight markers (lane M, labelled in kDa), purified ~35 kDa cMCR-1 protein (lane 1) and cMCR-1 crystals (lane 2) are shown. (**c**) Western blot analysis of the final purified cMCR-1 and cMCR-1 crystals using an anti-6×His antibody. Purified ~35 kDa cMCR-1 protein (lane 1) and cMCR-1 crystals (lane 2) are shown.

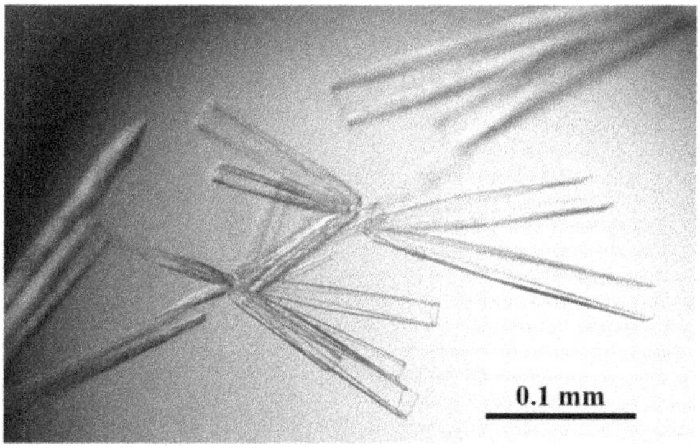

Figure 2. Crystals of cMCR-1.

Diffraction data for the cMCR-1-D-xylose complex was collected to 1.8 Å resolution (Figure 3) and on its basis, the cMCR-1-D-xylose co-crystals fell under space group $P2_12_12_1$, possessing unit-cell parameters a = 51.6 Å, b = 73.1 Å, c = 82.2 Å, α = 90°, β = 90°, γ = 90°. The asymmetric unit contained a single cMCR-1 molecule complexed with D-xylose. The data set of X-ray diffraction had a resolution range from 43.71 Å to 1.82 Å with 3.5% R_{merge} and 99.0% completeness. To elucidate the structure of cMCR-1-D-xylose complex, we employed molecular replacement method using cMCR-1 monomer (PDB entry 5GRR [7]) as a search model and obtained a clear solution. We confirmed the occurrence of a single protein molecule in the asymmetric unit by cross-rotation and translation-function calculations; the corresponding solvent content was 29.13%. Initial structure refinement using *PHENIX* [14] yielded a model (Figure 4a) with an R_{work} of 13.9% and an R_{free} of 17.8%.

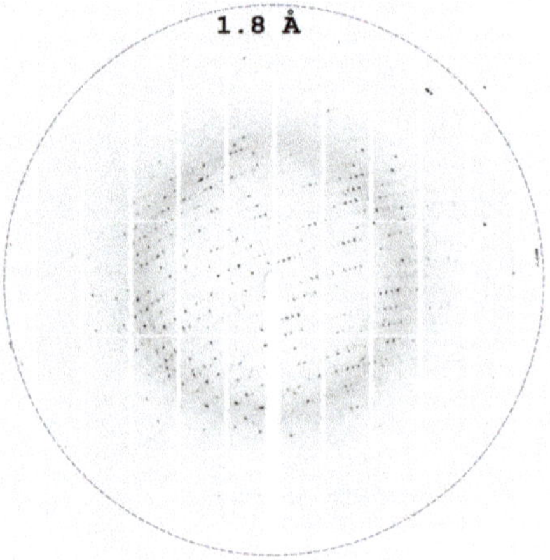

Figure 3. Representative X-ray diffraction pattern of cMCR-1-D-xylose co-crystal.

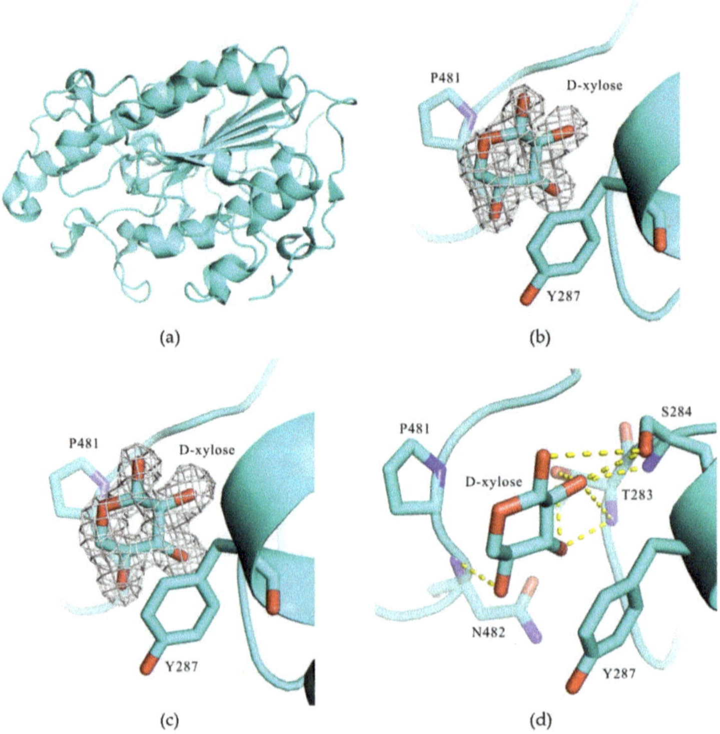

Figure 4. *Cont.*

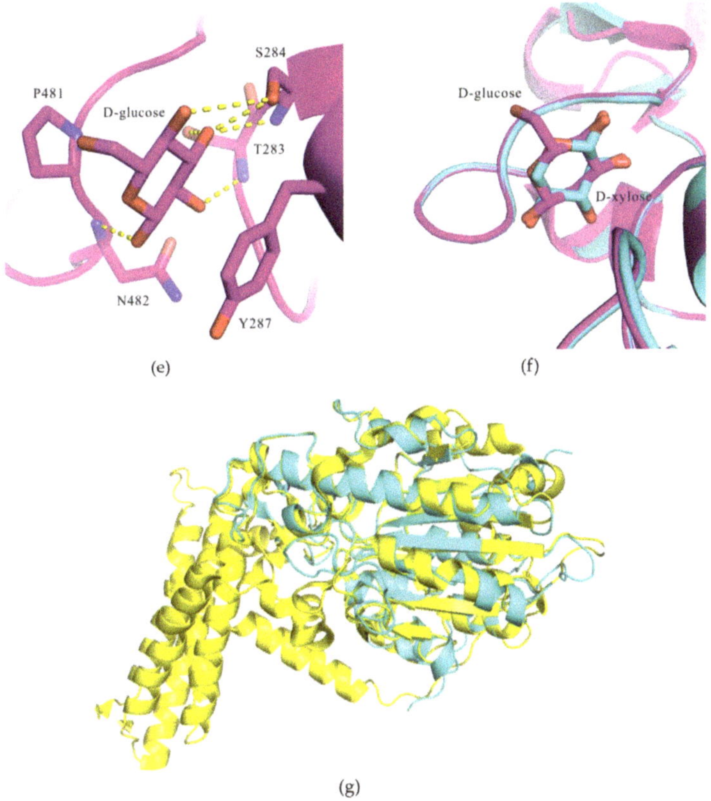

Figure 4. Structure of the cMCR-1-D-xylose complex. (**a**) The ribbon diagram showing the overall structure of cMCR-1. (**b**) The Fo-Fc electron-density map contoured at 3.0 σ depicting the D-xylose molecule. The map was calculated using the model omitting the D-xylose molecule after rounds of refinement. (**c**) The 2Fo-Fc electron-density map contoured at 1.0 σ depicting the D-xylose molecule. (**d**) Interaction between cMCR-1 and the D-xylose molecule. (**e**) Interaction between cMCR-1 and the D-glucose molecule. (**f**) Superposition of cMCR-1-D-xylose (cyan) and cMCR-1-D-glucose (magenta). (**g**) Superposition of cMCR-1-D-xylose (cyan) and EptA (yellow). All figures were prepared using *PyMOL* (Schrödinger).

As shown in the Fo-Fc map (contoured at 3.0 σ level) (Figure 4b) and 2Fo-Fc map (contoured at 1.0 σ level) (Figure 4c), a D-xylose molecule bound in the putative lipid A-binding pocket of cMCR-1. According to the PDB structure validation report, real space correlation coefficient (RSCC) and real space r-value (RSR) are 0.94 and 0.13, respectively, for the ligand D-xylose. The D-xylose, Pro481, and Tyr287 formed a sandwich structure with the D-xylose molecule in the middle (Figure 4d). Obviously, the hydrophobic stacking interaction played a crucial role in D-xylose recognition.

The structure of cMCR-1-D-xylose complex is similar with that of cMCR-1-D-glucose complex (PDB entry 5YLF [9]). Meanwhile, there are still many differences between the two structures. First, alignment of the two structures (Figure 4f) showed that both D-xylose and D-glucose bound to the same pocket of cMCR-1 and formed a π-π-conjugated interaction with Pro481 and Tyr287 of cMCR-1, but the skeletons of D-xylose and D-glucose were in the opposite positions. Next, we analyzed the cMCR-1-D-xylose and cMCR-1-D-glucose interactions by using

PISA server (http://www.ebi.ac.uk/pdbe/pisa). The accessible surface area (ASA) = 269.67 Å^2, buried surface area (BSA) = 194.27 Å^2, solvation energy effect ($\Delta^i G$) = −2.37 kcal/mol between D-xylose and cMCR-1, and the ASA = 303.17 Å^2, BSA = 204.82 Å^2, $\Delta^i G$ = −2.97 kcal/mol between D-glucose and cMCR-1. It suggested that the interaction between D-glucose and cMCR-1 is greater than that between D-xylose and cMCR-1. The D-xylose and D-glucose molecules also bound to cMCR-1 through a large number of hydrogen bonds. The O1, O2, O3, and O4 atoms of D-xylose hydrogen-bonded to Ser284 OG, Thr283 OG1, N/Ser284 N, OG, Thr283 OG1, N, and Asn482 N, respectively (Figure 4d). The O1, O2, O3, and O4 atoms of D-glucose hydrogen-bonded to Asn482 N, Thr283 N, Ser284 N, OG/Thr283 OG1, and Ser284 OG, respectively (Figure 4e).

We also conducted a comparison of cMCR-1-D-xylose complex with phosphoethanolamine transferase A (EptA) from *Neisseria meningitides* (PDB entry 5FGN [17]), the only structure of a full-length phosphoethanolamine transferase so far. The structure of cMCR-1-D-xylose complex can be well superimposed with the structure of EptA catalytic domain, with a Cα root-mean-square deviation of 2.0 Å as revealed by Dali server (http://ekhidna.biocenter.helsinki.fi/dali_server/start) [18] (Figure 4g). Anandan et al. [17] have shown that detergent dodecyl-β-D-maltoside (DDM) could bind in a substrate pocket of EptA, and the pocket bound by DDM was probably the phosphoethanolamine (PEA) binding pocket near the putative lipid A-binding pocket.

In conclusion, our finding that D-xylose could bind in the putative lipid A-binding pocket of cMCR-1 is interesting, which might provide a clue for MCR-1 inhibitor development. In vitro inhibitory assay is currently in progress to confirm if D-xylose could inhibit colistin resistance mediated by MCR-1.

Acknowledgments: Our study has been aided by an open grant from the Key Laboratory of Prevention and Control Agents for Animal Bacteriosis (Ministry of Agriculture) (KLAEMB-2017-07), a grant from the Research and Innovation Initiatives of WHPU (2018Y11), and National Science and Technology Major Project (2017ZX10201301-003-003).

Author Contributions: Zhao-Xin Liu and Chi Zeng designed the experiments. Zhao-Xin Liu performed the experiments. Zhao-Xin Liu, Chi Zeng, and Zhenggang Han analyzed the data. Xiao-Li Yu and Guoyuan Wen provided technical support. Zhao-Xin Liu, Chi Zeng, and Zhenggang Han completed the paper.

Conflicts of Interest: No interest conflict exists among the authors.

References

1. Logan, L.K.; Weinstein, R.A. The epidemiology of carbapenem-resistant *Enterobacteriaceae*: The impact and evolution of a global menace. *J. Infect. Dis.* **2017**, *215*, S28–S36. [CrossRef] [PubMed]
2. Poirel, L.; Jayol, A.; Nordmann, P. Polymyxins: Antibacterial activity, susceptibility testing, and resistance mechanisms encoded by plasmids or chromosomes. *Clin. Microbiol. Rev.* **2017**, *30*, 557–596. [CrossRef] [PubMed]
3. Liu, Y.Y.; Wang, Y.; Walsh, T.R.; Yi, L.X.; Zhang, R.; Spencer, J.; Doi, Y.; Tian, G.; Dong, B.; Huang, X.; et al. Emergence of plasmid-mediated colistin resistance mechanism MCR-1 in animals and human beings in China: A microbiological and molecular biological study. *Lancet Infect. Dis.* **2016**, *16*, 161–168. [CrossRef]
4. Hu, Y.Y.; Wang, Y.L.; Sun, Q.L.; Huang, Z.X.; Wang, H.Y.; Zhang, R.; Chen, G.X. Colistin resistance gene *mcr-1* in gut flora of children. *Int. J. Antimicrob. Agents* **2017**, *50*, 593–597. [CrossRef] [PubMed]
5. Hinchliffe, P.; Yang, Q.E.; Portal, E.; Young, T.; Li, H.; Tooke, C.L.; Carvalho, M.J.; Paterson, N.G.; Brem, J.; Niumsup, P.R.; et al. Insights into the mechanistic basis of plasmid-mediated colistin resistance from crystal structures of the catalytic domain of MCR-1. *Sci. Rep.* **2017**, *7*, 39392. [CrossRef] [PubMed]
6. Hu, M.; Guo, J.; Cheng, Q.; Yang, Z.; Chan, E.W.C.; Chen, S.; Hao, Q. Crystal structure of *Escherichia coli* originated MCR-1, a phosphoethanolamine transferase for colistin resistance. *Sci. Rep.* **2016**, *6*, 38793. [CrossRef] [PubMed]
7. Ma, G.; Zhu, Y.; Yu, Z.; Ahmad, A.; Zhang, H. High resolution crystal structure of the catalytic domain of MCR-1. *Sci. Rep.* **2016**, *6*, 39540. [CrossRef] [PubMed]
8. Stojanoski, V.; Sankaran, B.; Prasad, B.V.; Poirel, L.; Nordmann, P.; Palzkill, T. Structure of the catalytic domain of the colistin resistance enzyme MCR-1. *BMC Biol.* **2016**, *14*, 81. [CrossRef] [PubMed]

9. Wei, P.; Song, G.; Shi, M.; Zhou, Y.; Liu, Y.; Lei, J.; Chen, P.; Yin, L. Substrate analog interaction with MCR-1 offers insight into the rising threat of the plasmid-mediated transferable colistin resistance. *FASEB J.* **2018**, *32*, 1085–1098. [CrossRef] [PubMed]
10. Otwinowski, Z.; Minor, W. Processing of X-ray diffraction data collected in oscillation mode. *Methods Enzymol.* **1997**, *276*, 307–326. [CrossRef] [PubMed]
11. Battye, T.G.; Kontogiannis, L.; Johnson, O.; Powell, H.R.; Leslie, A.G. *iMOSFLM*: A new graphical interface for diffraction-image processing with *MOSFLM*. *Acta Crystallogr. D Biol. Crystallogr.* **2011**, *67*, 271–281. [CrossRef] [PubMed]
12. McCoy, A.J.; Grosse-Kunstleve, R.W.; Adams, P.D.; Winn, M.D.; Storoni, L.C.; Read, R.J. *Phaser* crystallographic software. *J. Appl. Crystallogr.* **2007**, *40*, 658–674. [CrossRef] [PubMed]
13. Emsley, P.; Cowtan, K. *Coot*: Model-building tools for molecular graphics. *Acta Crystallogr. D Biol. Crystallogr.* **2004**, *60*, 2126–2132. [CrossRef] [PubMed]
14. Adams, P.D.; Afonine, P.V.; Bunkóczi, G.; Chen, V.B.; Davis, I.W.; Echols, N.; Headd, J.J.; Hung, L.W.; Kapral, G.J.; Grosse-Kunstleve, R.W.; et al. *PHENIX*: A comprehensive Python-based system for macromolecular structure solution. *Acta Crystallogr. D Biol. Crystallogr.* **2010**, *66*, 213–221. [CrossRef] [PubMed]
15. Joosten, R.P.; Long, F.; Murshudov, G.N.; Perrakis, A. The PDB_REDO server for macromolecular structure model optimization. *IUCrJ* **2014**, *1*, 213–220. [CrossRef] [PubMed]
16. Chen, V.B.; Arendall, W.B.; Headd, J.J.; Keedy, D.A.; Immormino, R.M.; Kapral, G.J.; Murray, L.W.; Richardson, J.S.; Richardson, D.C. *MolProbity*: All-atom structure validation for macromolecular crystallography. *Acta Crystallogr. D Biol. Crystallogr.* **2010**, *66*, 12–21. [CrossRef] [PubMed]
17. Anandan, A.; Evans, G.L.; Condic-Jurkic, K.; O'Mara, M.L.; John, C.M.; Phillips, N.J.; Jarvis, G.A.; Wills, S.S.; Stubbs, K.A.; Moraes, I.; et al. Structure of a lipid A phosphoethanolamine transferase suggests how conformational changes govern substrate binding. *Proc. Natl. Acad. Sci. USA* **2017**, *114*, 2218–2223. [CrossRef] [PubMed]
18. Holm, L.; Laakso, L.M. Dali server update. *Nucleic Acids Res.* **2016**, *44*, W351–W355. [CrossRef] [PubMed]

crystals

MDPI

Brief Report

Refolding, Characterization, and Preliminary X-ray Crystallographic Studies on the *Campylobacter concisus* Plasmid-Encoded Secreted Protein Csep1^{p} Associated with Crohn's Disease

Mohammad Mizanur Rahman [1], Bradley Goff [2], Li Zhang [3,*] and Anna Roujeinikova [1,2,*]

1. Infection and Immunity Program, Monash Biomedicine Discovery Institute, Department of Microbiology, Monash University, Clayton, Victoria 3800, Australia; mohammad.mizanur.rahman@monash.edu
2. Department of Biochemistry and Molecular Biology, Monash University, Clayton, Victoria 3800, Australia; bgof1@student.monash.edu
3. School of Biotechnology and Biomolecular Sciences, University of New South Wales, Sydney, New South Wales 2052, Australia

* Correspondence: l.zhang@unsw.edu.au (L.Z.); anna.roujeinikova@monash.edu (A.R.); Tel.: +61-39-902-9194 (L.Z. & A.R.)

Received: 19 September 2018; Accepted: 14 October 2018; Published: 16 October 2018

Abstract: Colonization of *Campylobacter concisus* in the gastrointestinal tract can lead to the development of inflammatory bowel disease (IBD). Plasmid-encoded *C. concisus*-secreted protein 1 (Csep1^{p}) was recently identified as a putative pathogenicity marker associated with active Crohn's disease, a clinical form of IBD. Csep1^{p} shows no significant full-length sequence similarity to proteins of known structure, and its role in pathogenesis is not yet known. This study reports a method for extraction of recombinantly expressed Csep1^{p} from *Escherichia coli* inclusion bodies, refolding, and purification to produce crystallizable protein. Purified recombinant Csep1^{p} behaved as a monomer in solution. Crystals of Csep1^{p} were grown by the hanging drop vapour diffusion method, using polyethylene glycol (PEG) 4000 as the precipitating agent. A complete data set has been collected to 1.4 Å resolution, using cryocooling conditions and synchrotron radiation. The crystals belong to space group $P6_2$ or $P6_4$, with unit cell parameters a = b = 85.8, c = 55.2 Å, $\alpha = \beta = 90$, and $\gamma = 120°$. The asymmetric unit appears to contain one subunit, corresponding to a packing density of 2.47 Å^3 Da^{-1}.

Keywords: *Campylobacter consisus*; Crohn's disease; circular dichroism; protein crystallization; Csep1^{p}

1. Introduction

Campylobacter concisus is a gram-negative, spiral shaped, flagellated bacterium that, albeit being part of the normal human oral microflora [1], is also found in the intestinal tract, where its presence is associated with inflammatory bowel disease (IBD) [2]. *C. concisus* is genetically heterogeneous, with two distinct genomospecies (GS) that show differential propensities for translocation to and persistence in the human gastrointestinal tract [3–5]. In a genome study of 104 *C. concisus* isolates from 41 individuals, Kirk et al. (2018) [3] found GSII strains predominantly in gut mucosal samples (a result supported by a previous qPCR-based study [5]), and GSI strains were found to be overrepresented in oral samples. Translocation to and colonization of the gastrointestinal tract by *C. concisus* has been implicated as a potential etiological factor underlying inflammatory bowel disease (IBD); however, the molecular mechanisms underlying *C. concisus* pathogenesis in this respect remain poorly defined [2,6].

C. concisus has been shown to interact with host cells in a multifaceted manner. It is able to induce apoptosis and perturb the production of proteins associated with occluding junctions in the intestinal epithelium [7], modulate cell responses to bacterial lipopolysaccharides [8], and induce the production of pro-inflammatory cytokines [9]. By searching for homologues of the genes and proteins that had been previously implicated in colonization and virulence of other bacterial species, Kaakoush et al. [10] identified a set of 25 putative virulence factors present in the genome of the *C. concisus* strain 13826, including invasin InvA, adhesin CadF, two genes that encode zonula occludens toxins (ZOT), and phospholipase A (PldA). It was noted by Kaakoush et al. [10] that the presence of these virulence factors indicates that *C. concisus* may be capable of attaching to and invading host cells though a mechanism that targets occludens junctions, providing a starting point from which to investigate the enteric pathogenicity of *C. concisus*.

ZOT have been shown to compromise the intestinal barrier by interfering with proteins on the occludens junctions [11], whereas PldA has been shown to damage the membrane of mammalian cells [12]. Neither of the genes encoding these proteins, however, has a demonstrated association with IBD. Further, ZOT have been identified considerably more frequently in GSI *C. concisus* strains, which exhibit a reduced ability to persist in the GI tract [3]. As such, the usefulness of these gene products as molecular markers of pathogenicity in *C. concisus* with respect to IBD development is limited. Liu et al. [13] have recently identified a putative pathogenicity marker—the *C. concisus csep1-6bpi* gene. This gene was disproportionately present in oral *C. concisus* strains obtained from patients with active Crohn's disease (a clinical form of IBD), and were not detected in oral *C. concisus* strains isolated from both Crohn's patients in remission and healthy controls, suggesting that the gene may play a role in the development of Crohn's disease [13]. The *csep1-6bpi* gene is present in pICON plasmid or chromosome of GSII *C. concisus* strains, encoding *C. concisus*-secreted protein 1 (Csep1) that contains an N-terminal secretion signal (amino-acid residues 1-21); this protein was confirmed to have been secreted, being detected in the bacterial culture supernatant [13].

In this paper, we report expression, refolding from inclusion bodies, purification, and characterization of the pICON plasmid-encoded *C. concisus*-secreted protein 1 ($Csep1^{P}$), the novel putative virulence factor of *C. concisus*. This protein shows no significant full-length homology to proteins of known structure. The availability of pure recombinant $Csep1^{P}$ and determination of its atomic structure will greatly facilitate investigation into its function and its role in the *C. concisus* pathogenesis.

2. Materials and Methods

2.1. Gene Cloning and Overexpression

To express the protein in *E. coli*, the nucleotide sequence encoding $Csep1^{P}$ (locus tag CCS77_2074 on the pICON plasmid, GenBank ID CP021643.1) minus the signal peptide sequence was codon optimized, synthesized, and cloned into the pET151/D-TOPO vector (Invitrogen, Waltham, MA, USA) by GenScript. This expression vector contains an N-terminal His_6 tag followed by a tobacco etch virus (TEV) protease cleavage site. The recombinant protein used for biophysical assays and crystallization contained residues 22–222, plus an additional GIDPFT sequence at the N-terminus, as a cloning artifact originating from the TEV protease cleavage site. The vector was introduced into *E. coli* BL21 DE3 (Novagen, Merck Group, Darmstadt, Germany), and cells were then cultured with shaking in Luria–Bertani medium supplemented with 50 mg/L ampicillin at 310 K to an OD_{600} of 0.8. Overexpression of $Csep1^{P}$ was induced with 1 mM of isopropyl β-D-1-thiogalactopyranoside. Cells were then grown for a further 4 h at 310 K and harvested by centrifugation at 4800× *g* for 15 min at 277 K.

2.2. *Solubilization of Inclusion Bodies*

The cells were re-suspended in buffer A (20 mM Tris-HCl pH 8.0, 150 mM NaCl, and 1 mM phenylmethanesulfonyl fluoride (PMSF)), lysed using an EmulsiFlex-C5 cell disruption system (Avestin, Ottawa, ON, Canada) and centrifuged at 10,000× *g* for 30 min at 277 K. SDS-PAGE analysis of the proteins in the resultant pellet and supernatant indicated that Csep1^{P} is predominantly expressed in inclusion bodies (IBs). The IBs were solubilized by following the procedure described in [14], with some modifications. Briefly, the pellet was washed twice with buffer B (10 mM Tris-HCl pH 8.0, 0.2 mM PMSF, and 1% w/v Triton X-100) and once with buffer C (10 mM Tris-HCl pH 8.0 and 0.2 mM PMSF). After centrifugation at 10,000× *g* for 30 min, the supernatant was discarded and the IBs were solubilized in buffer D (10 mM Tris-HCl pH 8.0, 10 mM dithiothreitol (DTT), 8 M urea, and 0.2 mM PMSF) by axial rotation for 120 min at 277 K. The denatured protein solution was then clarified by centrifugation at 30,000× *g* for 30 min at 277 K, and protein concentration was determined using the Bradford assay [15]. Protein solution was aliquoted, snap-frozen in liquid nitrogen, and stored at 193 K.

2.3. *Refolding and Purification*

Refolding of recombinant Csep1^{P} was performed by diluting 70 mg of denatured protein into 250 mL of buffer E (3 M urea, 10 mM Tris-HCl pH 8.0, 0.4 M L-arginine monohydrochloride, 2 mM oxidized L-glutathione, and 20 mM reduced L-glutathione) followed by a 24-h incubation at 227 K with vigorous stirring. The sample was then dialyzed against 7.5 L of buffer F (10 mM Tris-HCl, pH 8.0) at 277 K, with four buffer changes over the period of 24 h. After that, Tris–HCl pH 8.0, NaCl, and imidazole were added to the sample to final concentrations of 20, 500, and 20 mM, respectively. The protein solution was loaded onto a 5 mL Ni-nitrilotriacetic acid (NTA) sepharose affinity column (GE Healthcare, Chicago, IL, USA) pre-equilibrated with buffer G (20 mM Tris–HCl pH 8.0, 500 mM NaCl, and 20 mM imidazole). The column was then washed with seven column volumes of the same buffer to remove unbound proteins, and Csep1^{P} was eluted with buffer H (20 mM Tris–HCl pH 8.0, 500 mM NaCl, and 500 mM imidazole). The N-terminal His_6 tag was cleaved off using His-tagged TEV protease (Invitrogen) during overnight dialysis against buffer I (20 mM Tris-HCl pH 8.0, 150 mM NaCl, 2 mM DTT, and 1% v/v glycerol at 277 K. NaCl and imidazole were then added to the sample to final concentrations of 500 and 20 mM, respectively. The uncleaved protein, His_6 tag, and TEV protease were removed over the Ni-NTA affinity column. The flow-through fractions were pooled, concentrated to 2 mL in an Amicon Ultracel 10 kDa cutoff concentrator, and loaded onto the Superdex 75 HiLoad 26/60 gel-filtration column (GE Healthcare, Chicago, IL, USA), pre-equilibrated with a buffer containing 20 mM Tris-HCl pH 8.0 and 150 mM NaCl (during the initial purification) or with buffer G (during the purification for crystallization) at a flow rate of 4 mL/min. The peak fractions of the eluate were pooled, and protein purity was assessed by SDS-PAGE. Tandem mass spectrometry analysis of the tryptic digest peptides obtained from the protein band cut out from the gel was performed using the Monash Biomedical Proteomics Facility. The oligomeric state of Csep1^{P} in solution was determined by calculating the molecular weight (MW) based on the retention volume, using the calibration plot for the Superdex 75 HiLoad 26/60 column: $V_{retention}$ (mL) = 631.3–104.3 × log MW [16].

2.4. *Thermal Shift Assay*

A thermal shift assay of protein stability in different buffers was performed using a Rotor-Gene Q Real time PCR instrument (QIAGEN, Hilden, Germany). Purified Csep1^{P} in buffer G was concentrated to 12 mg/mL (1 mM) and diluted into a range of different buffers, containing 10 × SYPRO Orange reagent (Sigma-Aldrich, 5000× stock, catalogue number S5692, St. Louis, MI, USA) to a concentration of 10 μM (volume = 25 μL). The samples were thermally denatured by heating from 35 °C to 90 °C at a ramp rate of 0.5 °C/min. Protein denaturation was monitored by following the SYPRO Orange fluorescence emission (λ_{ex} 530 nm/λ_{em} 555 nm). GraphPad Prism was used to fit the denaturation data to a derivation of the Boltzmann equation for the two-state unfolding model, in order to obtain the midpoint of denaturation (the melting temperature T_m) [17]. All experiments were performed in triplicate.

2.5. Circular Dichroism Spectroscopy

Prior to circular dichroism (CD) experiments, purified Csep1P was buffer-exchanged into 10 mM sodium phosphate pH 7.4. Far-UV CD spectra were recorded at a protein concentration of 0.24 mg/mL at 298 K, using a JASCO J-815 spectropolarimeter over a wavelength range from 200–250 nm with a scan rate of 20 nm/min. Spectra were recorded in triplicate and averaged. The secondary structure content was calculated using the BeStSel server [18].

2.6. Crystallization

Prior to crystallization, Csep1P was concentrated to 8 mg/mL and centrifuged at 277 K for 30 min at 13,200× *g* to clarify the solution. Initial screening for crystallization conditions was performed by the hanging-drop vapor-diffusion method, using an automated Phoenix crystallization robot (Art Robbins Instruments, Sunnyvale, CA, USA) and commercial screens Crystal Screen HT and PEG/Ion HT (Hampton Research, Aliso Viejo, CA, USA), JBS HTS1 and 2 (Jena Bioscience, Jena, Germany), and JCSG+ Suite (Qiagen, Hilden, Germany). The preliminary crystallization droplets contained 100 nL of protein solution mixed with 100 nL of reservoir solution and equilibrated against 50 μL of reservoir solution in a 96-well plate. After one day, crystals appeared in many different conditions. The condition containing 200 mM ammonium acetate, 100 mM sodium acetate trihydrate (pH 4.6), and 30% w/v polyethylene glycol (PEG) 4000 was chosen for optimisation. The refinement of this condition yielded monocrystals using 25% w/v PEG 4000, 100 mM ammonium acetate, 80 mM sodium acetate trihydrate (pH 4.6) as the reservoir solution, and 8 mg/mL of protein (drop size was 2 μL protein solution plus 2 μL reservoir solution, suspended over 500 μL reservoir solution).

2.7. Data Collection and Processing

For data collection, the Csep1P crystal was briefly soaked in a cryoprotectant solution containing 36% w/v PEG 4000, 100 mM ammonium acetate, 80 mM sodium acetate trihydrate (pH 4.6), and 10% v/v glycerol, and flash-cooled by plunging the crystal into liquid nitrogen. An X-ray diffraction data set was collected to 1.4 Å resolution on the MX2 beamline of the Australian Synchrotron (Figure 1). The data were processed and scaled using XDS [19] and AIMLESS [20] from the CCP4 suite [21]. The space group was determined with POINTLESS [22], and the Matthews coefficient was calculated using MATTHEWS_COEF [23] from the CCP4 software package. Data collection statistics are presented in Table 1.

Table 1. Data collection and processing statistics. Values in parentheses are for the highest resolution shell.

Diffraction Source	MX2 beamline, Australian Synchrotron
Detector	EIGER X 16M
Wavelength (Å)	1.07
Temperature (K)	100
Total oscillation span (°)	90
Mosaicity (°)	0.11
Space group	$P6_2$ or $P6_4$
Unit cell parameters	
a, b, c (Å)	85.8 85.8 55.2
α, β, γ (°)	90 90 120

Table 1. *Cont.*

Diffraction Source	MX2 beamline, Australian Synchrotron
Resolution range (Å)	28.08–1.40 (1.42–1.40)
Observed reflections	189,480 (2515)
Unique reflections	43,067 (1401)
Mean I/σ(I)	15.1 (1.0)
Completeness (%)	94 (61)
Multiplicity	4.4 (1.8)
R_{merge} [1]	0.030 (0.288)
$CC_{(1/2)}$ [2] (%)	99 (64)

[1] $R_{merge} = \frac{\left(\sum_h \sum_i |I_{hi} - \langle I_h \rangle|\right)}{\sum_h \sum_i |I_{hi}|}$, where I_{hi} is the intensity of the *i*th observation of reflection *h*. [2] $CC_{(1/2)}$ is the Pearson correlation coefficient calculated between two random half data sets.

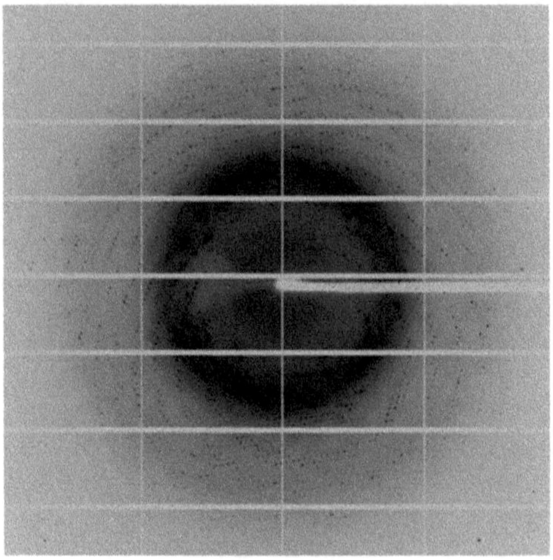

Figure 1. A representative oscillation image of the data collected from the Csep1^{P} crystal, using an EIGER X 16M pixel detector on the MX2 station at the Australian Synchrotron, Victoria, Australia. The edge of the detector corresponds to the resolution of 1.45 Å.

3. Results and Discussion

3.1. Cloning, Overexpression, Refolding, and Purification

An N-terminally His_6-tagged expression construct for recombinant Csep1^{P} lacking the signal peptide (residues 1–21) was created by ligating the synthetic, codon-optimized gene into the pET151/D-TOPO vector. Expression of Csep1^{P}-His_6 in *Escherichia coli* BL21 DE3 cells upon induction of T7 polymerase predominantly resulted in protein deposition in inclusion bodies (IBs). We isolated approximately 127 mg of protein in the form of washed IBs from 1 liter of bacterial culture. The recombinant Csep1^{P} was recovered from IBs by following the previously published refolding procedure, which involves diluting denatured protein into a buffer containing 10 mM Tris-HCl (pH 8.0), 10 mM dithiothreitol, 8 M urea, and 0.2 mM protease inhibitor phenylmethanesulfonyl fluoride [14]. Approximately 25 mg of tagged protein was obtained from 70 mg of IBs, corresponding to a soluble protein yield of 45 mg per 1 liter of culture (prior to purification).

Csep1^{P} was purified to higher than 95% homogeneity—based on Coomassie blue staining of the SDS–PAGE gel (Figure 2)—by affinity chromatography, followed by tag removal and gel filtration.

The recombinant protein comprised residues 22–222 of Csep1P, as well as six additional residues (GIDPFT) at the N-terminus, originating from the TEV cleavage site. The protein migrated on SDS–PAGE with an apparent molecular weight (MW) of 24 kDa (Figure 2). This value is very close to the MW calculated from the amino-acid sequence (24.34 kDa). The protein identity was confirmed by tandem mass spectrometry (MS) analysis of the tryptic digest peptides obtained from the protein band cut out from the gel. The search for peptides matching the Csep1P sequence against an *E. coli* background proteome, allowing for semi-tryptic specificity, three missed cleavages, and limited modifications, definitely identifying Csep1P based on 3102 spectra with 98% sequence coverage (Figure S1).

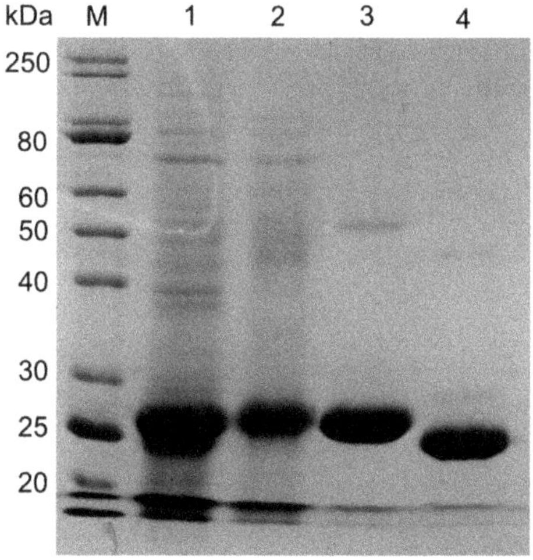

Figure 2. SDS-PAGE Coomassie Blue-stained gel (15%) of samples taken during the purification of recombinant Csep1P. Lane M: molecular weight ladder; Lane 1: inclusion bodies (IBs) after washing with buffer B; Lane 2: urea-solubilized IBs containing Csep1P-His$_6$; Lane 3: 15 µg of refolded Csep1P-His$_6$ after an affinity chromatography step; Lane 4: 15 µg of purified Csep1P with the His$_6$-tag removed.

3.2. *Protein Buffer Optimization*

Our initial attempt to concentrate the protein in a buffer containing 20 mM Tris-HCl pH 8.0 and 150 mM NaCl for crystallization experiments resulted in extensive precipitation, indicating limited solubility under those experimental conditions. We hypothesized that the protein would be more soluble in a buffer that increases its thermodynamic stability. We have therefore assessed the Csep1P stability in different buffers using a thermal shift assay. In this assay, a mixture of the purified protein and fluorescent dye is exposed to a temperature gradient; the dye's fluorescence increases when it binds to the protein's hydrophobic core, which becomes gradually exposed upon thermal denaturation. The protein unfolding curve is measured by following changes in the fluorescence. The melting temperature (T_m) value, which corresponds to the temperature at which 50% of the protein has denatured, provides a measure of the thermal stability of the protein.

We first screened nine different buffers commonly used in crystallography, in the pH range between 3.0 and 8.0, each at a concentration of 100 mM and in the presence or absence of 150 mM NaCl. Analysis of the protein unfolding curves (Figure S2a) indicated that under very acidic conditions (pH 3.0–4.0), the protein's fold is highly unstable, even at room temperature; the fluorescent signal started at a high value and decreased, rather than increasing, with temperature. No meaningful T_m value could be ascribed to those curves. Protein stability under moderately acidic conditions (pH 4.6) was also significantly lower than that in the pH range between 6.0 and 8.0, as evidenced by the lower

T_m value (Figure 3a). Furthermore, the presence of 150 mM NaCl had a favorable effect on the protein's stability, resulting in up to 2 °C increase in the T_m value when compared to the respective buffer without salt (Figure 3a), suggesting that higher ionic strength promotes stabilization of this protein fold. As Csep1^P in 100 mM HEPES pH 7.0, 150 mM NaCl showed the highest T_m value (64.6 °C) in this screen, we attempted to concentrate the sample in this buffer; however, that also resulted in protein precipitation.

We then took into account the observation that during the purification procedure, the protein could be concentrated to 10–20 mg/mL in a buffer containing 20 mM Tris-HCl (pH 8.0), 20 mM imidazole, 500 mM NaCl, 2 mM DTT, and 1% v/v glycerol. Since imidazole and glycerol are known to increase the stability of some proteins, we produced a screen designed to systematically test the effect of glycerol and imidazole at increasing salt concentrations (150, 250, and 500 mM NaCl) on Csep1^P stability. In addition, we tested the effect of charged amino acids L-Arg and L-Glu at 30 mM on T_m, as the Arg–Glu mix has also been shown to increase the solubility of proteins [24]. The results of the thermal shift assay using this screen are shown in Figure 3b. No significant differences in T_m were observed between the conditions with and without glycerol. In contrast, imidazole, the Arg–Glu mix, and higher NaCl concentrations each showed a stabilizing effect, as judged by an increase in the respective T_m value (Figure 3b), with 500 mM NaCl having a more pronounced effect than 150 or 250 mM. We have selected three conditions corresponding to the highest observed T_m values: (1) 100 mM Tris-HCl (pH 8.0), 30 mM Arg, 30 mM Glu, and 500 mM NaCl (T_m = 66.0 °C); (2) 100 mM Tris-HCl (pH 8.0), 10 mM imidazole, and 500 mM NaCl (T_m = 65.8 °C); and (3) 100 mM HEPES (pH 7.0), 10 mM imidazole, and 500 mM NaCl (T_m = 65.8 °C). We tested protein solubility in these buffers by concentrating the dilute protein, buffer-exchanged into the respective buffer by dialysis up to the solubility limit. The highest concentration of ~20 mg/mL was achieved in condition 2; similar levels were achieved in a slightly modified condition 2, containing 20 mM Tris-HCl (pH 8.0), 20 mM imidazole, and 500 mM NaCl (buffer G, see "Materials and Methods"). Therefore, to streamline the purification procedure while retaining protein solubility at high levels, the final gel-filtration and concentration steps were performed in buffer G.

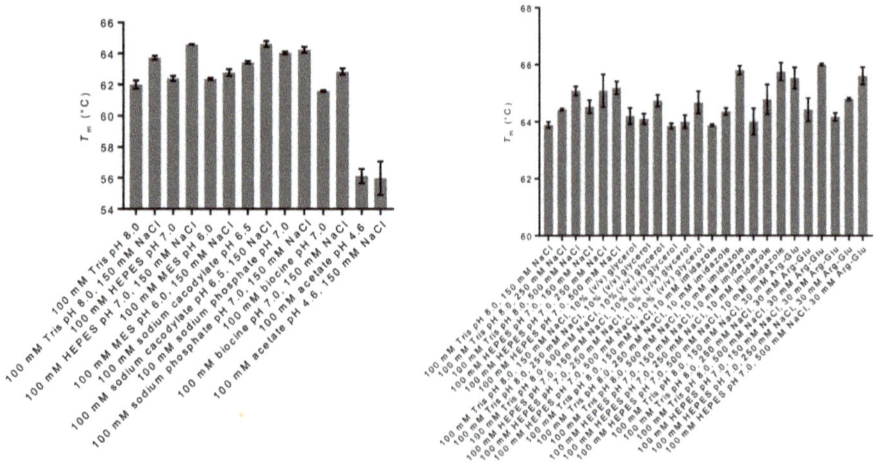

(a) (b)

Figure 3. Comparison of melting temperature (T_m) of Csep1^P in different buffers. (**a**) Effect of pH and presence or absence of 150 mM NaCl; (**b**) Effect of additives (glycerol, imidazole or Arg-Glu mix) and increasing NaCl concentration. Results are means ± S.D. for three independent replicates.

3.3. *Stoichiometry and Secondary Structure Content of Csep1*^{p}

When subjected to size-exclusion chromatography on a calibrated gel-filtration column, the protein eluted as a single, symmetrical peak with a retention volume of 176 mL (Figure S3), which corresponds to an apparent MW of approximately 23 kDa. This indicates that Csep1^{p} is monomeric under the tested buffer conditions.

To ascertain the fold integrity of the prepared Csep1^{p} sample prior to crystallization trials, we estimated its secondary structure content using circular dichroism (CD). Analysis of the CD spectrum (Figure 4) yielded values of 38% and 20% for α-helix and β-sheet content, respectively. These values are close to those predicted from the primary sequence analysis (α = 38%, β = 19%) using the Jpred4 server [25], confirming that Csep1^{p} extracted from IBs is folded.

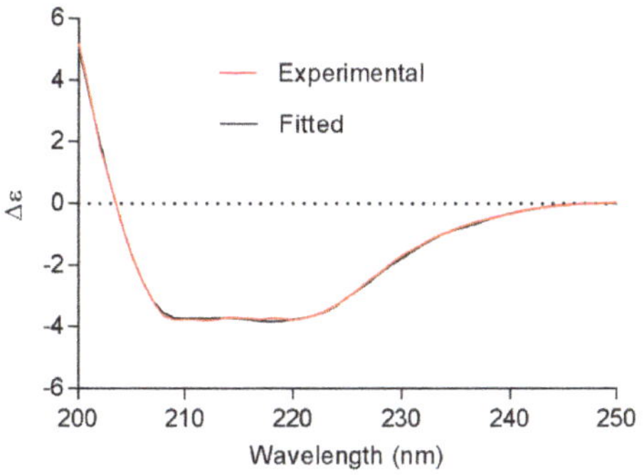

Figure 4. Circular dichroism (CD) spectrum of recombinant Csep1^{p}.

3.4. *Crystallization and Preliminary X-ray Analysis*

To initiate a study of the structure/function relationship of Csep1^{p}, we undertook robotic crystallization trials using commercially available screens, and optimized the preliminary hits manually to produce monocrystals of suitable size and diffraction quality. The best crystals were obtained using PEG 4000 as a precipitant and a protein concentration of 8 mg/mL. The crystals typically appeared after two days (Figure 5). An X-ray diffraction data set was collected for a single cryo-cooled crystal, using beamline MX2 at the Australian Synchrotron to a resolution of 1.4 Å. Auto-indexing of the diffraction data using XDS [19] was consistent with a trigonal or hexagonal crystal system. The data could be scaled using AIMLESS [20] in the hexagonal system, and analysis using POINTLESS [22] showed systematic absences along the 00*l* axis, with reflections only present when *l* = 3n, which suggested that the crystals belong to space group $P6_2$ or its enantiomorph $P6_4$. The average I/σ(I) value was 15.1 for all reflections (resolution range 28.08–1.40 Å) and 1.0 in the highest resolution shell (1.42–1.40 Å). A total of 189,480 measurements were made of 43,067 independent reflections. Data processing gave an R_{merge} of 0.03 for intensities (0.288 in the 1.42–1.40 Å resolution shell). The data was 94% complete, with 61% completeness in the highest resolution shell (Table 1). Analysis of the data using PHENIX Xtriage [26] detected no signs of twinning. Calculation of the Matthews coefficient and solvent content for one molecule in the asymmetric unit gave values of 2.47 $Å^3$ Da^{-1} and 50%, respectively, which lies in the range observed for protein crystals [27]. Determination of the structure will require heavy-atom derivatization or selenomethionine substitution to allow experimental phasing using multiple isomorphous replacement or single- or multi-wavelength anomalous dispersion methods.

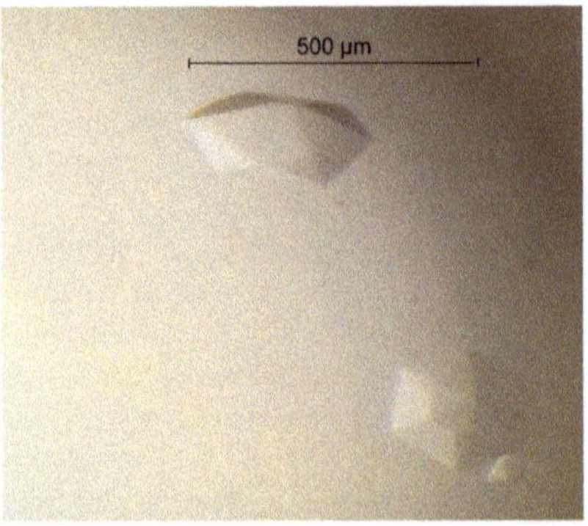

Figure 5. Crystals of Csep1P from a pICON plasmid.

Supplementary Materials: The following are available online at http://www.mdpi.com/2073-4352/8/10/391/s1. Figure S1: Csep1P sequence coverage by the tryptic digest peptides identified in the sample using tandem mass spectrometry analysis. Figure S2: Normalised thermal unfolding (melting) curves of Csep1P in different buffers, measured by following changes in the fluorescence of SYPRO Orange. (a) Effect of pH and presence or absence of 150 mM NaCl; (b) effect of additives (glycerol, imidazole, or Arg–Glu mix) and increasing NaCl concentration. Figure S3: Gel filtration chromatogram. The minor peak at the void volume (~85 mL) corresponds to a small amount of non-specific, large aggregates.

Author Contributions: L.Z. and A.R. conceived and coordinated this study; M.M.R. and A.R. designed the experiments; M.M.R. and B.G. performed the experiments; M.M.R. and A.R. analyzed the data; all authors wrote or edited the paper; all authors read and approved the final manuscript.

Acknowledgments: We thank Danuta Maksel and Geoffrey Kwai Wai Kong at the Monash Macromolecular Crystallisation Facility for their assistance in setting up robotic crystallization screens. Part of this research was undertaken on the MX2 beamline of the Australian Synchrotron (AS), Victoria, Australia. We also thank the AS staff for their assistance with data collection.

Conflicts of Interest: The authors declare that they have no competing interests.

References

1. Zhang, L.; Budiman, V.; Day, A.S.; Mitchell, H.; Lemberg, D.A.; Riordan, S.M.; Grimm, M.; Leach, S.T.; Ismail, Y. Isolation and detection of *Campylobacter concisus* from saliva of healthy individuals and patients with inflammatory bowel disease. *J. Clin. Microbiol.* **2010**, *48*, 2965–2967. [CrossRef] [PubMed]
2. Zhang, L.; Lee, H.; Grimm, M.C.; Riordan, S.M.; Day, A.S.; Lemberg, D.A. *Campylobacter concisus* and inflammatory bowel disease. *World J. Gastroenterol.* **2014**, *20*, 1259–1267. [CrossRef] [PubMed]
3. Kirk, K.F.; Meric, G.; Nielsen, H.L.; Pascoe, B.; Sheppard, S.K.; Thorlacius-Ussing, O.; Nielsen, H. Molecular epidemiology and comparative genomics of *Campylobacter concisus* strains from saliva, faeces and gut mucosal biopsies in inflammatory bowel disease. *Sci. Rep.* **2018**, *8*, 1902. [CrossRef] [PubMed]
4. Mahendran, V.; Octavia, S.; Demirbas, O.F.; Sabrina, S.; Ma, R.; Lan, R.; Riordan, S.M.; Grimm, M.C.; Zhang, L. Delineation of genetic relatedness and population structure of oral and enteric *Campylobacter concisus* strains by analysis of housekeeping genes. *Microbiology* **2015**, *161*, 1600–1612. [CrossRef] [PubMed]
5. Wang, Y.; Liu, F.; Zhang, X.; Chung, H.K.L.; Riordan, S.M.; Grimm, M.C.; Zhang, S.; Ma, R.; Lee, S.A.; Zhang, L. *Campylobacter concisus* genomospecies 2 is better adapted to the human gastrointestinal tract as compared with *Campylobacter concisus* genomospecies 1. *Front. Physiol.* **2017**, *8*, 543. [CrossRef] [PubMed]
6. Zhang, L. Oral *Campylobacter* species: Initiators of a subgroup of inflammatory bowel disease? *World J. Gastroenterol.* **2015**, *21*, 9239–9244. [CrossRef] [PubMed]
7. Nielsen, H.L.; Nielsen, H.; Ejlertsen, T.; Engberg, J.; Günzel, D.; Zeitz, M.; Hering, N.A.; Fromm, M.; Schulzke, J.-D.; Bücker, R. Oral and fecal *Campylobacter concisus* strains perturb barrier function by apoptosis induction in HT-29/B6 intestinal epithelial cells. *PLoS ONE* **2011**, *6*, e23858. [CrossRef] [PubMed]

8. Ismail, Y.; Mahendran, V.; Octavia, S.; Day, A.S.; Riordan, S.M.; Grimm, M.C.; Lan, R.; Lemberg, D.; Tran, T.A.; Zhang, L. Investigation of the enteric pathogenic potential of oral *Campylobacter concisus* strains isolated from patients with inflammatory bowel disease. *PLoS ONE* **2012**, *7*, e38217. [CrossRef] [PubMed]
9. Man, S.M.; Kaakoush, N.O.; Leach, S.T.; Nahidi, L.; Lu, H.K.; Norman, J.; Day, A.S.; Zhang, L.; Mitchell, H.M. Host attachment, invasion, and stimulation of proinflammatory cytokines by *Campylobacter concisus* and other non-*Campylobacter jejuni Campylobacter* species. *J. Infect. Dis.* **2010**, *202*, 1855–1865. [CrossRef] [PubMed]
10. Kaakoush, N.O.; Man, S.M.; Lamb, S.; Raftery, M.J.; Wilkins, M.R.; Kovach, Z.; Mitchell, H. The secretome of *Campylobacter concisus*. *FEBS J.* **2010**, *277*, 1606–1617. [CrossRef] [PubMed]
11. Mahendran, V.; Liu, F.; Riordan, S.M.; Grimm, M.C.; Tanaka, M.M.; Zhang, L. Examination of the effects of *Campylobacter concisus* zonula occludens toxin on intestinal epithelial cells and macrophages. *Gut Pathog.* **2016**, *8*, 18. [CrossRef] [PubMed]
12. Istivan, T.S.; Coloe, P.J.; Fry, B.N.; Ward, P.; Smith, S.C. Characterization of a haemolytic phospholipase A(2) activity in clinical isolates of *Campylobacter concisus*. *J. Med. Microbiol.* **2004**, *53*, 483–493. [CrossRef] [PubMed]
13. Liu, F.; Ma, R.; Tay, C.Y.A.; Octavia, S.; Lan, R.; Chung, H.K.L.; Riordan, S.M.; Grimm, M.C.; Leong, R.W.; Tanaka, M.M.; et al. Genomic analysis of oral *Campylobacter concisus* strains identified a potential bacterial molecular marker associated with active Crohn's disease. Emerg. *Microbes Infect.* **2018**, *7*, 64. [CrossRef] [PubMed]
14. Liu, Y.C.; Roujeinikova, A. Expression, refolding, purification and crystallization of the sensory domain of the TlpC chemoreceptor from *Helicobacter pylori* for structural studies. *Protein Expr. Purif.* **2015**, *107*, 29–34. [CrossRef] [PubMed]
15. Bradford, M.M. A rapid and sensitive method for the quantitation of microgram quantities of protein utilizing the principle of protein-dye binding. *Anal. Biochem.* **1976**, *72*, 248–254. [CrossRef]
16. Aydin, I.; Dimitropoulos, A.; Chen, S.H.; Thomas, C.; Roujeinikova, A. Purification, crystallization and preliminary X-ray crystallographic analysis of the putative *Vibrio parahaemolyticus* resuscitation-promoting factor YeaZ. *Acta Crystallogr. Sect. F Struct. Biol. Cryst. Commun.* **2011**, *67*, 604–607. [CrossRef] [PubMed]
17. Orwig, S.D.; Lieberman, R.L. Biophysical characterization of the olfactomedin domain of myocilin, an extracellular matrix protein implicated in inherited forms of glaucoma. *PLoS ONE* **2011**, *6*, e16347. [CrossRef] [PubMed]
18. Micsonai, A.; Wien, F.; Bulyaki, E.; Kun, J.; Moussong, E.; Lee, Y.H.; Goto, Y.; Refregiers, M.; Kardos, J. BeStSel: A web server for accurate protein secondary structure prediction and fold recognition from the circular dichroism spectra. *Nucleic Acids Res* **2018**, *46*, W315–W322. [CrossRef] [PubMed]
19. Kabsch, W. XDS. *Acta Crystallogr. D Biol. Crystallogr.* **2010**, *66*, 125–132. [CrossRef] [PubMed]
20. Evans, P.R.; Murshudov, G.N. How good are my data and what is the resolution? *Acta Crystallogr. D Biol. Crystallogr.* **2013**, *69*, 1204–1214. [CrossRef] [PubMed]
21. Winn, M.D.; Ballard, C.C.; Cowtan, K.D.; Dodson, E.J.; Emsley, P.; Evans, P.R.; Keegan, R.M.; Krissinel, E.B.; Leslie, A.G.; McCoy, A.; et al. Overview of the CCP4 suite and current developments. *Acta Crystallogr. D Biol. Crystallogr.* **2011**, *67*, 235–242. [CrossRef] [PubMed]
22. Evans, P. Scaling and assessment of data quality. *Acta Crystallogr. D Biol. Crystallogr.* **2006**, *62*, 72–82. [CrossRef] [PubMed]
23. Kantardjieff, K.A.; Rupp, B. Matthews coefficient probabilities: Improved estimates for unit cell contents of proteins, DNA, and protein–nucleic acid complex crystals. *Protein Sci.* **2003**, *12*, 1865–1871. [CrossRef] [PubMed]
24. Golovanov, A.P.; Hautbergue, G.M.; Wilson, S.A.; Lian, L.-Y. A simple method for improving protein solubility and long-term stability. *J. Am. Chem. Soc.* **2004**, *126*, 8933–8939. [CrossRef] [PubMed]
25. Drozdetskiy, A.; Cole, C.; Procter, J.; Barton, G.J. JPred4: A protein secondary structure prediction server. *Nucleic Acids Res.* **2015**, *43*, W389–W394. [CrossRef] [PubMed]
26. Zwart, P.H.; Grosse-Kunstleve, R.W.; Lebedev, A.A.; Murshudov, G.N.; Adams, P.D. Surprises and pitfalls arising from (pseudo)symmetry. *Acta Crystallogr. D Biol. Crystallogr.* **2008**, *64*, 99–107. [CrossRef] [PubMed]
27. Matthews, B.W. Solvent content of protein crystals. *J. Mol. Biol.* **1968**, *33*, 491–497. [CrossRef]

crystals

MDPI

Article

Crystallization and Crystallographic Analysis of a *Bradyrhizobium Elkanii* USDA94 Haloalkane Dehalogenase Variant with an Eliminated Halide-Binding Site

Tatyana Prudnikova [1,2,†], Barbora Kascakova [1,†], Jeroen R. Mesters [3], Pavel Grinkevich [1], Petra Havlickova [1], Andrii Mazur [1,2], Anastasiia Shaposhnikova [1,2], Radka Chaloupkova [4], Jiri Damborsky [4,5], Michal Kuty [1,2] and Ivana Kuta Smatanova [1,2,*]

1. Faculty of Science, University of South Bohemia in Ceske Budejovice, Branisovska 1760, 37005 Ceske Budejovice, Czech Republic
2. Center of Nanobiology and Structural Biology, Institute of Microbiology of the Czech Academy of Sciences, Zamek 136, 37333 Nove Hrady, Czech Republic
3. Institute of Biochemistry, University of Lübeck, Ratzeburger Allee 160, 23538 Lübeck, Germany
4. Loschmidt Laboratories, Department of Experimental Biology and RECETOX, Faculty of Science, Masaryk University, Kamenice 5/A4, 62500 Brno, Czech Republic
5. International Clinical Research Center, St. Anne's University Hospital Brno, Pekarska 53, 65691 Brno, Czech Republic

* Correspondence: ivanaks@seznam.cz or talianensis@gmail.com

† These authors contributed equally to this work.

Received: 27 June 2019; Accepted: 19 July 2019; Published: 23 July 2019

Abstract: Haloalkane dehalogenases are a very important class of microbial enzymes for environmental detoxification of halogenated pollutants, for biocatalysis, biosensing and molecular tagging. The double mutant (Ile44Leu + Gln102His) of the haloalkane dehalogenase DbeA from *Bradyrhizobium elkanii* USDA94 (DbeAΔCl) was constructed to study the role of the second halide-binding site previously discovered in the wild-type structure. The variant is less active, less stable in the presence of chloride ions and exhibits significantly altered substrate specificity when compared with the DbeAwt. DbeAΔCl was crystallized using the sitting-drop vapour-diffusion procedure with further optimization by the random microseeding technique. The crystal structure of the DbeAΔCl has been determined and refined to the 1.4 Å resolution. The DbeAΔCl crystals belong to monoclinic space group C121. The DbeAΔCl molecular structure was characterized and compared with five known haloalkane dehalogenases selected from the Protein Data Bank.

Keywords: Haloalkane dehalogenase; halide-binding site; random microseeding

1. Introduction

Hazardous halogenated compounds are an important class of environmental pollutants. An obvious critical step in the potential biodegradation pathway is the dehalogenation process [1,2]. Haloalkane dehalogenases (HLDs) play an essential role in biodegradation of the halogenated pollutants. HLDs are predominantly bacterial enzymes that belong to the superfamily of α/β-hydrolases and catalyze the hydrolytic conversion of a wide range of halogenated aliphatic compounds, and therefore play an important role in bioremediation [3] and industrial biocatalytic processes [4]. An aliphatic alcohol, a halide and a hydrogen cation are released during the enzymatic dehalogenation of haloalkanes by HLDs. The tertiary structures of HLDs are composed of a conserved α/β-hydrolase core domain and an α-helical cap domain [5]. The core domain is responsible for the catalytic reaction of the enzyme and the cap domain is essential for substrate specificity and recognition [6]. A deep cleft is situated between these two domains, allowing the solvent to access the buried active site. The active site is composed of two halide-anion stabilizing residues and the catalytic triad consisting of a nucleophile, a base and an acid [7]. HLDs can be divided into three subfamilies, HLD-I, HLD-II and HLD-III, according to the composition of the catalytic residues and the anatomy of the cap domain [4].

A novel HLD DbeA from *B. elkanii* USDA94, a member of HLD-II subfamily [4], was structurally and biochemically characterized [7]. The structure of DbeA wild type was determined to 2.2 Å resolution and displays a typical topology of the α/β-hydrolases (EC 3.8.1.5). The unique feature of the DbeA structure is the presence of two halide-binding sites, both fully occupied by chloride anions [7]. The first halide-binding site is located in the protein active site and is involved in substrate binding and stabilization of halogen ion produced during dehalogenation reaction. DbeA active site consists of five catalytic residues: two halide stabilizing residues (Trp104 and Asn38) and three amino acids essential for the catalytic activity of the enzyme [2,4]: the nucleophile Asp103, the catalytic base His271, and the catalytic acid Glu127. The second halide-binding site in DbeA is unique and has never been observed within HLD structures deposited in the PDB [8]. The second halide-binding site, which is buried in the protein core domain and located approximately 10 Å far from the first halide-binding site, is formed by five amino-acid residues: Ile44, Gln274, Gln102, Gly37 and Thr40 [7]. Superposition of the DbeA structure with other related HLD-II members revealed the presence of two unique amino acids in the second halide-binding site: Gln102 instead of a typical His and Ile44 as a substitution of an ordinary Leu, thereby sufficiently increasing the cavity volume to accommodate the second halide ion. The variant DbeAΔCl (Ile44Leu+Gln102His) was constructed and biochemically characterized to elucidate the role of the second halide-binding site in structure and function of DbeA [7].

Removal of the second halide binding site in DbeA significantly changed the substrate specificity of DbeAΔCl and reduced the catalytic activity by an order of magnitude towards most of the tested substrates [7]. Wild-type DbeA is more active, its melting temperature rises with an increasing concentration of chloride salts and the binding energy for chloride ions is higher when compared with the DbeAΔCl variant [7]. It was suggested that the chloride anion bound in a vicinity of second binding site may increase basicity of catalytic histidine and consequently accelerate the nucleophilic addition of water to the alkyl-enzyme intermediate [7]. In previous attempts, the crystallization of DbeAΔCl was unsuccessful. The obtained crystals were very unstable, sensitive to mechanical stress and poorly diffracted X-rays to about 10 Å resolution. It took several years to grow crystals with an improved diffraction quality. Here, we report the successful crystallization, structure determination and further characterization of DbeAΔCl variant.

2. Materials and Methods

2.1. Gene Synthesis, Cloning, Expression and Protein Purification

The recombinant gene dbeAΔCl-His6 (Ile44Leu + Gln102His) was synthesized artificially (Entelechon, Regensburg, Germany) according to the DbeA sequence [7] (Table 1). The restriction endonucleases NdeI and XhoI (Fermentas, Burlington, Canada) and T4 DNA ligase (Promega, Madison, USA) were applied to transfer the synthesized gene into the expression vector pET-21b (Novagen, Madison, USA). In order to overexpress DbeAΔCl in *E. coli* BL21(DE3) cells, the final genes were transcribed by T7 RNA polymerase, which is expressed by the isopropyl β-D-1-thiogalactopyranoside (IPTG)-inducible lac UV5 promoter. Cells containing the plasmid were cultured in Luria broth medium at 310 K. When the culture reached an optical density of 0.6 at a wavelength of 600 nm, gene expression (at 293 K) was induced by the addition of 0.5 mM IPTG. The cells were subsequently harvested and disrupted by sonication using a Soniprep 150 (Sanyo Gallenkamp PLC, Loughborough, England). The supernatant was collected after centrifugation at 100,000 g for 1 h. The crude extract was further purified on a HiTrap Chelating HP 5 ml column charged with Ni^{2+} ions (GE Healthcare, Uppsala, Sweden). The His-tagged enzyme was bound to the resin in the presence of 20 mM potassium phosphate buffer pH 7.5, 0.5 M sodium chloride, 10 mM imidazole. Unbound and non-specifically bound proteins were washed out by a buffer containing 37.5 mM imidazole. The target enzyme was eluted with a buffer containing 300 mM imidazole. The active fractions were pooled and dialyzed overnight against 50 mM Tris-HCl pH 7.5. The DbeAΔCl enzyme was stored at 277 K in 50 mM Tris-HCl pH 7.5 buffer prior to analysis. The DbeAΔCl production information is summarized in Table 1.

Table 1. Production specifics for DbeAΔCl.

Source Organism	***Bradyrhizobium Elkanii* USDA94**
DNA source	Artificially synthesized DNA
Transport vector	pMA
Expression vector	pET-21b
Expression host	*E. coli* BL21(DE3)
Complete amino acid sequence of DbeAΔCl	MTISADISLHHRAVLGSTMAYRETGRSDAPHVLFLHGNPTSSYL WRNIMPLVAPVGHCIAPDLIGYGQSGKPDISYRFFDQADY LDALIDELGIASAYLVAHDWGTALAFHLAARRPQLVRGLA FMEFIRPMRDWSDFHQHDAARETFRKFRTPGVGEAMILDN NAFVERVLPGSILRTLSEEEMAAYRAPFATRESRMPTLML PRELPIAGEPADVTQALTAAHAALAASTYPKLLFVGSPGA LVSPAFAAEFAKTLKHCAVIQLGAGGHYLQEDHPEAIGRS VAGWIAGIEAASAQRHAALEHHHHHH

2.2. Crystallization

The freshly isolated and purified DbeAΔCl protein was crystallized at a concentration of 30 mg.ml^{-1} in 50 mM Tris–HCl buffer pH 7.5 by the sitting-drop vapour-diffusion procedure [9]. For initial screening several commercial precipitants kits were used: JBScreen Classic Kits № 1-10 and Wizard I-III (Jena Bioscience GmbH, Jena, Germany), Morpheus®HT-96, JCSG-plus™ HT-96, PACT premier™ HT-96 and Structure Screen 1 + 2 HT-96 kit (Molecular Dimensions Ltd (MDL), Suffolk, UK), Crystal Screen kit, PEGRx HT™ and PEG/Ion HT™ (Hampton Research (HR), Aliso Viejo, USA) and Axygen I-VIII crystallization kits (Axygen Biosciences, Union City, USA). The CombiClover 96 well plates (MDL, Suffolk, UK) for manual screening experiments as well as Swissi polystyrene MRC 2-drop plate (MDL, Suffolk, UK) were utilized for the initial screening on an Oryx3 robot (Douglas Instruments Ltd, Hungerford, UK) for the DbeAΔCl protein.

The hanging drop crystallization trials were carried out in Limbro 24 well plates (HR, Aliso Viejo, USA). The Douglas Instruments, USA Vapour Batch 96 well plates were used to perform the microbatch under oil crystallization [10]. Macroseeding experiments [11] were carried out by lowering the protein concentration to 20–25 mg.ml^{-1}. The counter-diffusion crystallization was performed in single glass capillaries with inner diameters ranging from 0.1 to 0.4 mm (HR, Aliso Viejo, USA) using a three-layer configuration [12].

2.3. Data Collection, Processing and Structure Solution

X-ray diffraction data at 100 K were collected to the 1.4 Å resolution at the BESSY II electron storage ring on beamline MX 14.1 of the Helmholtz-Zentrum Berlin (Berlin-Adlershof, Germany; [13]). 2500 images were processed with the graphical user interface XDSAPP [14] for running XDS [15]. Phasing by molecular replacement was performed using MOLREP [16] and the structure of DbeA as the template (PDB code 4k2a; [7]). One molecule was found in the asymmetric unit of DbeAΔCl. Structure refinement and model building was performed using isotropic and anisotropic refinement protocols in REFMAC5 [17] and Coot [18] from the CCP4 package [19], respectively. The quality of the protein models was confirmed with MolProbity [20,21] and wwPDB [22] validation servers. The structure of the DbeAΔCl has been deposited to the Protein Data Bank under accession code 6s42.

Figures with the structure were prepared using the program PyMOL [23]. The complete information about the data collection, processing, and refinement statistics are provided in Table 2.

Table 2. Data collection and crystallographic statistics.

X-ray Diffraction Data Collection Statistics	
Space group	C121
Cell parameters (Å, °)	a = 128.95, b = 63.95, c = 46.05; $\alpha = \gamma = 90$, $\beta = 106.27$
Wavelength (Å)	0.918
Resolution (Å)	1.39
Number of unique reflections	68,322
Redundancy	2.18 (2.20)
Completeness (%)	96.13 (92.17)
R_{merge} #	4.9 (26.3)
Average I/σ(I)	12.72 (3.19)
Wilson B (Å^2)	20.8
Refinement Statistics	
Resolution range (Å)	41.96–1.4 (1.43–1.39)
No. of reflections in working set	64,907 (4,609)
R value (%) ##	13.98
R_{free} value (%) ###	15.22
RMSD bond length (Å)	0.006
RMSD angle (°)	1.635
No. of atoms in AU	2,823
No. of protein atoms in AU	2,333
No. of water molecules in AU	470
No. of iodide ions in AU	8
No. of chloride ions in AU	3
Mean B value (Å^2)	13.42
Ramachandran Plot Statistics	
- Residues in favoured regions (%)	97.2
- Residues in allowed regions (%)	100
PDB code	6s42

The data in parentheses refer to the highest-resolution shell. # $R_{merge} = (|I_{hkl} - \langle I \rangle|)/I_{hkl}$, where the average intensity ⟨I⟩ is taken over all symmetry equivalent measurements and I_{hkl} is the measured intensity for any given reflection; ## R-value = $||F_o| - |F_c||/|F_o|$, where F_o and F_c are the observed and calculated structure factors, respectively; ### R_{free} is equivalent to R value but is calculated for 5% of the reflections chosen at random and omitted from the refinement process.

3. Results and Discussion

The crystallization procedure previously successfully used for growing of the DbeAwt protein [24] was applied to prepare crystals of freshly purified DbeAΔCl protein. Initially, only a light amorphous precipitation was observed. Further optimization of the crystallization conditions was carried out by variation of the protein and precipitant concentrations (PEG and salt) and drop protein-reservoir ratio composition. This optimization did not improve the results. Additional screening was conducted by applying further commercial crystallization kits: JBScreen Classic Kits № 1-10 and Wizard I-III (Jena Bioscience GmbH, Jena, Germany). Again, only very small and thin needle crystals or a heavy amorphous precipitation were observed. The next optimization step was based on moving the system closer to the metastable zone based on the phase transition diagram by decreasing the concentration of the crystallization drop components. The reduction of the DbeAΔCl concentration and variation of the precipitant concentrations yielded small microcrystals (Figure 1a) and a two-dimensional (2D) single needle crystals (Figure 1b) within a period of 1-3 weeks. The microcrystals were grown within three weeks from precipitant consisting of 28% (w/v) PEG 4000, 0.2 M Li_2SO_4 in 0.1 M Tris pH 8.2 buffer and a 30 mg.ml^{-1} protein concentration. The small needle crystals were observed after 10 days at a 10 to

30 mg.ml^{-1} protein concentration in precipitant composed of 12– 17% (w/v) PEG 3350, 115– 125 mM $MgCl_2$ in 100 mM Tris-HCl 7.5 buffer.

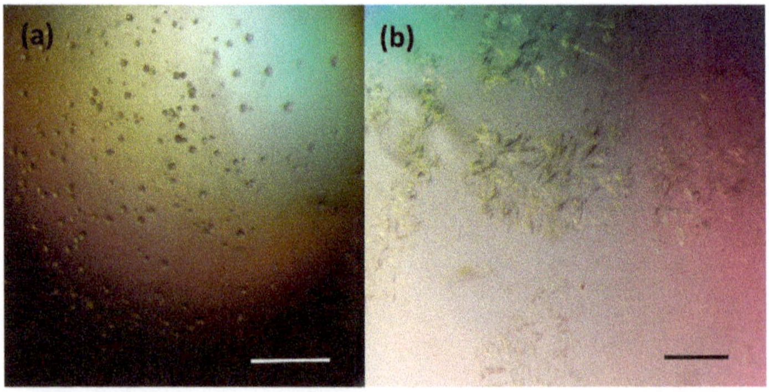

Figure 1. Results of initial crystallization experiments of DbeAΔCl protein from *B. elkanii* USDA94: (**a**) microcrystals and (**b**) small 2D needle crystals. The scale bar represents 100 μm.

In order to improve the crystal quality, Morpheus®HT-96, JCSG-plus™ HT-96, PACT premier™ HT-96 and Structure Screen 1 + 2 HT-96 kit (Molecular Dimensions Ltd (MDL), Suffolk, UK), Crystal Screen kit, PEGRx HT™ and PEG / Ion HT™ (Hampton Research (HR), Aliso Viejo, USA) and Axygen I-VIII crystallization kits (Axygen Biosciences, Union City, USA) were applied. Finally, small 3D crystals (Figure 2a) with a dimension of about 15 × 5 × 35 μm were grown from a solution containing 26.57% (w/v) hexanediol at 295 K over 10 days. The crystals diffracted X-rays to a maximum resolution of 8–10 Å. The quality of these crystals did not allow to record good diffraction data and additional strategies were needed to improve the size and shape of obtained crystals.

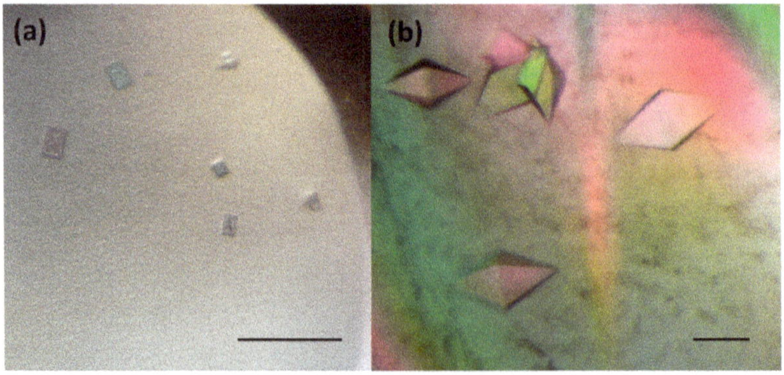

Figure 2. DbeAΔCl crystals used for diffraction analysis: (**a**) small 3D crystals grown at 26.5% (w/v) hexanediol and (**b**) big 3D crystals developed by random seeding experiments. The scale bar represents 100 μm.

Further optimization using the precipitant solution mentioned above was pursued by setting up the experiments at a lower temperature (193K), by application of the Additive Screen (HR) and by variation of the protein concentration from 10 to 40 $mg.ml^{-1}$, however, without success. The same conditions were tested in a hanging drop vapour-diffusion, microbatch under oil, counter-diffusion and macroseeding procedure. All these experiments did not significantly improve the diffraction quality of the crystals. Next, the small crystals (Figure 2a) were used for random microseeding experiments [25] with application of two commercial crystallization kits: PACT premier™ HT-96 and Structure Screen 1 + 2 HT-96 kit (MDL), which finally resulted in the appearance of 3D crystals with an average dimensions 70 × 75 × 100 μm within a period of two weeks (Figure 2b). These crystals, later used for X-ray data collection experiments, were obtained by lowering the protein concentration to 20 $mg.ml^{-1}$ and applying the following precipitant: 0.2M sodium iodide and 20% (w/v) PEG 3350.

Crystals of DbeAΔCl diffracted X-rays to 1.4 Å resolution and belong to the monoclinic base-centered space group C121. The diffraction data allowed localizing 294 amino-acid residues fitting to the one molecule in the asymmetric unit. The overall shape of DbeAΔCl is like a block with 2,823 non-hydrogen atoms (470 water molecules, 8 iodide anions, 3 chloride anions and hexanediol molecule) and corresponds to the canonical architecture of HLDs of the α/β hydrolase fold superfamily (Figure 3a). The structural organization of DbeAΔCl displays two compact domains: an α/β hydrolase core domain and a helical cap domain with the active site located between them. The cap domain (residues 134–214) consists of five α-helices (α4, α5, α5´, α6, α7 and α8) and six loops, together forming a lid, protecting the active site cavity. The core domain (residues 4–133 and 215–298) consists of a central twisted eight-stranded β-sheet with the β2 strand running antiparallel. The β-sheet region is flanked by six α-helices: two elements (α1 and α2) cover one side and the remaining four (α3, α9, α10 and α11) the other side [2,26] (Figure 3a). The protein displays a monomer as the biological unit according to the analysis of crystal contacts between molecules in the unit cell and crystal packing. The exploration of macromolecular interfaces by PDBePISA server [27] underpins the monomeric nature of the protein.

The DbeAΔCl active site displays a substrate-binding pocket typical for all haloalkane dehalogenases. The enzyme's active site cavity contains the catalytic triad consisting of Asp103, His 271 and Glu127. The nucleophile Asp103 is located at the turn between β-strand β5 and helix α3. The catalytic base His271 is positioned on the loop joining β8 and α11. The catalytic acid Glu127 is located behind β-strand β6. Inspecting the electron density map, one iodide anion and one molecule of hexanediol as components of precipitant cocktail were identified near the DbeAΔCl active site. The iodide ion is mainly stabilized by interactions with the N atoms of two halide-binding residues: Asn34 $N^{\delta 2}$ and Trp104 $N^{\epsilon 1}$ with distances of 3.72 Å and 3.47 Å, respectively. Further coordination is realized with the N atom of the pyrrolidine ring of Pro205 at a distance of 3.64 Å and an O atom of bound hexanediol at 3.68 Å distance (Figure 3b).

The substitutions Ile44Leu and Gln102His introduced into the structure of DbeA result in the elimination of the second halide-binding site (Figure 3c). The site is occupied by a water molecule in the mutant enzyme. Apparently, insufficient space is left for the positioning of the second halide anion as observed in the DbeAwt structure. The atoms of three residues: Thr40 $O^{\gamma 1}$, His102 $N^{\delta 1}$ and Gly37 N, with hydrogen bond distances of 2.74 Å, 2.79 Å, and 2.94 Å, respectively, coordinated water molecule W71 that was modelled instead of the halide anion (Figure 3c). Further coordination is realized by water molecule W11 at a distance of 2.73 Å. Water molecule W11 is situated between two halide-binding sites, 6.73 Å away from the iodide ion in the canonical active site of the protein. Water W11 is stabilized by interaction with the $O^{\delta 2}$ atom of catalytic nucleophile Asp103 at 2.71 Å distance.

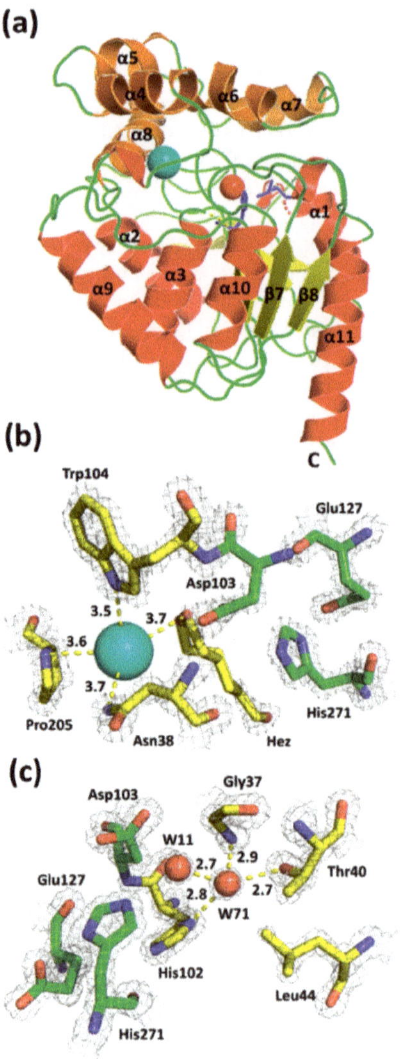

Figure 3. The overall structure of DbeAΔCl (**a**), close-up view of the canonical DbeAΔCl active site (**b**) and second halide-binding site (**c**). The $2F_o$-F_c electron-density map contoured at 2σ is shown in grey (**a**) Cα ribbon trace shows elements of the protein secondary structure. The α-helices are coloured red for the main domain and brown for the cap domain; β-strands are coloured yellow; loops are shown in green; iodide ion is presented as a cyan sphere, water molecule (W71) is shown as a red sphere; the two point substitutions Ile44Leu + Gln102His introduced into DbeA are highlighted as blue sticks. (**b**) The iodide ion in the active site is presented as a cyan sphere with coordination interaction distances in Å and highlighted by yellow dashed lines; hexanediol (Hez) (shown in two alternative conformations) and amino acids Asp103, Trp104, Pro205 coordinating the iodide ion are shown as sticks with carbon atoms coloured yellow. Carbon atoms of catalytic triad are highlighted in green. (**c**) The water molecules W11 and W71 are presented as red spheres; amino acids coordinating water molecule W71 are shown as sticks with carbon atoms coloured in yellow with and interactions with distances in Å shown by yellow dashed lines. Water molecule W71 is located in second-halide binding site.

The sequence of DbeAΔCl from *B. elkanii* (PDB ID code: 6s42) was aligned (Figure 4) and compared with five known sequences of HLDs deposited in the PDB: DbeA wild type from *B. elkanii* (PDB ID code: 4k2a [7]), DhaA from *Rhodococcus species* (PDB ID code: 1bn6 [1]), DhlA from *Xanthobacter autotrophicus* (PDB ID code: 1cij [28]), LinB from *Sphingomonas paucimobilis* (PDB ID code: 1cv2 [29]) and DmbA from *Mycobacterium tuberculosis* (PDB ID code: 2qvb [30]). The reason for selection of these dehalogenases is that DbeAwt is the same protein without two point mutations occurring in DbeAΔCl with 99.3% sequence identity, DhaA as HLD with highest sequence identity shows 50.7% and DhlA as HLD with lowest sequence identity displays 26.5% in comparison to DbeAΔCl. Alignment of DbeAΔCl with different kind of haloalkane dehalogenase LinB from the same substrate specificity group (SSG-I) demonstrates 47.4% sequence identity similar to DmbA as another type of HLD from different SSG-III (44.8%) [31].

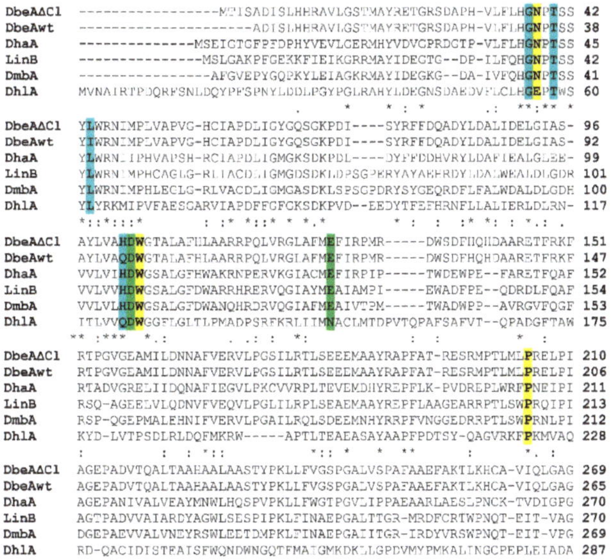

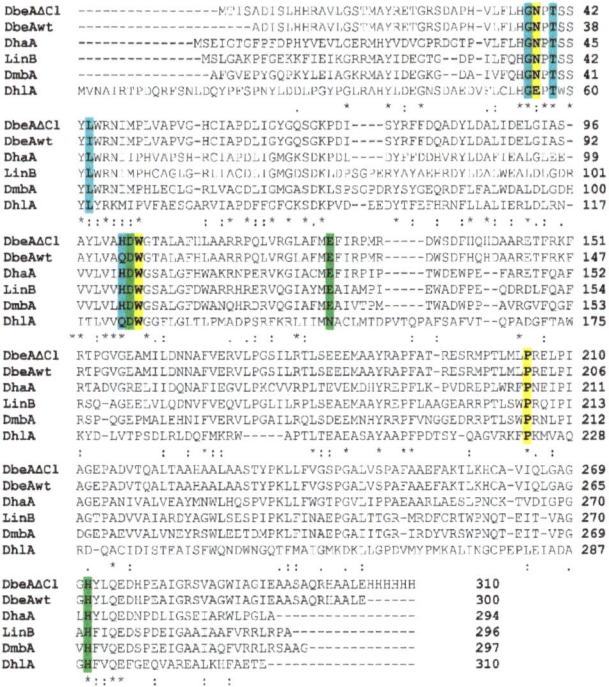

Figure 4. Multiple sequence alignment of DbeAΔCl with five Haloalkane dehalogenases (HLDs) deposited in PDB. Amino acids from canonical HLD active site are highlighted in yellow, catalytic triad residues are highlighted in green and second halide binding site residues are highlighted in cyan. Sequence alignment was performed by ClustalW [32].

The molecular structures of DbeA, DhaA, LinB, DmbA, and DhlA were superposed with the DbeAΔCl enzyme's Cα atoms with root mean square deviations of 0.335, 1.018, 0.985, 1.127 and 2.112 Å, respectively. 3D-superpositions of the DbeAΔCl structure with the closest similarity model (DhaA) and the lowest identity model (DhlA) are shown in Figure 5a.

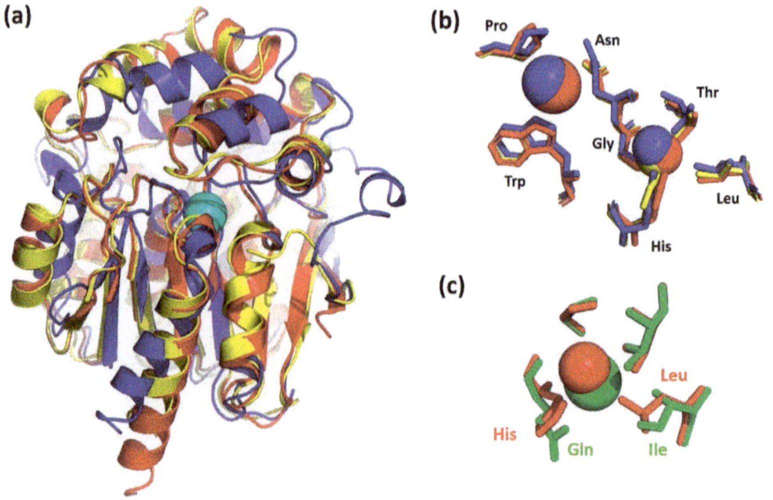

Figure 5. Structural comparison of DbeAΔCl with homologous dehalogenases. (**a**) DbeAΔCl secondary structure superposition with DhaA and DhlA. Cα ribbon trace shows elements of the protein secondary structures. The DbeAΔCl is coloured in red; DhaA is coloured in blue; DhlA is shown in yellow; iodide ion (cyan sphere) is placed in the canonical active site of DbeAΔCl. (**b**) Superposition of DbeAΔCl, DhaA and DhlA active sites. Amino acids of DbeAΔCl are shown as red sticks; amino acids of DhaA are shown as blue sticks; amino acids of DhlA are shown as yellow sticks. The iodide ion and water molecule W11 bound to DbeAΔCl are represented as red spheres; bromide ion and water are shown as blue spheres for the DhaA structure and the single water molecule in the DhlA is shown as yellow sphere. (**c**) Superposition of second halide binding site residues of mutant and wild type DbeA. DbeAΔCl amino acids are shown as red sticks; amino acids of DbeAwt are shown as green sticks. Water W71 is shown as red sphere and Cl^- ion is shown as green sphere.

In general, the secondary structure elements of the core domain are better conserved compared to the cap domain for the most dehalogenases. Significant differences in cap domains define substrate specificity and variability [1]. The considerable divergence among the superimposed structures was also observable at the relatively disordered N- and C-terminal parts of the proteins (Figure 5a). The position of the active site residues was well conserved among HLDs with some differences in DhlA structure. DbeAΔCl, DbeA, LinB, DhaA and DmbA belong to the HLD-II subfamily with Asp-His-Glu catalytic triad and Asn-Trp halide binding residues [4] whereas DhlA belongs to the HLD-I subfamily with an Asp-His-Asp + Trp-Trp catalytic pentad composition. The positions of the halide-binding residues Asn38 and Trp104 in the DbeAΔCl structure is conserved in comparison to the rest of HLD structures. The position of the catalytic triad (Asp103; His271 and Glu127) is well conserved among the dehalogenases from HLD-II subfamily. The side chain of the catalytic acid Asp in DhlA is situated slightly deeper in the active site cavity in comparison to HLD-II structures.

Superposition of the iodide ion at the DbeAΔCl active site with halide ions in the vicinity of halide-stabilizing residues of the rest structures reveals some structural differences. DhlA and DmbA contain halide ions inside the active site: Br^- (DhlA) and Cl^- (DmbA) with shifts in 0.47 and 0.39 Å from the DbeAΔCl iodide ion position, respectively. However, there is only a water molecule in the LinB structure (0.27 Å away from iodide ion of DbeAΔCl) and an empty space in the DhaA structure (the closest water molecule is located 6.48 Å away from the iodide ion position).

Water molecule W71, which substitutes the halide ion at the second halide-binding site of the DbeA wild type structure, was found at the canonical place where only a water molecule is present in all the HLDs with a shift of 1.39 Å compared to the DhaA structure, 0.80 Å to the DhlA, 0.55 Å to the

LinB and 0.43 Å to the DmbA, and 1.28 Å away compared to the second chloride anion in the DbeA structure. Superposition of halide binding sites of the DbeAΔCl structure with halide binding sites of the closest relative (DhaA) and the lowest identity HLD (DhlA) is shown in the Figure 5b.

4. Conclusions

After many crystallization experiments and optimization cycles, we found that two point mutations deeply buried in the structure have a significant influence on the crystallization of the macromolecule. Finally, random microseeding experiments with a seed stock prepared from small 3D crystals obtained at a protein concentration of 20 mg.ml^{-1} and 0.2 M sodium iodide plus 20% (w/v) PEG 3350 as the precipitant solutions produced crystals of the double mutant DbeAΔCl from *B. elkanii* USDA94. These crystals were of sufficient quality for X-ray diffraction experiments. The structure of DbeAΔCl was solved using molecular replacement and refined to 1.4 Å resolution. Overall, the structure is very similar to other HLDs structures of the α/β hydrolase fold superfamily (EC 3.8.1.5). The substitutions Ile44Leu and Gln102His resulted in a space reduction of the second halide-binding site and thus incapability of DbeAΔCl to bind a second halide anion as compared to the wild type structure. Instead of a chloride anion, a water molecule (W71) was found in the site of which the consequences are, DbeAΔCl is less active and less stable in the presence of chloride salts when compared with the DbeAwt enzyme [7]. The DbeAΔCl structure of B. *elkanii* was aligned and compared with five molecular structures of haloalkane dehalogenases selected from the PDB. The elements of the secondary structure and the catalytic pentad are well conserved. Superposition of the iodide ion at the DbeAΔCl active site with the other structures reveals some structural differences. The water molecule W71 located at the compromised second halide-binding site of DbeAΔCl was coordinated at the canonical place as compared to other HLDs.

Author Contributions: T.P., J.R.M. and I.K.S. designed the experiments and solved the structure. T.P., B.K., P.G., J.R.M. and M.K. analyzed the data. T.P. and B.K. wrote the manuscript. R.C., J.D., A.M., P.H. and A.S. provided technical support.

Funding: This work was supported by the Grant Agency of the Czech Republic 17-24321S; DAAD mobility grant DAAD-16-09; ERDF project CZ.02.1.01/0.0/0.0/15_003/0000441; Ministry of Education, Youth and Sports of the Czech Republic (CZ.1.05/2.1.00/01.0024, CZ.1.05/2.1.00/01.0001, and LM2015055); GAJU 17/2019/P.

Acknowledgments: The diffraction data were collected on the beam line MX14.1 at the BESSY II electron storage ring operated by the Helmholtz-Zentrum Berlin. We would particularly like to acknowledge the help and support of Manfred S. Weiss during data collection. Also, we would like to thank Stefan A. Kolek (Douglas Instruments Ltd, Hungerford, UK) for providing crucial information about the random seeding experiment during the 2014 FEBS-Instruct practical course PC14-005 at Nove Hrady, Czech Republic.

Conflicts of Interest: No interest conflict exists among the authors.

References

1. Newman, J.; Peat, T.S.; Richard, R.; Kan, L.; Swanson, P.E.; Affholter, J.A.; Holmes, I.H.; Schindler, J.F.; Unkefer, C.J.; Terwilliger, T.C. Haloalkane dehalogenases: Structure of a Rhodococcus enzyme. *Biochemistry* **1999**, *38*, 16105–16114. [CrossRef] [PubMed]
2. Janssen, D.B.; Dinkla, I.J.T.; Poelarends, G.J.; Terpstra, P. Bacterial degradation of xenobiotic compounds: Evolution and distribution of novel enzyme activities. *Environ. Microbiol.* **2005**, *7*, 1868–1882. [CrossRef] [PubMed]
3. Prokop, Z.; Sato, Y.; Brezovsky, J.; Mozga, T.; Chaloupkova, R.; Koudelakova, T.; Jerabek, P.; Stepankova, V.; Natsume, R.; van Leeuwen, J.G.; et al. Enantioselectivity of haloalkane dehalogenases and its modulation by surface loop engineering. *Angew. Chem. Int. Ed. Engl.* **2010**, *49*, 6111–6115. [CrossRef] [PubMed]
4. Chovancova, E.; Kosinski, J.; Bujnicki, J.M.; Damborsky, J. Phylogenetic analysis of haloalkane dehalogenases. *Proteins* **2007**, *62*, 305–306. [CrossRef] [PubMed]
5. Holmquist, M. Alpha/beta-hydrolase fold enzymes: Structures, functions and mechanisms. *Curr. Protein Pept. Sci.* **2000**, *1*, 209–235. [CrossRef] [PubMed]

6. Prokop, Z.; Oplustil, F.; DeFrank, J.; Damborsky, J. Enzymes fight chemical weapons. *Biotechnol. J.* **2006**, *1*, 1370–1380. [CrossRef]
7. Chaloupkova, R.; Prudnikova, T.; Rezacova, P.; Prokop, Z.; Koudelakova, T.; Daniel, L.; Brezovsky, J.; Ikeda-Ohtsubo, W.; Sato, Y.; Kuty, M.; et al. Structural and functional analysis of a novel haloalkane dehalogenase with two halide-binding sites. *Acta Cryst.* **2014**, *70*, 1884–1897. [CrossRef]
8. Berman, H.M.; Westbrook, J.; Feng, Z.; Gilliland, G.; Bhat, T.N.; Weissig, H.; Shindyalov, I.N.; Bourne, P.E. The Protein Data Bank. *Nucleic Acids Res.* **2000**, *28*, 235–242. [CrossRef]
9. Ducruix, A.; Giegé, R. *Crystallization of Nucleic Acids and Proteins*; Oxford University Press: Oxford, UK, 1999. [CrossRef]
10. Chayen, N.E.J. Crystallization with oils: A new dimension in macromolecular crystal growth. *J. Cryst. Growth* **1999**, *196*, 434–441. [CrossRef]
11. Bergfors, T.M. *Protein Crystallization: Techniques, Strategies and Tips*; International University Line: La Jolla, CA, USA, 1999. [CrossRef]
12. Gavira, J.A.; Jesus, W.; Camara-Artigas, A.; Lopez-Garriga, J.; Garcia-Ruiz, J.M. Crystallization and diffraction patterns of the oxy and cyano forms of the Lucina pectinata haemoglobins complex. *Acta Cryst.* **2006**, *62*, 196–199. [CrossRef]
13. Gerlach, M.; Mueller, U.; Weiss, M.S. The MX beamlines BL14.1-3 at BESSY II. *JLSRF* **2016**, *2*, 1–6. [CrossRef]
14. Sparta, K.M.; Krug, M.; Heinemann, U.; Mueller, U.; Weiss, M.S. XDSAPP2.0. *J. Appl. Cryst.* **2016**, *49*, 1085–1092. [CrossRef]
15. Kabsch, W. Automatic processing of rotation diffraction data from crystals of initially unknown symmetry and cell constants. *J. Appl. Cryst.* **1993**, *26*, 795–800. [CrossRef]
16. Vagin, A.; Teplyakov, A. MOLREP: An automated program for Molecular Replacement. *J. Appl. Cryst.* **1997**, *30*, 1022–1025. [CrossRef]
17. Murshudov, G.N.; Skubak, P.; Lebedev, A.A.; Pannu, N.S.; Steiner, R.A.; Nicholls, R.A.; Winn, M.D.; Long, F.; Vagin, A.A. REFMAC5 for the refinement of macromolecular crystal structures. *Acta Cryst.* **2011**, *67*, 355–367. [CrossRef]
18. Emsley, P.; Lohkamp, B.; Scott, W.G.; Cowtan, K. Features and development of Coot. *Acta Cryst.* **2010**, *66*, 486–501. [CrossRef]
19. Winn, M.D.; Ballard, C.C.; Cowtan, K.D.; Dodson, E.J.; Emsley, P.; Evans, P.R.; Keegan, R.M.; Krissinel, E.B.; Leslie, A.G.W.; McCoy, A.; et al. Overview of the CCP4 suite and current developments. *Acta Cryst.* **2011**, *67*, 235–242. [CrossRef]
20. Chen, V.B.; Arendall, W.B.; Headd, J.J.; Keedy, D.A.; Immormino, R.M.; Kapral, G.J.; Murray, L.W.; Richardson, J.S.; Richardson, D.C. MolProbity: All-atom structure validation for macromolecular crystallography. *Acta Cryst.* **2010**, *66*, 12–21. [CrossRef]
21. Hintze, B.J.; Lewis, S.M.; Richardson, J.S.; Richardson, D.C. Molprobity's ultimate rotamer-library distributions for model validation. *Proteins* **2016**, *84*, 1177–1189. [CrossRef]
22. Gore, S.; Velankar, S.; Kleywegt, G.J. Implementing an X-ray validation pipeline for the Protein Data Bank. *Acta Cryst.* **2012**, *68*, 478–483. [CrossRef]
23. Schrodinger, L.L.C. The PyMOL Molecular Graphics System, Version 2.0. 2019. Available online: https://pymol.org/2/ (accessed on 23 July 2019).
24. Prudnikova, T.; Mozga, T.; Rezacova, P.; Chaloupkova, R.; Sato, Y.; Nagata, Y.; Brynda, J.; Kuty, M.; Damborsky, J.; Kuta-Smatanova, I. Crystallization and Preliminary X-ray Analysis of a Novel Haloalkane Dehalogenase DbeA from Bradyrhizobium elkani USDA94. *Acta Cryst.* **2009**, *65*, 353–356. [CrossRef]
25. Shaw Stewart, P.D.; Kolek, S.A.; Briggs, R.A.; Chayen, N.E.; Baldock, P.F.M. Random Microseeding: A Theoretical and Practical Exploration of Seed Stability and Seeding Techniques for Successful Protein Crystallization. *Cryst. Growth Des.* **2011**, *11*, 3432–3441. [CrossRef]
26. Ollis, D.L.; Cheah, E.; Cygler, M.; Dijkstra, B.; Frolow, F.; Franken, S.M.; Harel, M.; Remington, S.J.; Silman, I.; Schrag, J.; et al. The alpha/beta hydrolase fold. *Protein Eng.* **1992**, *5*, 197–211. [CrossRef] [PubMed]
27. Krissinel, E.; Henrick, K. Inference of macromolecular assemblies from crystalline state. *J. Mol. Biol.* **2007**, *372*, 774–797. [CrossRef] [PubMed]
28. Pikkemaat, M.G.; Ridder, I.S.; Rozeboom, H.J.; Kalk, K.H.; Dijkstra, B.W.; Janssen, D.B. Crystallographic and kinetic evidence of a collision complex formed during halide import in haloalkane dehalogenase. *Biochemistry* **1999**, *38*, 12052–12061. [CrossRef] [PubMed]

29. Marek, J.; Vevodova, J.; Smatanova, I.K.; Nagata, Y.; Svensson, L.A.; Newman, J.; Takagi, M.; Damborsky, J. Crystal structure of the haloalkane dehalogenase from Sphingomonas paucimobilis UT26. *Biochemistry* **2000**, *39*, 14082–14086. [CrossRef]
30. Mazumdar, P.A.; Hulecki, J.C.; Cherney, M.M.; Garen, C.R.; James, M.N.G. X-ray crystal structure of Mycobacterium tuberculosis haloalkane dehalogenase Rv2579. *Biochim. Biophys. Acta.* **2008**, *1784*, 351–362. [CrossRef]
31. Koudelakova, T.; Bidmanova, S.; Dvorak, P.; Pavelka, A.; Chaloupkova, R.; Prokop, Z.; Damborsky, J. Haloalkane dehalogenases: Biotechnological applications. *Biotechnol. J.* **2013**, *8*, 32–45. [CrossRef]
32. Larkin, M.A.; Blackshields, G.; Brown, N.P.; Chenna, R.; McGettigan, P.A.; McWilliam, H.; Valentin, F.; Wallace, I.M.; Wilm, A.; Lopez, R.; et al. Clustal W and Clustal X version 2.0. *Bioinformatics* **2007**, *23*, 2947–2948. [CrossRef]

MDPI
St. Alban-Anlage 66
4052 Basel
Switzerland
Tel. +41 61 683 77 34
Fax +41 61 302 89 18
www.mdpi.com

Crystals Editorial Office
E-mail: crystals@mdpi.com
www.mdpi.com/journal/crystals

www.ingramcontent.com/pod-product-compliance
Lightning Source LLC
LaVergne TN
LVHW071949080526
838202LV00064B/6714
* 9 7 8 3 0 3 9 2 1 4 0 3 7 *